T0181168

NATURAL LAWS IN SCIENTIFIC PRACTICE

NATURAL LAWS IN SCIENTIFIC PRACTICE

MARC LANGE

OXFORD
UNIVERSITY PRESS
2000

OXFORD
UNIVERSITY PRESS

Oxford New York
Athens Auckland Bangkok Bogotá Buenos Aires Calcutta
Cape Town Chennai Dar es Salaam Delhi Florence Hong Kong Istanbul
Karachi Kuala Lumpur Madrid Melbourne Mexico City Mumbai
Nairobi Paris São Paulo Singapore Taipei Tokyo Toronto Warsaw

and associated companies in
Berlin Ibadan

Published by Oxford University Press, Inc.
198 Madison Avenue, New York, New York 10016

Oxford is a registered trademark of Oxford University Press

Library of Congress Cataloging-in-Publication Data
Lange, Marc 1963–
Natural laws in scientific practice / Marc Lange.
p. cm.
Includes bibliographical references and index.
ISBN 978-0-19-533133-2
1. Science—Philosophy. 2. Philosophy and science.
3. Nature. I. Title.
Q175.L2442 2000
501—dc21 99-38157

Printed in the United States of America
on acid-free paper

Praise the Lord, for he hath spoken
Worlds his mighty voice obeyed;
Laws, which never shall be broken,
For their guidance he hath made.

English Hymnal (1796),
Foundling Hospital Collection, no. 535

Preface

This books concerns the concept of a law of nature as it figures in science. Certain intuitions suggest that laws of nature bear special relations to counterfactual conditionals, inductive confirmations, and scientific explanations. But these intuitions have been notoriously difficult to cash out precisely, which has led some philosophers to suggest recently that the concept of a natural law be dropped altogether from our accounts of scientific reasoning.

Much of this book is devoted to a careful unpacking of these intuitions, with a view to determining what role, if any, the concept of a natural law should be expected to play in our understanding of scientific practice. In chapters 2 and 3, I work toward giving a precise explication of the special relation that laws bear to counterfactuals, a relation that reflects the laws' special kind of necessity ("physical necessity"). Although an accident (a contingent nonlaw) may be capable of "supporting" a broad range of counterfactuals, there is a sharp distinction (as I show) between the laws and the accidents in their relations to counterfactuals. There is a sense in which the laws collectively possess a maximal degree of invariance under counterfactual suppositions, though this sense must be specified carefully. I also explain what *multiple* grades of physical necessity would amount to, and the precise sense in which these various grades would fall in the range between logical and conceptual necessity, on the one hand, and no necessity at all on the other.

It has often been held that an accident's predictive accuracy cannot be confirmed by its instances in the same manner as a law's can by its instances. In chapters 4 and 5, I try to uncover the intuitions motivating this thought, and to see whether they make sense in the face of serious challenges prompted by current philosophical accounts of theory confirmation. Ultimately, I use the results of chapters 2 and 3 to develop a precise conception of "inductive projection" and then to show that confirmation of this sort is unavailable to a hypothesis that is believed to lack physical necessity. I use these ideas to offer a novel response to Hempel's "paradox of the ravens" (and, later, to Goodman's "grue" problem). Although I argue that the physical necessities in a given possible world are not fixed by the non-nomic facts there, I emphasize that empirical evidence bears upon the correctness of counterfactual conditionals for precisely the same reasons as it bears upon the correctness of predictions regarding actual, as yet unexamined cases. I consider why it is especially valuable in science to identify which facts are physically necessary, and argue that lawhood's scientific significance results partly from the special epistemic status of "inductive strategies."

Laws of nature have traditionally been regarded as constituted by or at least manifested in exceptionless regularities. In chapter 6, however, I argue that this view is mistaken; rather, a law of nature is associated with an inference rule that is "reliable"—accurate enough for certain purposes, though perhaps not perfectly truth-preserving. In fact, I argue that this conception is needed in order to capture the kind of necessity that laws possess—that is, to vindicate the intuition that a natural law would still have held under certain counterfactual suppositions. I argue that a generalization may be violated in a given possible world and yet be associated with a law there, as long as those violations remain "offstage"—outside the range of facts with which we are concerned in contemplating that world. I argue that some natural laws have provisos limiting their application to certain practical purposes (in connection with which certain approximations can be tolerated) or to cases where there is no information contrary to the prediction that the law would support. In chapter 8, I argue that provisos of this kind are associated with claims such as "the human being has 46 chromosomes." These claims state laws of certain biological fields (such as cardiology, developmental biology, and neurology) and express rules for default reasoning. But a law of, say, cardiology is at the same time an accident of, say, evolutionary biology or fundamental physics. I use my account of the laws' distinctive relation to counterfactuals in order to make precise the concept of a law of some particular scientific discipline. Though these views are radical, they are motivated by the traditional conception of the laws as standing in special relations to counterfactuals, explanations, and inductions.

This account of a given law's relation to certain purposes not only puts us in a position to understand the sense in which there may be laws of "special sciences," such as cardiology or island biogeography, but also enables us to understand how, in a given science, there may be various laws of different degrees of accuracy and different scopes. In a typical pattern, there is a fairly broad law that is accurate enough for certain applications (such as the ideal gas laws, van der Waals's law, or Hooke's law), and then more accurate laws with narrower scopes (such as a law with higher order correction factors for a particular species of gas or a particular kind of spring). If the various more accurate laws are combined, the result is a physical necessity with the same broad scope as the original, modestly accurate law. But this physical necessity is not itself a law; it is not the result of any single inductive strategy because the higher order correction factors needed for one kind of gas do nothing to suggest the higher order correction factors needed for another kind of gas. That is, we could not justly arrive at this physical necessity by observing only gases of one kind; we would have to observe samples of each kind of gas in the physical necessity's scope.

I argue that the laws are associated with the results of the best set of inductive strategies for us to carry out. (This is as close as I come to giving a reductive analysis of the concept of a natural law.) I argue that this account captures the important (and often neglected) distinction between laws of nature and physically necessary nonlaws. This distinction is crucial to the sense in which laws are "universal" or "general," as well as to the laws' relation to the natural kinds. In the course of cashing out the concept of the "best inductive strategies" in chapter 7, I investigate the notion of a hierarchy of laws governing laws. A symmetry principle, for instance, is a law governing laws, as are principles expressing the sorts of laws in which, say, a chemical category must figure in order to qualify as a distinct chemical species. The metalaws' relation to the laws they govern reproduces the relation in which the latter laws stand to the non-nomic facts that they govern.

In chapter 8, I examine the laws' relation to scientific explanations. I argue neither that every scientific explanation must appeal to a law nor that every law is explanatorily potent. However, I do explain why a fact's lawhood can make a difference to its explanatory power and how the fact that *h* is a law can explain why *h* obtains. I devote a good deal of attention to understanding the sense in which an explanation in a macro science (such as evolutionary biology or island biogeography) contributes something that cannot be supplied, even in principle, by a complete explanation at the most fundamental microphysical level. I argue that the autonomy of macro sciences results ultimately from the fact that the macro laws' range of invariance (i.e., the range of counterfactual suppositions under which the laws of that macro discipline would still have held, in virtue of the fact that they are laws of that field) includes some counterfactual suppositions under which the fundamental microphysical laws would *not* still have held.

In discussing these topics, I presuppose no familiarity with the philosophical issues surrounding the concept of a natural law. I begin from scratch—first (in chapter 1) motivating the intuitive distinctions between laws and accidents in terms of counterfactuals, confirmations, and explanations, then posing various difficulties for these intuitions, and finally, refining the initial proposals into more adequate accounts of the phenomena. However, I do not purport to present a neutral survey of the literature on natural law or to offer a comprehensive introduction to the various reductive accounts of natural law that have been proposed. Although I believe that I have made this book accessible to a motivated undergraduate audience, the issues are, of course, very difficult ones. Those discussions that are particularly technical, or that treat side issues, I have relegated to appendices or notes. Some discussions in the main text (such as parts of chapter 2, section 1) are inevitably denser than others. But even there, I believe that a reader could read for the main ideas, and then go on to read the following sections without loss of comprehension, returning later to examine the details of the argument.

My purpose throughout is to reveal the functions that the concept of a natural law performs in an account of scientific reasoning, and the constraints that scientific practice imposes on a philosophical account of what a natural law is.

Acknowledgments

My greatest philosophical debts are to my teachers Bas van Fraassen and Robert Brandom. Without them, nothing like this book would have been written. I am especially grateful to them, as well as to Philip Kitcher, for a good deal of practical advice, encouragement, and moral support through the years.

Larry Bonjour very kindly read the penultimate draft of this book. On two separate occasions, John Carroll offered very valuable comments on earlier drafts. Henry Hodes and anonymous referees for Oxford University Press, *Philosophy of Science, The Philosophical Review, Nous, Analysis, The Journal of Philosophy*, and *Philosophy and Phenomenological Research* at various times gave me many useful suggestions that later found their way into the manuscript. At earlier stages in my thinking, Jim Woodward, Alex Rosenberg, Carl Hoefer, Alan Hajek, and many other friends and colleagues allowed me to draw on their expertise. In graduate school, my conversations with Ken Gemes greatly influenced the initial course of this project. To all, my thanks.

The University of Washington Philosophy Department (Ken Clatterbaugh, chair) provided me with a very congenial environment in which to finish this work, for which I am very grateful.

I thank my children, Rebecca and Abe, for the wonderful ways they have distracted me during the writing of this book. Most of all, I thank Dina Eisinger, my partner in all things, who along with stirring my heart has lovingly accommodated my work. Her thoughts on matters philosophical and exigetical have greatly improved this book. I dedicate it, with love, to her.

Some passages from various papers of mine have managed to make it into this book. I thank the publishers of these papers for permission to reprint those passages here: "Natural Laws and the Problem of Provisos," *Erkenntnis* 38 (March 1993): 233–48, with kind permission from Kluwer Academic Publishers; "Are There Natural Laws Concerning Particular Biological Species?" *The Journal of Philosophy* 92 (August 1995): 430–51; "Inductive Confirmation, Counterfactual Conditionals, and Laws of Nature," *Philosophical Studies* 85 (January 1997): 1–36, with kind permission from Kluwer Academic Publishers; "Laws, Counterfactuals, Stability, and Degrees of Lawhood," *Philosophy of Science* 66 (June 1999): 243–67; and "Why Are the Laws of Nature So Important to Science?" *Philosophy and Phenomenological Research* 59 (September 1999): 625–52.

Contents

List of Symbols

$\blacksquare h$	it is a law that h
$\Box h$	it is physically necessary that h
$\Rightarrow$	logical entailment
$\Leftrightarrow$	logical biconditional
$>$	subjunctive conditional ("$p > q$" means "were p the case, then q would be the case")
$\supset$	material conditional
&	and
$\vee$	or
$\neg$	not
$\in$	is an element of
$\notin$	is not an element of
$\cup$	union
$\subset$	is a proper subset of
$\subseteq$	is a subset of
$\not\subset$	is not a proper subset of
Con(Γ,p)	p is logically consistent with every member of set Γ
U	the set of non-nomic claims (i.e., claims purporting to describe non-nomic facts)

Λ	the logical closure in U of the facts $s \in$ U where ■s holds (i.e., the logical closure in U of the laws governing non-nomic facts)
U^+	the set of claims that can be formed from ■, □, and the members of U, as long as any expression in the scope of ■ or □ belongs to U
Λ^+	the logical closure in U^+ of the facts of the form ■s and ¬■t where $s,t \in$ U
U*	the set of subjunctive conditionals having antecedents and consequents in U
C_P	"were there an *F* possessing property *P*, then all *FP*s would be *G*"
Θ	the set of exactly the hypotheses $h_i \in$ U that are currently salient on the inductive strategies that we are carrying out
Ψ	the logical closure in U^+ of the meta-laws (i.e., of the laws governing laws governing non-nomic facts)

NATURAL LAWS IN SCIENTIFIC PRACTICE

• ONE

Introduction

1. The strategy of this book

We are accustomed to thinking of the universe as governed by various "laws of nature" and of science as aiming to discover these laws. What is it that science thereby aims to discover? What is it for there to be a certain "law of nature"?

I approach these questions by examining scientific practice—in particular, the special roles played in science by beliefs about the laws (I introduce these roles in section 2 below). What difference does it make to scientific reasoning that a certain fact is believed to be a natural law? To account for these differences, what must it *be* for that fact to be a law?

1.1. My strategy compared to others

Philosophical work on the concept of a natural law has recently taken one of two approaches.

Some philosophers, such as David Armstrong (1978, 1983) and David Lewis (1973, 1986b, 1994), pursue *reductive* projects in an attempt to ascertain what it is for something to be a natural law—for example, what makes it the case that "All gold objects are electrically conductive" expresses a law whereas "All gold objects are smaller than one cubic mile," albeit true (let us assume), does not.[1] In order to succeed, these accounts must first identify a claim c such that $c \Leftrightarrow \blacksquare p$ is a conceptual truth (where "$\blacksquare p$" means "It is a law that p"). They must then argue that the fact expressed by c is ontologically more basic than p's lawhood. Without this second step, they have failed to show that the above biconditional achieves a genuine reduction of p's lawhood—that the fact expressed by c is *responsible for making* it the case that $\blacksquare p$. Reductive analyses are motivated, often in large part, by the aim of fitting the natural laws into some general metaphysical picture. Such a picture is needed to support the contention that certain sorts of facts are ontologically prior to facts about the laws. For Armstrong, these purportedly more basic facts concern universals; for Lewis, they concern the distribution among spacetime points of various natural, intrinsic, local properties.

Other philosophers, such as Bas van Fraassen (1989), are skeptics about natural law. They argue that the concept of a natural law should play no role in our account of scientific practice. They offer reasons why various proposed reductive analyses do not succeed even on their own terms. More fundamentally, they also argue that the features of scientific practice that the concept of a natural law might be thought useful for explicating (which I introduce in section 2 below) either are not genuine features of scientific practice or are better explicated without invoking this concept.

This book takes neither of these approaches. It addresses questions that must be answered prior to developing a reductive account, arriving at answers that suggest many proposed reductive analyses to be mistaken. And it argues against the skeptical approach.

My project has two components. Its first aim is to explicate various respects in which it makes a difference to science that some fact is a matter of natural law. Our beliefs about the laws make a difference to the ways in which we regard evidence as inductively bearing upon certain facts. Our beliefs about the laws also make a difference to our beliefs regarding counterfactual conditionals and scientific explanations. Many philosophers have held that such differences exist. But it has proved notoriously difficult to capture these intuitive differences precisely and plausibly. When proposed explications of these differences have been offered, cases have then been found in which claims believed to express nonlaws appear to perform the very functions that, according to these explications, are reserved for the claims believed to express laws. Skeptics like van Fraassen use these cases to argue that there is no work for the concept of a natural law to perform in an account of scientific reasoning. (I briefly review some of these difficulties in the next section.)

During the course of this book, I elaborate various intuitions suggesting that our beliefs about the laws make a difference to our beliefs about certain other matters. Once I have refined these intuitions, I reconcile them with various apparent counterexamples and their intuitive motivations. In this way, I argue that scientific practice includes phenomena that the concept of a natural law should be used to save. This is the best way I know to argue against skepticism regarding natural law.

1.2. How metaphysics can mislead

The first aim of my project, then, is to identify the respects in which the laws' scientific roles differ from the roles played by facts that are not laws. I must then explain *why* the laws bear these special relations to counterfactuals, explanations, and inductions. This is my project's second (and final) component.

Armstrong, Lewis, and others who seek to give a reductive analysis of natural law also take up this task. They defend their analyses $c \Leftrightarrow \blacksquare p$ by arguing that scientists can use p to perform the special functions reserved for laws exactly when scientists believe c. This approach, whatever particular analysis it offers, tends to result in accounts having two deficiencies. Both of these deficiencies can be traced to the fact that a large part of the motivation for any reductive analysis of natural law must ultimately be some broader metaphysical picture into which natural law is being made to fit.

The first deficiency is that these accounts, on my view, tend to trade away fidelity to the laws' scientific roles in exchange for coherence with some attractive broad metaphysical view. In other words, these accounts are too willing to bite the bullet: If laws appear to be used a certain way in scientific reasoning, but this feature cannot easily be accommodated by the metaphysical picture, then understandably, the apparent feature tends to be discounted, on the strength of the motivations for the metaphysical picture.

For instance, I shall argue that it is difficult to accommodate the laws' scientific roles if we insist on regarding each law in a given possible world as associated straightforwardly with a regularity there. Difficulties also arise if we hold that any two possible worlds are alike in their laws as long as they are alike in their "non-nomic" facts—roughly speaking, in the facts that the laws govern (as Newton's laws of motion purport to govern facts about moving particles).[2] Other difficulties arise if we hold that the laws would have been different had certain facts obtained that are "physically possible"—roughly speaking, that do not themselves "break" the actual laws. I believe that Lewis, for instance, underestimates the difficulties that these views encounter partly because he regards these various views as making important contributions to his general campaign on behalf of "Humean supervenience." This is the idea that laws, chances, causal relationships, and so on, are nothing but patterns that supervene on the distribution of natural, intrinsic, localized properties.

For example (as I discuss in chapter 2), Lewis takes the idea that the laws would have been different had certain particular states of affairs p obtained—where p does not itself break the actual laws—to be closely tied to the idea that laws are at bottom nothing but regularities in particular facts.[3] Whether the laws would have been different had p obtained depends for Lewis on whether the laws are different in the possible world where p obtains (the "p-world") that is *most similar* to the actual world. That a given p-world has the same laws as the actual world requires that it be in some degree similar to the actual world. But this similarity can be outweighed—because laws are nothing but regularities in particular facts—by differences from the actual world in other matters of particular fact. In that event, the laws are different in the closest *p*-world. As Lewis says, it seems inevitable that similarity in laws can be traded off against certain similarities in particular fact if a similarity in laws is at bottom nothing but a similarity in particular fact.

I try not to be too much concerned with how well some account I propose coheres with some otherwise attractive metaphysical program. I hope thereby to increase the chance that I am making my philosophical theory fit the facts of scientific practice rather than construing those facts to fit my theory. Accordingly, I refrain from appealing either to a Humean or to a necessitarian picture to support my explications of the laws' scientific roles. Philosophical modesty compels me to admit that I am not sufficiently confident of either picture to use it to warrant setting aside certain counterfactuals, explanations, and inductions that otherwise seem correct.

Do not misunderstand me: I will be appealing to various intuitions about laws. And these intuitions will be metaphysical, at least in the sense that various broad metaphysical programs (such as Lewis's) will be unable to accommodate some of them. But the intuitions I shall invoke (such as the distinction between laws and initial conditions) will be strictly *about laws*; they are not driven by broader metaphysical agendas. They are therefore less likely, I believe, to cut the concept of a natural law to fit a Procrustean bed.

1.3. *How metaphysical appeals can be found unilluminating*

There is a second deficiency to which reductive analyses are prone. They maintain that when $\blacksquare p$ holds, then *c* holds, which enables *p* to play certain special functions in science. This explanation of the laws' special roles tends to rely heavily on the associated metaphysical picture. We need this picture in order to understand the alleged law-making fact *c*—in particular, to render *c* intelligible *independent of the concept of a natural law*. In that way, when we use *c* to derive *p*'s power to perform certain functions, our derivation *avoids presupposing* what it is intended to explain: that *p*'s *lawhood* makes *p* capable of performing these functions.

But someone who is unhappy with the metaphysical picture cannot use it to supply *c* with some content independent of the concept of a law and independent of the roles that scientific practice takes the laws as capable of performing. She might appeal to the concept of a law in order to understand *c*. But then to show that *c* makes *p* capable of playing certain roles, she must note that $c \Leftrightarrow \blacksquare p$ and that $\blacksquare p$ makes *p* capable of playing these roles, thereby appealing to the fact that she was trying to explain. Alternatively, she might appeal to these roles to cash out what it is for *c* to hold. But then the fact that *c* holds ends up coming too close to what it is intended to explain—that *p* can play these roles—turning that explanation into a mere stipulation.

Consequently, it may be difficult for someone who does not accept the metaphysical picture to extract an illuminating explanation of why our beliefs about the laws make a difference to our beliefs about counterfactuals, explanations, and inductions.

For example, suppose that laws are analyzed somehow as "natural necessities." Any analysis (I shall argue) must explain why the laws would still have held had there obtained any "physically possible" state of affairs—for instance (presuming it to be a law that all gold objects are electrically conductive), why it is that had the object in my hand been made of gold, then it would have been electrically conductive. If the explanation is that natural necessities are just that—necessary—and so would still have obtained had other things been different, then at best we have the first step in an explanation. We do not have a complete explanation yet, since we need some account of what it is for a fact to possess "natural necessity." This should tell us in *what ways* other things could have been different without changing the laws. We might try to manage without such an account by proposing that *p*'s "natural necessity" amounts to nothing more than that *p* would still have obtained had any physically possible situation obtained. But then *p*'s natural necessity becomes what it was intended to explain. (Moreover, this analysis of "natural necessity" does not work if "physical possibility" is itself understood in terms of natural necessity.) Alternatively, if we say that *p*'s "natural necessity" is just *p*'s lawhood, then we are right back where we started; to understand why *p*'s natural necessity ensures that *p* would still have obtained had any physically possible situation obtained, we must appeal to lawhood's relation to counterfactuals, which is what we were trying to explain. So an appeal to natural necessity either is explanatorily unilluminating—as empty as using "dormitive virtue" to explain why opium is soporific—or else requires a substantial account of what natural necessity is.

The trouble with this approach is not merely that such an account is difficult to give. It is that if we are indebted to some metaphysical picture for funding our explanations of scientific practice, then someone who is unwilling to buy into that metaphysical picture will find our explanations unilluminating. For instance, Armstrong uses his account of universals in order to elaborate the notion of a "natural necessity": a natural necessity is a connection of "nomic necessitation" among universals. Armstrong (1983, p. 103) then posits that a connection of nomic necessitation would still have obtained had unactualized physical possibilities obtained, and so a given law would still have been a law had any physical possibility been realized. Armstrong is prepared to regard "nomic necessitation" as a primitive relation, presumably on the strength of the other arguments he offers for his metaphysical picture. But someone who is unprepared to grant Armstrong this unexplained explainer will regard Armstrong's explanation as unilluminating—as having done little more than replace the question, "Why would the laws still have obtained had the object in my hand been made of gold?" with the question, "Why would the relations of nomic necessitation still have obtained had the object in my hand been made of gold?"[4]

Likewise, Armstrong purports to explain why the lawhood of "All gold objects are electrically conductive" allows the electrical conductivity of the object in my hand to be explained by the fact that it is gold. Armstrong contends that when one fact explains another, the relevant universals (e.g., goldness, electrical conductivity) stand in a relation of nomic necessitation. But in accounting for why a given scientific explanation presupposes a certain natural law, Armstrong's proposal prompts the question, "Why does that scientific explanation presuppose a certain relation of nomic necessitation among universals? What is a relation of nomic necessitation that it carries explanatory significance?" Armstrong (1983) says only: "To posit such a necessitation *unifies* the given phenomenon (which is a mark of explanation), and leads to further predictions extending beyond what is observed" (p. 102). But this is very thin; even if explanation is construed as unification of some sort, we must be told why a relation of nomic necessitation is precisely the kind of relation that can bind things together into the relevant sort of unity. We must also be told why certain accidents can sometimes supply the relevant unification and why certain laws cannot, though they constitute relations of nomic necessitation (see section 2.3 below).

In short, the concept of nomic necessitation is doing all of the work in these accounts. Therefore, someone who does not find the motivation for Armstrong's metaphysical picture sufficient to justify using "nomic necessitation" to do the heavy philosophical lifting will regard Armstrong as merely stipulating, in an ad hoc manner, that "nomic necessitation" possesses whatever characteristics (e.g., explanatory import, invariance under the right class of counterfactual suppositions) are needed to match the various special roles that laws play in scientific work.

I intend to leave myself less vulnerable to such a charge. Although I also appeal to a conceptual truth of the form $c \Leftrightarrow \blacksquare p$ in explaining the laws' relations to counterfactuals, explanations, and inductions, I do not take c to be ontologically prior to $\blacksquare p$. I refer to c not as a "reductive analysis" of $\blacksquare p$, but as the "root commitment" we undertake in believing $\blacksquare p$. From c, it follows (I contend) that p can perform all of the special functions that set the laws apart. In particular, I show that in undertaking this root commitment (i.e., in believing that c), we commit ourselves to regarding p (the claim thereby believed to express a law) as having been confirmed in a special manner. I show that this commitment, in turn, requires us to accept each of the counterfactual conditionals in a special range, and that these beliefs are responsible for the special relation between our beliefs about the laws and our beliefs about scientific explanations. In this derivation, I presuppose neither some special role played by the laws nor some potentially question-begging metaphysical picture. (It is not, then, some general Humean presupposition that natural necessities are conceptually impossible or epistemically inaccessible that leads me to avoid any reference to them in my explanations of scientific practice. Indeed, I argue that no reduction of the laws to Humean facts is possible.) My argument for $c \Leftrightarrow \blacksquare p$ is that it best explains the laws' capacity to carry out the special functions that science calls upon them to perform.

1.4. How I shall proceed

My explanation of the laws' scientific roles supplies no reductive analysis of lawhood. But it imposes severe constraints on what a law of nature could be. Roughly, I argue that to account for scientific practice, a reductive analysis

cannot require that every natural law be associated straightforwardly with a regularity;

cannot regard the laws as supervening on the "non-nomic" facts;

must recognize that some of the (contingent) logical consequences of the laws—that is, some of the "physical necessities"—are not laws themselves;

must regard the physical necessities as supervening on the facts regarding which subjunctive conditionals "were *p* the case, then *q* would be the case" are correct, where *p* and *q* are "non-nomic" claims;

must entail that the laws would still have been laws had any unrealized physical possibility been realized;

must recognize that there are laws *of* particular scientific disciplines, that this concept (rather than the concept of a law *simpliciter*) is what principally figures in scientific reasoning, and that the laws of one scientific field may be among the nonlaws of another;

must permit there to be multiple "grades" of physical necessity, that is, multiple grades of necessity "between" logical or conceptual necessity and no necessity at all; and

must make the laws independent of our beliefs about them and yet bound up with certain aspects of our purposes and justificatory practices.

So, although I shall offer no reduction of lawhood to something ontologically more basic, I shall ultimately say a good deal about what lawhood must be for it to function as it does in science.

When $c \Leftrightarrow \blacksquare p$ is proposed as a reductive analysis, *c* must be considered ontologically more basic than $\blacksquare p$. I operate under no such restriction. I am free, for instance, to include beliefs about counterfactual conditionals in the root commitment without having to argue that a counterfactual conditional's correctness is ontologically prior to the lawhood of various facts. Accordingly, I shall not hold that the correctness of certain counterfactual conditionals makes certain facts into laws rather than that the laws make certain counterfactual conditionals correct.

Counterfactual conditionals are notoriously sensitive to the context in which they appear. For example, context influences whether we should say, "Had Babe Ruth been playing professional baseball this year, then he would have hit a great many home runs, even against modern pitching, because he was such an outstanding hitter," or whether we should say, "Had Babe Ruth been playing professional baseball this year, then he would have hit only a dozen or so homers, since after all, the Bambino would now have been about a hundred years old." Such context sensitivity—as well as the obvious fact that counterfactuals concern *unrealized* states of affairs—makes it difficult to say whether counterfactuals purport to state facts, and so have truth-values, or whether they are asserted "correctly" or "incorrectly" in some other sense. Science apparently treats counterfactual conditionals in much the same way as it treats noncounterfactual claims. For instance, as we shall see, empirical evidence confirms counterfactuals in the course of confirming more straightforward descriptions of the world. Apparently, science presupposes counterfactual conditionals to have a kind of objective correctness or incorrectness (in a given context), and science aims to discover which counterfactual conditionals are correct and which are not (in various contexts). In keeping with philosophical modesty, I also assume that counterfactuals have some sort of objective correctness. My arguments do not depend on any particular account of what their "correctness" consists in.

Since I offer no account of what makes counterfactuals correct, I cannot determine whether the laws are laws in virtue of certain counterfactuals holding or whether certain counterfactuals obtain partly in virtue of which facts are laws. Hence, even though I identify the precise relation to counterfactuals that distinguishes the laws *and specify this relation without using the concept of a law,* I cannot present this result as a reduction of the laws to ontologically more basic truths. Still, this result shows that *p*'s lawhood, insofar as it goes beyond *p*'s truth, is a matter of objective fact, as long as counterfactuals possess a kind of objective correctness.

Even without laboring under the restrictions imposed by some general metaphysical picture of which facts are ontologically more basic than which others, we will find challenge enough in identifying a root commitment that explains why beliefs about the laws play their special roles in scientific practice. Ultimately, I explain why it is so important for science to discover whether some fact is a law, and also rule out various proposed reductive analyses of law—because, as I show below, none of them is capable of accounting for the laws' *precise roles* in science.

My purpose, then, is to see what can be learned about what natural laws *are* solely from what they *must be* in order to function as they do in science. It might be objected that by concentrating on the laws' scientific roles, I focus on epiphenomena rather than on what is central to lawhood, and so am misled in various ways. I am prepared to concede that perhaps there are other notions of natural law besides the one used in science. Perhaps these other notions involve natural necessities or Humean regularities. I believe that the scientific concept of natural law is remote from natural necessities and Humean regularities (at least as typically understood). So if a philosopher (in discussing, say, free will and determinism, or the role of the mental or the moral in the natural order) depicts natural laws as essentially involving such things, then she is operating with a concept of natural law that does not appear in science. She should not look to scientific practice to validate the concept of natural law to which she is appealing.

2. Three roles played by laws in science

Many proposed reductive analyses fail to explicate correctly the laws' special roles in scientific reasoning. Often (though not always) they characterize these roles in excessively vague terms—for example, laws "support counterfactuals," possess "explanatory power," and are "confirmed by their instances." The features setting laws apart are too subtle to be captured in this crude way. If the roles allegedly reserved for claims believed to express laws are stated in such general terms, then (as skeptics about law have pointed out) these same roles can be performed even by claims believed to express facts that are not laws.

2.1. Laws versus accidents

It is a commonplace that one of the goals of natural science is to discover the laws of nature. But not everything that science discovers, or even considers an important discovery, is considered to be a law. For example, it was recently discovered that the dinosaurs' extinction was caused by Earth's collision with a large rocky body. But this discovery did not reveal some previously unknown law of nature.

Let's contrast some laws (according to current science, at least) with some facts that are not laws. We are all familiar with Newton's law of gravitation, the various conservation laws, and the gas laws.[5] Each of these is "general" or "universal" in a vague intuitive sense that does not apply to certain facts that are not laws, such as that there is a mountain on Venus at 38 degrees 20 minutes north latitude, 63 degrees 14 minutes west longitude. So to improve our grip on the intuitive difference between a law and a fact that does not rise to the level of a law, we should contrast the above laws with mere facts that are likewise intuitively "general." These have been termed "accidental generalizations," "universal coincidences," or "historical accidents on the cosmic scale."[6] (Contingent facts that are not laws are also sometimes referred to as "initial conditions" or "boundary conditions.")[7] Some of the accidental generalizations discussed in the philosophical literature (and there presumed to be true, at least for the sake of argument) are as follows:

"All solid gold cubes are smaller than one cubic mile."[8]

"All of the coins in my pocket today are made of silver."[9]

"All persons now in this room are less than thirty years old."[10]

"All dodos have a white feather in their tails."[11]

"Everyone who drinks from this bottle wears a necktie."[12]

"All mountains in the United Kingdom are less than five thousand feet in height."[13]

"All moas die before age fifty."[14]

"All of the screws currently in Smith's car are rusty."[15]

Perhaps some of these accidental generalizations are not intuitively "general" in just the same sense as Newton's law of gravitation, the various conservation laws, and the gas laws. Some include indexicals ("currently," "today," "my") or demonstratives ("this"). Some refer to particular locations ("my pocket," "the United Kingdom"), particular times ("today," "now"), or particular objects ("this bottle," "Smith's car"). Many philosophers have suggested that law-statements include neither indexicals nor demonstratives nor ineliminable references to particular times, places, events, objects, persons, nations, and so on. (I examine this view in appendix 2 to this chapter.) But at least "All solid gold cubes are smaller than one cubic mile" expresses an accident possessing the same intuitive generality as the laws I mentioned initially.

I shall now introduce some apparent differences between the scientific roles of laws and accidents. I shall also note some of the difficulties encountered in cashing out these intuitions, which have suggested to some that the concept of a law contributes nothing to our understanding of scientific reasoning. We shall see.

2.2. Laws and counterfactual conditionals

Suppose I carelessly brush against the ceramic handle of a pot on a hot stove. Had the handle instead been made of copper, it would have been thermally conductive, and so I would have burned myself. This counterfactual holds because it is a law that copper is thermally conductive. Likewise, if some 10-kilogram object is acted upon by no forces, then given Newton's second law of motion ($F = ma$), it follows that had the object remained 10 kilograms but been acted upon by a net force of 10 newtons, its acceleration would have been 1 meter per second per second.

In contrast, although all of the coins in my pocket today are made of silver, a penny would still *not* have been made of silver had it been in my pocket today, since the regularity involving the coins in my pocket today is merely accidental. Similarly (to use Popper's example), consider the moa, an extinct bird that lived only in New Zealand. Suppose that all moas died before age 50; although the moa had a hardy constitution, New Zealand happened to be heavily populated by a certain virus, which can kill moas while they are still young. On this scenario, "All moas die before age 50," although true, is an accidental generalization. Had a moa lived in an environment free of the virus, it might have lived beyond age 50.

These examples suggest that a fact's lawhood is reflected in which counterfactuals are correct. Although many philosophers defend some such view, there are many obstacles to making it precise and plausible. These difficulties, if stubborn enough, would suggest that our beliefs about the laws make no difference to our beliefs about counterfactuals—that the best explanation of examples such as those above makes no appeal to any distinction between laws and accidents.

Here is a small sample of these difficulties. Sometimes an accident appears to behave just like a law in connection with certain counterfactuals. For instance, anthropologists believe it to be an accident that any person of entirely Native-American heritage is blood type O or blood type A. Research has suggested that all Native Americans are descended from a very small band that crossed the Siberia-Alaska land bridge, and as it happened, allele B was not represented in that company. Thus, anthropologists believe that if an additional person of entirely Native-American heritage were born today, then that person's blood would be type O or type A. Likewise, if we believe that "All [fifty] of the pears on the tree are now ripe" is an accidental generalization, then typically, we believe that had there been another pear on the tree, it would also have been ripe. It is no coincidence that all of the actual pears ripened at the same time, considering that they all experienced roughly the same environmental conditions. Similarly, the function relating a certain car's maximum speed on a dry, flat road to its gas pedal's distance from the floor is not a law but can apparently support some counterfactuals; it has, Trygve Haavelmo (1944) says (in a celebrated passage in which he draws an analogy to the kinds of relations discovered in econometrics), "invariance with respect to certain hypothetical changes" (p. 29). For example, when we have depressed the gas pedal to one inch from the floor, this function tells us what the car's maximum speed would have been had we instead depressed the gas pedal to one-half inch from the floor. This change in the gas pedal would not have altered the internal mechanism of the car, which is responsible for this function (and is a matter of accidental fact). For that matter, this function tells us what the car's maximum speed would have been had the driver been wearing a differently colored shirt. This just goes to show that *any* accidental truth is invariant with respect to *certain* hypothetical changes.

Thus, the intuitive difference between laws and accidents in their relations to counterfactuals cannot adequately be captured by the sort of characterization often found in the philosophical literature, such as this:

> [L]aws must not only apply to the existing physical world but must also cover physical situations which, though non-existent, are permitted by the laws of nature. . . . Mere accidental regularities, however, do not extend to physically non-existing situations. . . . Thus, only true laws support counterfactuals, while accidental regularities do not. (Weinert 1995, pp. 18–19)

Any account of natural law that so characterizes the distinction between laws and accidents is doomed to fail; the phenomenon that it purports to save does not exist.

It might initially be supposed that even though an accident would still have obtained under *some* range of counterfactual circumstances, that range is *narrower* than the range under which a law would still have held—in other words, that an accident's range of invariance is a proper subset of a law's. This seems to be Haavelmo's suggestion in his discussion of "the degree of permanence of economic laws." Haavelmo (1944, pp. 28–29) contends that the "degree of autonomy" of the accidental relation between a certain car's maximum speed on a dry, flat road and its gas pedal's distance from the floor is smaller than the "degree of autonomy" of the laws of thermodynamics because "the range of hypothetical changes" under which the laws are invariant "contains completely another subclass," namely, the range across which the accident holds constant.

However, this is an incorrect way to understand the difference between laws and accidents in their relation to counterfactuals. There can be a counterfactual supposition under which a given accident would still have obtained but a given law would not. For instance, suppose that, as it happens, all of the wiring now lying on the table is made of copper. Had copper been electrically insulating, then all of the wires on the table would have been useless. Here we have a counterfactual antecedent under which a certain natural law (that all copper objects are electrically conductive) would obviously not still have obtained, whereas a certain accidental generalization (that all of the wires now on the table are made of copper) would still have held.[16] There are many similar examples. Suppose that President Clinton has a policy of putting only dimes in a certain pocket, and suppose that he happens never to violate this policy. So, "All of the items that are ever in Clinton's pocket are dimes" is true, and this generalization has a certain range of invariance: had you handed Clinton a quarter and asked him to put it in this pocket, he would have refused. But its range of invariance is limited: perhaps if you had handed Clinton a check for an extraordinarily large campaign contribution and asked him to put *that* in the pocket, he would have done so; this may fall outside of the generalization's range of invariance. Now the Lorentz force law implies that a charge q moving with speed v perpendicular to the local magnetic field of magnitude B experiences a force of magnitude qvB. Consider what would have happened if we had taken a given quarter and asked Clinton to pocket it, and if one of the electrons in the quarter once long ago had been accelerated to beyond the speed of light. Now such an acceleration presumably falls outside the range of invariance of the Lorentz force law; the law breaks down under this physically impossible circumstance (i.e., where v exceeds c). However, none of this would have made any difference to Clinton; the quarter would still have been worth only 25 cents. Had there been such a quarter and had we asked Clinton to put it in his dime pocket, he would have refused.[17] My point is that the Clinton generalization is preserved (in a certain conversational context) under this counterfactual supposition, whereas the Lorentz force law is not. For that matter, if a body on Jupiter last year had accelerated despite experiencing no net force and we had asked Clinton to pocket a quarter, then he would still have refused but Newton's first "law" of motion would not still have obtained. A contingent nonlaw's range of invariance may extend to some circumstances that fall outside of a given law's range of invariance. So it cannot be that a law's range of invariance is automatically broader than an accident's in that the latter is a proper subset of the former.

Each of the counterfactual suppositions I have just discussed is logically inconsistent with some law. Accordingly, it might be suggested that a given law's range of invariance includes every counterfactual supposition that is logically consistent with the laws, whereas these are not all included in an accident's range of invariance. But even if something like this is true (as I consider more closely in chapter 2), it is not an entirely satisfactory way of distinguishing the laws' especially intimate relation to counterfactuals. The source of my dissatisfaction is *not* that if the laws' special range of invariance is itself delimited by reference to the laws, then whether some truth is a law cannot be fixed simply by which counterfactual conditionals are correct; which counterfactual antecedents are logically consistent with the laws—and so what the laws are—would first somehow have to be fixed. I see no reason why this is a problem. Why should we have expected the laws to be determined simply by the correctness of various counterfactuals? Admittedly, we should have expected this had we thought that lawhood simply *is* invariance under a certain range of counterfactual antecedents. The laws must then not be needed to designate that range, on pain of circularity. But, as discussed above, I do not contend that some fact's invariance under a certain range of counterfactual antecedents *makes* that fact a law. Perhaps it is the other way around: the lawhood of various facts is responsible for the correctness of various counterfactual conditionals.[18]

Instead, here is why I am dissatisfied with elaborating the laws' special relation to counterfactuals in terms of the laws' invariance under every counterfactual antecedent that is consistent with the laws: because this does not explain why the laws' relation to counterfactuals is so *special*. For *any* set of truths (except the set of all truths), even a set that includes some accidents, there is arguably *some* range of counterfactual suppositions under which it is invariant. For the laws, one such range consists of those antecedents that are consistent with the laws. Perhaps another set of truths is invariant under every counterfactual antecedent consistent with "George Washington was the first president of the United States." What is so special about the first sort of invariance that it gives the laws an especially intimate relation to counterfactuals? Unless we already have some ground for regarding the laws as special—and so for privileging the range of counterfactual antecedents consistent with the laws—we cannot regard the laws' invariance under *these* counterfactual antecedents as giving the laws a *special* relation to counterfactuals. It may give the laws a *unique* relation to counterfactuals; perhaps it genuinely sets the laws apart from the accidents. But this fact would have no great significance, since it would require that the relevant range of counterfactual antecedents be gerrymandered precisely to suit the laws. Suppose we had begun instead with a certain range of counterfactual antecedents that are logically consistent with a given accidental generalization, encompassing some that are logically inconsistent with laws. Then we might have found the accident rather than a given law to be invariant under all of *these* counterfactual antecedents.

In chapters 2 and 3, I reconcile all of these points with the original intuition that the laws bear an especially intimate relation to counterfactuals. The relation that every law but no accident bears to counterfactuals reveals a sense in which the laws *collectively* possess *uniquely* a *maximal* range of invariance. This explains why there is a sense of necessity ("physical necessity") associated with the laws but none associated with any accident—as we recognize when we say that a 10-kilogram object acted upon by a net force of 10 newtons *must* accelerate by 1 meter per second per second and that a perpetual motion machine is *impossible*, whereas a moa *could* have lived beyond age 50 (had it not been infected by the virus). In other words, I explain why an accident does *not* possess some grade of necessity inferior to that of the laws whereas the laws possess a grade of necessity inferior to that of the logical truths, even though any accident would still have held under *some* counterfactual suppositions and there are some logically possible circumstances under which a law would *not* still have held. I also explicate the sense in which physical necessity lies "between" logical (or conceptual) necessity and no necessity at all, and what it would mean for there to be *multiple* grades of necessity between these.

2.3. Laws and scientific explanations

A "scientific explanation" is a certain kind of answer to a why-question, such as, "Why did the dinosaurs go extinct?" Intuitively, whether some fact *p* is a law or an accident can affect *p*'s role in scientific explanations. Carl Hempel and Paul Oppenheim (1948) incorporated this intuition into their deductive-nomological (D-N) model of scientific explanation. They maintained that every scientific explanation is a "covering-law" explanation—that is, works by virtue of a law that "covers" the fact being explained. For example, we might explain why a certain powder burns with green flames, rather than with flames of any other color, by noting that the powder is a copper salt and that it is a law of nature that all copper salts, when ignited, burn with green flames.[19] That every explanation requires a covering law would account for why we cannot explain why my wife and I have two children by citing the fact that all of the families on our block have two children—since this last is an accident.

However, I reject the view that every explanation requires a covering law—that "a law can, whereas an accidental generalization cannot, serve as the basis for an explanation" (Hempel 1966, p. 56; cf. Carnap 1966, p. 7). For instance, that all of the coins in my pocket are made of silver explains why it was that I ended up with a silver coin when I selected a coin at random from my pocket. Likewise (recalling Haavelmo's example from the preceding section), a given car's maximum speed on a dry, flat road could be explained by the distance of its gas pedal from the floor and the relation between this distance and its maximum speed. This relation is accidental (since it depends on how the car happens to be constructed). Analyses of natural law that purport to account for an accidental generalization's inability to explain a fact it "covers" are doomed to failure, since the phenomenon for which the analyses strive to account does not exist.

The same goes for analyses of natural law that purportedly account for why the lawhood of "All *F*s are *G*" makes an object's *F*ness relevant to explaining its *G*ness. For instance, it is a law that all pendulums of period T have a length equal to $gT^2/4\pi^2$, and yet (at least typically) a pendulum's period does not help to explain its length. Likewise, if we account for the laws' explanatory power by appealing to some property possessed only by laws, then we have to give an entirely different account of an accidental generalization's explanatory power—which seems implausible.

But although I recognize that the lawhood of "All *F*s are *G*" is neither necessary nor sufficient for an object's *F*ness to be explanatorily relevant to its *G*ness, I nevertheless disagree with those who contend that some fact's lawhood makes no difference at all to its explanatory significance. We would cease to accept certain factors as explanatory if we believed certain regularities to be accidental rather than physically necessary. For instance, suppose that all ignited samples of copper salt burn with green flames not because this is physically necessary, but because as it happens, no sample of a certain rare copper salt, whose flames are not green, is ever ignited. Then the fact that the given powder is a copper salt does not explain why it burns with green flames when it is ignited.

There is another respect in which a fact's lawhood apparently makes a difference to its explanatory role. We have observed many ignited samples of copper salts. Why have they *all* burned with green flames? Because it is a law that all copper salts, when ignited, burn with green flames. Does lawhood play a role in this explanation, or is the explanatory burden carried entirely by the fact that every actual combustion of any copper salt sample (whether it has been observed or not) involves green flames? Perhaps the fact that all ignited copper salt samples *we have examined* burn with green flames could be explained simply by the fact that all ignited copper salt samples *whatsoever* burn with green flames, just as we can explain why every coin that I selected randomly from my pocket turned out to be silver by appealing to the fact that all of the coins in my pocket were silver; no matter which copper-salt samples we examined, or which coins we picked from my pocket, the fact being explained would still have obtained. (Note the counterfactual! This is very suggestive.) But suppose we want to explain why all ignited samples of copper salt (examined *and* unexamined) burn with green flames. We can do so by appealing to the fact that *it is a law* that all ignited copper salts burn with green flames; no matter which *possible* samples of ignited copper salts had been actualized, all ignited samples of copper salt would have burned with green flames. Here, of course, lawhood is carrying the explanatory load; we cannot use the fact that all ignited copper salts burn with green flames to explain itself. Though we can explain why all ignited samples of copper salts burn with green flames by appealing to the fact that this is a law, we cannot explain why all of the coins in my pocket are silver by appealing to the fact that *it is an accident* that all of the coins in my pocket are silver.[20] So lawhood makes a difference to such explanations.

In chapter 8, I argue that lawhood's explanatory significance derives from the laws' relation to counterfactuals. Whether some fact is a law matters to its explanatory power only insofar as it reflects that fact's range of invariance; if an accidental truth has an appropriate range of invariance, then it may be just as capable as a law of serving as an explanatory generalization. Whether a given range of invariance is appropriate for a given scientific explanation depends on the why-question and its context. That every coin we selected randomly from my pocket is silver can be explained by the fact that every coin in my pocket is silver, since then (here comes the counterfactual) no matter which coins we had selected from my pocket, we would have gotten exclusively silver ones. On the other hand, it is not the case that "Every coin in my pocket is silver" would still have been true no matter which coins had been put into my pocket. That every coin in my pocket is silver therefore cannot explain why a given coin in my pocket is silver.

In chapter 8, I also discuss the explanations supplied by macro-level scientific disciplines, such as evolutionary biology, ecology, and economics, and their relation to the explanations provided by micro-level disciplines, such as particle physics. I argue that the macro disciplines are "autonomous"; they have their own laws, which have "analogs" neither among the laws of the micro-level discipline nor even (contrary to Fodor 1981) among the micro discipline's physical necessities involving "wildly disjunctive" properties. I argue that, as a result, macro-level explanations are not in any sense shorthand for micro-level explanations; it is not the case that macro explanations are convenient in practice but dispensable in principle. Rather, in virtue of the difference between macro and micro laws in their range of invariance, macro explanations contribute something that micro explanations cannot. Nevertheless, all of the why-questions that macro explanations answer can also be answered at the micro level.

2.4. Laws and induction

Suppose a scientist is entertaining the hypothesis "All *F*s are *G*," which she believes may express a law. She can test this hypothesis by examining various *F*s to ascertain whether or not they are *G*. By discovering an *F* to be *G* (i.e., by discovering an "instance" of the hypothesis), she may confirm the hypothesis "inductively." That is, roughly speaking, she may justly increase her confidence, regarding each *unexamined* case (i.e., unexamined *F*), that it conforms to the hypothesis (i.e., is *G*). If she assembles enough evidence that inductively confirms the hypothesis, she may become justified in using the hypothesis to make *predictions*.

History provides many instances of scientists inductively confirming hypotheses that they believe may state laws. One famous example is Robert

Boyle's inductive confirmation of the law that bears his name: that the pressure exerted by a "parcel of air" at a given temperature is inversely proportional to its volume. Boyle took a curved glass tube, closed at one end and open at the other, and poured in a quantity of mercury, entrapping some air in the closed end. He continued adding mercury until the levels of mercury on both sides of the curve were the same (see figure 1.1). At this point, the pressure exerted downward on the mercury column in the closed end by the trapped air must equal the pressure exerted downward on the mercury column in the open end by the atmosphere. Boyle measured the length (A) of tube occupied by the trapped gas and also measured (using a barometer) the atmospheric pressure, which he found to equal the downward pressure exerted by a column of mercury 29⅛ inches tall. Boyle then poured in more mercury, measured the diminished length (A) of tube now occupied by the trapped gas, and measured the extent (B) to which the new level of mercury in the open end exceeded the new level of mercury in the closed end. He then added more mercury and made new measurements of A and B. Since the trapped air supports B inches of mercury along with the atmosphere, the pressure (D) of that trapped air equals (B + 29⅛) inches of mercury. Boyle displayed his results in a table so that his readers could easily see the relation that they suggested (see figure 1.2). Boyle (in Birch 1772, p. 159) noted that

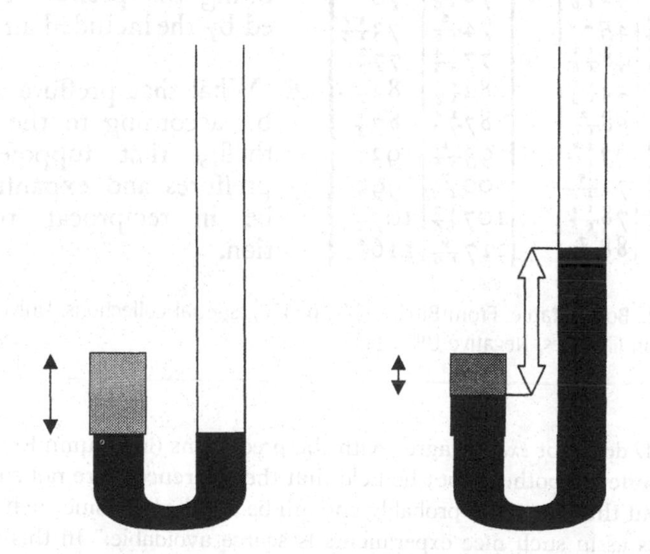

Figure 1.1. Boyle's apparatus. The left drawing displays the air (gray) trapped by mercury (black) in the enclosed end of Boyle's J-shaped tube, where the height of the mercury column in the open and closed ends of the tube is the same. Hence the pressure exerted on the trapped gas equals the atmospheric pressure. The solid arrow represents the length A at this point. The right drawing depicts the situation when more mercury has been added. The pressure exerted on the trapped gas now equals the atmospheric pressure along with the pressure contributed by the mercury in the additional length B of tube indicated by the open arrow.

A table of the condenfation of the air.

A	*A*	*B*	*C*	*D*	*E*
48	12	00	Added to $29\frac{1}{8}$ makes	$29\frac{2}{16}$	$29\frac{2}{16}$
46	$11\frac{1}{2}$	$01\frac{7}{16}$		$30\frac{9}{16}$	$33\frac{6}{16}$
44	11	$02\frac{13}{16}$		$31\frac{15}{16}$	$31\frac{12}{16}$
42	$10\frac{1}{2}$	$04\frac{6}{16}$		$33\frac{8}{16}$	$33\frac{1}{7}$
40	10	$06\frac{3}{16}$		$35\frac{5}{16}$	35- -
38	$9\frac{1}{2}$	$07\frac{14}{16}$		37	$36\frac{15}{19}$
36	9	$10\frac{2}{16}$		$39\frac{5}{16}$	$38\frac{7}{8}$
34	$8\frac{1}{2}$	$12\frac{8}{16}$		$41\frac{10}{16}$	$41\frac{2}{17}$
32	8	$15\frac{1}{16}$		$44\frac{3}{16}$	$43\frac{11}{16}$
30	$7\frac{1}{2}$	$17\frac{15}{16}$		$47\frac{1}{16}$	$46\frac{3}{5}$
28	7	$21\frac{3}{16}$		$50\frac{5}{16}$	50- -
26	$6\frac{1}{2}$	$25\frac{3}{16}$		$54\frac{5}{16}$	$53\frac{10}{13}$
24	6	$29\frac{11}{16}$		$58\frac{13}{16}$	$58\frac{2}{8}$
23	$5\frac{3}{4}$	$32\frac{3}{16}$		$61\frac{5}{16}$	$60\frac{18}{23}$
22	$5\frac{1}{2}$	$34\frac{15}{16}$		$64\frac{1}{16}$	$63\frac{6}{11}$
21	$5\frac{1}{4}$	$37\frac{15}{16}$		$67\frac{1}{16}$	$66\frac{4}{7}$
20	5	$41\frac{9}{16}$		$70\frac{11}{16}$	70- -
19	$4\frac{3}{4}$	45- -		$74\frac{2}{16}$	$73\frac{11}{19}$
18	$4\frac{1}{2}$	$48\frac{12}{16}$		$77\frac{14}{16}$	$77\frac{2}{3}$
17	$4\frac{1}{4}$	$53\frac{11}{16}$		$82\frac{12}{16}$	$82\frac{4}{17}$
16	4	$58\frac{2}{16}$		$87\frac{14}{16}$	$87\frac{3}{8}$
15	$3\frac{3}{4}$	$63\frac{15}{16}$		$93\frac{1}{16}$	$93\frac{1}{5}$
14	$3\frac{1}{2}$	$71\frac{5}{16}$		$100\frac{7}{16}$	$99\frac{6}{7}$
13	$3\frac{1}{4}$	$78\frac{11}{16}$		$107\frac{13}{16}$	$107\frac{7}{13}$
12	3	$88\frac{7}{16}$		$117\frac{9}{16}$	$116\frac{4}{8}$

AA. The number of equal fpaces in the fhorter leg, that contained the fame parcel of air diverfly extended.

B. The height of the mercurial cylinder in the longer leg, that compreffed the air into thofe dimenfions.

C. The height of the mercurial cylinder, that counterbalanced the preffure of the atmofphere.

D. The aggregate of the two laft columns *B* and *C*, exhibiting the preffure fuftained by the included air.

E. What that preffure fhould be according to the hypothefis, that fuppofes the preffures and expanfions to be in reciprocal proportion.

Figure 1.2. Boyle's table. From Birch (1772, p. 158). Special Collections, University of Washington Libraries, Negative UW18411.

column D does not *exactly* agree with the predictions (in column E) made by the suggested hypothesis, but he held that the differences "are not so considerable, but that they may probably enough be ascribed to some such want to exactness as in such nice experiments is scarce avoidable." In this way, he inductively confirmed his "law."

Of course, not every known law of nature "All *F*s are *G*" was originally ascertained exclusively or even partially by the discovery of instances that inductively confirmed it. Peter Achinstein (1971) puts the point well:

> [L]aws typically receive support from varied sources and not simply, and sometimes not at all, from finding [*F*s] which also satisfy [*G*]. This may be because of the difficulty of observing [*F*s] or in determining

> whether [*G*] is satisfied—owing to the fundamental, abstract, and quantitative character of the law. It may also be because of the difficulty of observing a large number and variety of instances—owing to the law's generality. Therefore, support often comes not just from instances. It may also come from theories. Maxwell derived his distribution law for molecular velocities from theoretical assumptions of kinetic theory. Support may come from analogies with other known facts. Priestly suggested the plausibility of an inverse-square law for electric charges by analogy with Newton's inverse-square law for gravitation. (p. 48)

On the other hand, many hypotheses needed only a single instance in order to receive sufficient inductive confirmation to warrant their acceptance. A chemist may need to measure the boiling point (at a given pressure) of just one sample of a new compound in order to justify believing that all samples of this compound (at the given pressure) boil at that temperature. Nancy Cartwright (1989) has discussed "one-shot experiments":

> It does not take a lifetime of associations to convince a reasonable person of electromagnetic induction; Oersted's single experiment was quite sufficient. . . . The bulk of experiments that support the gigantic edifice of 20th-Century physics are never repeated, and they involve no statistics. The trick of the outstanding experimenter is to set the arrangement just right so that the observed outcome means just what it is intended to mean. . . . [O]nce the genuine effect is achieved, that is enough. The physicist need not go on running the experiment again and again to lay bare a regularity before our eyes. A single case, if it is the right case, will do. (p. 92)

Even when a law was not originally discovered by being inductively confirmed by instances, we now regard its instances as inductive evidence for it. In a college physics lab, I had to conduct my own direct test of the Maxwell distribution law; clearly, I was expected to regard my observations as crude versions of the sort of observations that are capable of bearing inductively on the law.

Intuitively, it seems that by discovering an *F* to be *G*, we confirm that all *F*s are *G differently* depending on whether we believe that it may be a law. In particular, we can regard the observed instance as confirming the hypothesis "inductively"—roughly, as confirming, of any unexamined case, that it accords with the hypothesis—only if we believe that the hypothesis may state a law. For example, that all eight coins I have so far received in change today are pennies does not increase my confidence that the next coin I receive in change today will be a penny, for I believe that it would be nothing but a coincidence if all of the coins I receive in change today turned out to be pennies. Since I believe it to be merely coincidental that all of the coins I have already received in change today have been pennies, I do not regard this as any evidence that the next coin I receive will also be a penny. To justify believing that all of the coins I receive in change today are pennies, I must wait until the day is over and review every single coin I have received. Likewise, to justify believing the hypothesis that all presidents of the United States elected in a year ending with "0" between 1821 and 1979 died in office, I must check the record of each president so elected. That William Henry Harrison (elected in 1840) died in office does not confirm that James Garfield (elected in 1880) died in office. Intuitively, that is because I believe that if the hypothesis "All presidents . . . " is true, it is just an accident.

Apparently, our belief that a given hypothesis may state a law makes a difference to the way we regard instances as able to confirm it. Yet this intuition is very difficult to cash out precisely and plausibly. We sometimes regard instances of "All *F*s are *G*" as confirming unexamined *F*s to be *G* even if we believe that it is not physically necessary that all *F*s are *G*. For instance, as I mentioned in section 2.2, anthropologists contend that, as a matter of accidental fact, every person of entirely Native-American heritage has blood of type O or type A. Obviously, anthropologists reached this conclusion without having to examine every Native American who ever lives. It sufficed for anthropologists to check the blood types of a large representative sample of Native Americans. Likewise (following Salmon 1989, p. 49), we might confirm that unexamined cases accord with "All of the pears now on this tree are ripe" by tasting a few pears from the tree. We did not have to taste all of the pears on the tree in order to be justified in believing this generalization, even though we believe it to be accidental.

A reductive analysis of law had better not be defended on the grounds that it can explain why it is only those hypotheses that we believe may state laws that we are capable of confirming by instances to hold in unexamined cases—since this is not so. Nevertheless, I disagree with van Fraassen's (1989, pp. 134–38, 163–69) contention that no important kind of confirmation is available to a hypothesis only if we believe that it may state a law. In chapter 4, I elaborate carefully the intuitive motivations for believing that candidate laws alone can be confirmed "inductively." Then I argue that these intuitions cannot be captured by characterizing "inductive" confirmation merely as confirmation that unexamined cases agree with the hypothesis. I explain why the accidental generalizations in the examples I have just given (concerning Native Americans and pears) are intuitively not being confirmed "inductively" there. Having properly explicated the notion of "inductive" confirmation, I demonstrate that we cannot confirm a hypothesis "inductively" if we believe that it is an accidental generalization if it is true. In chapter 5, I explain why "inductive" confirmation is so important in science.

On my view, "inductive" confirmation is distinguished most fundamentally by *the reason why* the evidence bears on unexamined cases. The same reason in virtue of which the evidence bears upon *actual* unexamined cases also makes it bear upon various *counterfactual* (and perforce unexamined) cases. So in confirming *h* "inductively," we confirm various counterfactual conditionals—each counterfactual that intuitively would be "supported" by *h*. The same sort of account applies to accidents, explaining why they are taken to support certain counterfactuals. For example, by picking a pear from the tree and finding it to be ripe, we do not *inductively* confirm "All of the pears on this tree are now ripe," ultimately because we believe that this generalization lacks physical necessity. But perhaps for each *actual* unexamined pear on the tree, we confirm that it is ripe. The examined pear's ripeness might bear on an actual unexamined pear's ripeness because we believe that both pears have experienced roughly the same ripeness-inducing environmental conditions. This reason also applies to certain counterfactual pears. The examined pear's ripeness must then confirm the ripeness of those counterfactual pears that we believe would (had they existed) have experienced those same environmental conditions. Hence, an observed pear's ripeness confirms that had there been another pear on the tree, it would also have been ripe. My beliefs regarding the range of counterfactual circumstances under which "All of the pears on the tree are ripe" would still have been true are determined by the range of counterfactual pears whose ripeness I take my observations of actual pears to have confirmed.

Observations confirm counterfactual conditionals in just the way that they confirm predictions regarding actual cases that have yet to be examined. Our reason for taking some observation as bearing upon some actual unexamined case may also justify our taking it as bearing upon various counterfactual cases; we do not need some further, new kind of reason to justify this. Accordingly, wholesale skepticism regarding the possibility of knowing what would have happened under *unrealized* circumstances is incompatible with willingness to make predictions concerning *actual* unexamined events.

These three distinctive roles played by laws in science—in connection with counterfactuals, explanations, and inductions—are intended to reflect certain intuitive differences between laws and accidents. Any proposed reductive analysis of law (or proposal regarding the "root commitment" associated with belief in a law) must account for these roles and explain what they have to do with one another.

3. Coming attractions

According to certain philosophers, we describe no fact by calling "All *F*s are *G*" a law—beyond what we describe by deeming "All *F*s are *G*" true: a certain uniformity of nature. By holding "All *F*s are *G*" to be a law, we merely assert its truth and express our approval for its use in certain special roles. R. B. Braithwaite (1953), for instance, endorses this view:

> In common with most of the scientists who have written on philosophy of science from Ernst Mach and Karl Pearson to Harold Jeffreys, I agree with the principal part of Hume's thesis—the part asserting that universals of law are objectively just universals of fact, and that in nature there is no extra element of necessary connection. . . . [T]he difference between universals of law and universals of fact [lies] in the different roles they play in our thinking rather than in their objective content. (pp. 294–95; see also p. 304)

An account that fails to portray a claim's lawhood, over and above its truth, as a matter of fact may explain why the various special roles played by claims believed to express laws are bound up with one another; it may reveal why our approving of a claim's use in one of these special roles requires our approving of its use in each of the others. But such an account cannot show why we *know* more by directing all of these attitudes toward one truth than by directing them toward another, since no *knowledge* is involved in believing a claim to be a law, insofar as this goes beyond believing it to be true.

I do not endorse a view like Braithwaite's—as long as there is an objective sense of correctness for counterfactual conditionals, which philosophical modesty compels me to presume. The root commitment associated with belief in a law is belief in a certain fact that goes beyond any corresponding Humean regularity. (Indeed, I argue in chapter 6 that, for many laws, there *is* no corresponding Humean regularity.) This fact is objective in the sense that it does not depend on our beliefs. We could be mistaken about it. Yet this fact is irreducibly *normative*—a fact about how (in view of our concerns and justificatory practices) we *ought* to reason. In particular, the root commitment is to a belief regarding the rules of inference that result from the "inductive strategies" we ought to use. Let me roughly sketch this root commitment.

Consider Hooke's law, $F = -kx$, that a spring under tension or compression exerts a restoring force F proportional to its displacement x from its equilibrium length.[21] On my account, Hooke's law is associated with a rule of inference; the root commitment associated with belief in Hooke's law requires belief that in certain circumstances, an inference from, say, k and x to F, made in accordance with Hooke's law, is an acceptable step to take in the course of making certain kinds of empirical predictions. *Part* of what makes this step acceptable is that it belongs to a procedure yielding conclusions that are close enough to the truth for the relevant purposes. Why merely "close enough to the truth"? Even from true premises, Hooke's law does not always yield true conclusions. Of course, $F = -kx$ is not true in cases where the spring is compressed or stretched beyond its "elastic limit," or where the spring is heated, and so on—but Hooke's law should not be used in making predictions regarding such circumstances. More to the point is that even within the range in which it is appropriately used, $F = -kx$ is not true. It treats the restoring force as varying linearly with the body's distance from its equilibrium position, but for any actual spring, an equation specifying the exact force has smaller, nonlinear terms—in which x is raised to powers greater than one. So whether it is acceptable for scientists to infer by way of $F = -kx$ depends on the degree of approximation that their project can tolerate. It also depends on how they arrived at their premises and how they intend to use the conclusion that they would reach by way of $F = -kx$. Perhaps such a conclusion should not be plugged into certain other laws, themselves only approximately true, because this would yield conclusions too inaccurate for the purposes at hand. Hooke's law may specify an acceptable step to take in solving a given sort of problem, but only in the course of a particular sequence of steps.[22] (I discuss below whether a more "fundamental" law would exhibit similar features.)

Hooke's law, then, is associated with the inference rule that *in certain circumstances*, we are entitled to infer in accordance with $F = -kx$. We are merely *entitled*, not *committed*, to reason in accordance with $F = -kx$ in these circumstances, since the inferential path that includes Hooke's law is only one of the routes by which we could reach satisfactory conclusions. In coming to understand Hooke's law, a student learns the circumstances in which it is applicable. (In a statement of Hooke's law, they are often left implicit or expressed merely as the absence of "disturbing factors.") Because the circumstances in which $F = -kx$ is applicable have a partly pragmatic character, Hooke's law (I argue in chapter 6) is associated with no regularity—at least, none of the sort that is typically thought to be associated with every law.

Frequently, when scientists discover that for certain purposes, a given calculational procedure is successful in certain conditions, they have as yet no explanation of its success. To explain the extent of the law's accuracy, scientists may adopt a theory that introduces an "ideal" case conforming exactly to the law. For example, a statement of Boyle's law is accompanied by an implicit proviso, which limits the law's proper application to certain purposes and to gases at low temperatures and high pressures. In the course of explaining the extent of the law's accuracy, scientists defined an "ideal gas" as a gas of particles having zero radii and exerting no noncontact forces on one another. Note that the proviso to Boyle's law does not require that these ideal conditions obtain; the concept of an ideal gas was introduced to *explain* the extent of the law's accuracy, not to enable the law to be stated in the first place. Rather, the proviso's truth depends on the gas's low pressure and high temperature, as well as the pragmatic context. Even when the proviso is satisfied, the law's predictions are not exactly true.

Certain law-statements come with the implicit proviso "as long as there is no evidence to the contrary," again making it difficult to identify the law with a regularity. Such is the case with biological laws about particular species—for example, "The human being has 46 chromosomes"—as I discuss in chapter 8. While it is obviously not the case that all human beings have 46 chromosomes, the policy of presuming any given human being to have 46 chromosomes, in the absence of any reason to expect otherwise, is a good rule to use for certain purposes. In human medicine, it "supports" counterfactuals and its reliability is confirmed "inductively" in the manner of a law. That a claim of this sort expresses a rule of default reasoning accounts for the fact that "The human being has 46 chromosomes" does not entail "The human being with Down syndrome has 46 chromosomes." Not only in biology and other "special sciences," but also in the physical sciences (I suggest in chapter 6), there are laws that reflect the expectations we should have in the absence of information to the contrary.

A law, then, is associated with an inference rule that is "reliable"—sufficiently accurate for the relevant purposes—and its reliability may be the only regularity with which the law is associated. But the accidental generalizations, being truths, are no less "reliable." What sets the physical necessities apart from the accidents is that the former are reliable not only to what *actually* happens, but also to what *would have* happened under various unrealized circumstances. As I showed in section 2.2 above, it is crucial here to identify the precise range of unrealized circumstances under which *h* would still have held if and only if *h* is physically necessary. I devote chapters 2 and 3 to resolving this issue, and argue in chapter 5 that this range is ultimately responsible for the scientific importance of discovering the laws. I set aside these key questions until then.

What "reliability" requires of an inference rule depends on how much accuracy, to what sorts of facts, the relevant purposes demand. This point arises not only in connection with provisos, but also in connection with counterfactuals. The laws would have been no different had my alarm clock failed to awaken me this morning, or had I never been born, or had human beings never evolved because no rocky body crashed into Earth to wipe out the dinosaurs, or had the Sun never formed. Now when we are interested in what would have happened had some counterfactual circumstance been realized, we are often unconcerned with how that counterfactual circumstance itself would have managed to come to pass. For instance, suppose a politician is arguing (as some do, at the time of this writing) that Operation Desert Storm should have been continued until it resulted in the death or capture of Saddam Hussein. The politician might assert that had Julius Caesar rather than George Bush commanded Desert Storm, then (having learned from his campaigns in Gaul that there is no substitute for total victory) he would have persevered until he had captured or killed Saddam Hussein. In accepting this counterfactual, the politician does not consider—nor should he consider—how Caesar could have managed to command Desert Storm and still have led a Roman army in Gaul 2,000 years earlier. This is utterly irrelevant to the point he is making: that a lesson of Caesar's campaigns, had it been taken to heart, would have prompted a different strategy in the Gulf War. In the "possible world" called to mind in this context by the counterfactual supposition "Had Caesar commanded Desert Storm," the actual laws must be violated, since in that world, Caesar still participates in the Gallic Wars. Nevertheless, just as the laws would have been no different had I never been born, or had the dinosaurs never gone extinct, or had the Sun never formed, it seems intuitively that the laws would have been no different had Julius Caesar commanded Operation Desert Storm. Caesar would still have had to deal with the impediments to desert warfare that the laws present, and we are guided by our beliefs about the actual laws in predicting what would have happened had Caesar (armed with his Gallic experience) been in command.

This same issue arises in connection with many ordinary counterfactuals, not just in ostentatious cases like this. The counterfactual supposition holds in the given possible world without the causal antecedents apparently demanded by the actual laws, and yet those laws intuitively remain laws there. The conclusion I draw (in chapter 2) is that the laws governing a given possible world may be violated in that world and, let me emphasize, nevertheless remain the laws there, as long as those violations are of no concern to us. A law of a given possible world is associated with an inference rule that is "reliable" there—sufficiently accurate, for our purposes, concerning the facts of that world. What sort of accuracy suffices depends on what facts about that world we are interested in knowing. When we contemplate the "closest possible world" in which Caesar commands Desert Storm and our purpose is to consider how he would have prosecuted the Gulf War, we are unconcerned with how Caesar could have come to be conducting Desert Storm in the first place. The laws in this possible world (which, again, I argue, are the same as the actual laws) are accurate to the events with which we are concerned—but "offstage," in bringing about the counterfactual supposition, violations of the actual laws ("miracles") may occur. Violations of the laws also exist "offstage" in the actual world, as I explained above by reference to provisos. It might have been objected then that there are exceptions to laws of specialized sciences, such as biology, but not to laws of the physical sciences; *they* must be true. Had I responded with Hooke's and Boyle's laws, it might have been admitted that certain "derivative" or "approximate" laws of physics may be merely reliable, but the fundamental laws must be true. I might have responded with a law governing one of the fundamental kinds of forces, which could still hold even if it is false as a description of component forces because there are no *component* forces at all, only net forces. (The absence of any component electrostatic forces to conform to a regularity associated with the electrostatic-force law would not diminish that law's nomic status; after all, there are no component Coriolis forces, and yet there is a Coriolis-force law, which is used to explain the directions of cyclones and trade winds.) But a better reply, I think, is to recall that, in the possible world we invoked by the supposition "Had Caesar commanded Operation Desert Storm," the actual laws intuitively remain the laws and yet are violated, since only by virtue of this violation can Caesar command Desert Storm and still have participated in the Gallic Wars. That $\blacksquare h$ fails to entail h, albeit counterintuitive, is needed to account for our intuitions regarding the laws' relation to counterfactuals, a relation in which even the most fundamental laws stand.

Of course, the inference rule expressed by "The human beings has 46 chromosomes" is reliable only because of various historical accidents—accidents of human evolution—such as that there happened never to occur a certain mutation. While human medicine is uninterested in what would have happened had this mutation taken place and some alternative evolutionary history been realized, evolutionary biology deals precisely with phenomena that depend on the occurrence of mutations and the spread of new traits through populations. Consequently, this law of medicine is not a law of evolutionary biology; it is merely an accident of evolution. Reliability for medical purposes does not require that an inference rule have any degree of accuracy regarding what would have happened had the relevant accidents of human evolution been different. In disciplines such as cardiology, neurobiology, and developmental biology, various claims function as laws (e.g., in connection with counterfactuals), though none is a law of physics or evolutionary biology. Likewise, a law in physics is not a law in folk psychology. When an inference rule allowing the prediction of actions from desires, beliefs, and other conditions (ceteris paribus) is confirmed inductively, it is confirmed to be reliable not merely under the actual laws of physics, but even if they had been different in certain ways. From the viewpoint of folk psychology, it just happens to be the case that all of the physical realizations of some intentional state are drawn from a certain limited range (namely, the range consistent with the laws of physics); a necessity of physics is an accident of folk psychology.

Let me now return to Hooke's law and sketch the root commitment we undertake in believing it to be a law: we believe that the corresponding inference rule's reliability is the conclusion arrived at by one of the strategies belonging to the "best set of inductive strategies." To pursue this strategy, we observe the behavior of various springs under no disturbing influences. We see whether these data suggest some rule of a certain kind—say, underwriting inferences to any such spring's x (or F) from premises that include its F (or x). If so, these data are regarded as confirming the salient rule *inductively*. (Boyle, e.g., followed an inductive strategy in reviewing the table reproduced in section 2.4 above.) One reason that this strategy belongs to the *best* set of inductive strategies for us to pursue is that it requires very few observations to suggest a reliable rule (namely, $F = -kx$). Another reason is that on this strategy, data involving springs of any one kind—for example, iron Slinky-type springs—suffice to suggest a rule that turns out to be reliable regarding springs of any other kind—for example, balls of plastic wrap. It is therefore advantageous to pursue this broad strategy rather than one devoted to iron Slinky-type springs, another to balls of plastic wrap, and so on. To regard iron Slinky-type springs that accord with the salient rule as bearing indiscriminately upon springs of every other kind is to use our evidence most effectively.

Nevertheless, along with Hooke's law, there are also laws applying only to particular kinds of springs. Each of these laws includes nonlinear terms and yields more accurate predictions than Hooke's law. One might then wonder why Hooke's "law" qualifies as a law. Why aren't there just these various specialized laws? For that matter, why are there *different* nonlinear laws for springs of different kinds rather than a single complicated law that specifies the nonlinear terms for each kind of spring?

For many purposes, the linear term is enough to yield sufficiently accurate predictions of the restoring force. Though the size of the spring constant k (the coefficient of the linear term) depends on the spring's material and design, the spring constant of *any* spring can be expressed as (F_0/x_0) for small x_0. In contrast, the precise coefficients of the nonlinear terms depend on the material and construction of the spring in such a way that there are no simple expressions for these terms that apply to all springs.[23] Rather, there are different, relatively simple expressions for the nonlinear terms for rubber bands, panes of glass, balls of plastic wrap, iron Slinky-type springs, and so on. For this reason, in searching for rules that very accurately cover springs' restoring forces, it is better to treat the different kinds of springs separately. Observations of balls of plastic wrap are useful in suggesting such a rule covering all balls of plastic wrap, but will mislead us concerning the correct nonlinear terms for iron Slinky-type springs. In contrast, Hooke's law is quickly rendered salient as we accumulate evidence that accords tolerably well with it, even evidence drawn from examining only balls of plastic wrap. It is best, then, to pursue both the broad inductive strategy that results in Hooke's law as well as the narrower inductive strategies that result in various specialized rules.

On this view, a contingent logical consequence of a law-statement may fail to state a law (though it must, of course, be physically necessary) since it may not be the result of an inductive strategy in the best set. As explained above, an inference rule combining two of the specialized rules—"Use nonlinear terms . . . for iron Slinky-type springs, but . . . for balls of plastic wrap"—is not yielded by any single inductive strategy in the set. That some physical necessities are not laws is closely related to the fact that the union of two natural kinds may not itself be a natural kind. Although "All emeralds are green" and "All rubies are red" express laws, "All things that are emeralds or rubies are green if emeralds and red if rubies" expresses a physical necessity that is not a law. This rule fails to be rendered salient by examination of emeralds alone or rubies alone.

Similarly, consider the categories (beliefs, desires, actions, etc.) that figure in the laws of folk psychology. Presumably, they have nonaccidentally coextensive analogs in terms of neuroscience or elementary-particle physics; all of the realizations of a given intentional state are α_1 in hardware A or β_1 in hardware B or γ_1 in hardware C, and so on, where A, B, C, . . . are all of the hardware possible under the laws of physics. But this analog is highly disjunctive. When we pursue an inductive strategy in neuroscience or physics, these disjunctive analogs do not become salient; we are not justified in inductively confirming the reliability of an inference rule in neuroscience or elementary-particle physics that is analogous to an inference rule at which we arrive by pursuing one of the best inductive strategies in folk psychology. That is because instances of the disjunctive rule's reliability, drawn from some of these disjuncts, do not justify regarding the rule's implications for the other disjuncts as expressing what it would take for those cases to behave in the same way as the examined cases. In other words, no proper subset of the rule's disjuncts succeeds in suggesting its other disjuncts, just as the color of emeralds fails to suggest the color of rubies. That α_1 is always followed by α_2 in hardware A, and β_1 is always followed by β_2 in hardware B, does not suggest that γ_1 is always followed by γ_2 in hardware C. The folk-psychological categories, when "translated" into highly disjunctive combinations of physical categories, are gerrymandered, not natural kinds.[24] This arrangement is typical of laws in "special sciences." Whether a category counts as gerrymandered or salient depends on whether it is suggested when we pursue an inductive strategy, and so depends on our justificatory practices. This is a normative matter, not determined by the non-nomic facts (even supplemented by the correctness of various counterfactuals with non-nomic antecedents and consequents).

My concern throughout is to vindicate the traditional intuitive motivations for distinguishing laws from accidents—namely, the different relations they bear to counterfactuals, inductions, and explanations. (I argue that those traditional intuitions *demand* certain departures from what is typically taken for granted in discussions of natural law, such as that the laws in a given possible world correspond to exceptionless regularities there.[25]) The root commitment associated with belief in a law tells us something about what a natural law must be in order for it to function as scientific reasoning presumes.

Appendix 1: The variety of "laws of nature"

Scientists occasionally label some discovery a "law." In order to improve our grasp of the intuitive difference between a law and an accident, I now present a sample of the claims that are explicitly called "laws" in various scientific disciplines.[26] Another motivation for presenting this list is to combat the tendency to focus right away only on what I would call the fundamental laws of physics (see chapter 8). My approach to understanding lawhood is through understanding the special roles performed in scientific practice by claims believed to express laws. It is important to remain open to the possibility that in different scientific disciplines, these special roles are played by various different claims. Accordingly, a law of physiology may be an accident of physics.

Chemistry

the law of definite proportions (any compound always contains the same elements in the same proportions)

Faraday's first law of electrolysis (the mass of an element liberated or deposited on entering into reaction at an electrode during electrolysis is proportional to the quantity of electricity passing through)

the ideal gas laws

various reaction laws, which take forms such as "Chemical substances *W* and *X*, under . . . conditions, yield chemical substances *Y* and *Z*,"

and

"The rate of the . . . reaction varies with temperature according to the equation . . . "

Earth science

the law of superposition (in a sequence of sedimentary rocks, a deeper layer is older than a shallower layer)

the law of intrusion (an igneous intrusion is older than the rock which it intrudes)

Werner's law (the order of formations is never inverted)

the sixth-power law (the weight of the largest particles that can be moved by a stream varies with the sixth power of the stream velocity)

Ecology

Liebig's law of the minimum (the distribution of a species will be controlled by that environmental factor for which the organism has the narrowest range of adaptability or control)

the law of constant final yield (as individual plants in a population grow, the yield per unit area rapidly becomes independent of sowing or planting density)

the 3/2 thinning law (as individual plants in a population grow, the mean weight W to which a plant surviving to maturity grows is related to the density d as $W = kd^{-3/2}$)

Economics

Greshem's law (bad currency replaces good in circulation)

the laws of supply and demand

Epidemiology

Farr's law (an epidemic curve first ascends rapidly, then more slowly to a maximum, with a descent more rapid than the ascent)

Zeune's law (the proportion of cases of blindness is less in the temperate than in the frigid zone and increases in the torrid zone as the equator is approached)

Evolutionary biology

von Baer's laws (characters common to larger groups of species are likely to be more ancestral than characters common only to subgroups of that group; characters common to larger groups of species are likely to appear earlier in embryos than characters common only to subgroups)

Dollo's law (reversion to an ancestral peculiarity virtually never happens)

Wallace's law (every species has come into existence coincident both in time and space with a preexisting closely allied species)

Genetics

Mendel's laws

the Hardy-Weinberg law (if there are exactly two alleles (A and S) at a given locus in the genome of a sexually reproducing species, and if p

and q (= 1 − p) are respectively the fractions of these alleles in one generation, then in the absence of disturbing factors, the likelihoods of the genotypes AA, AS, and SS in the next generation are p^2, $2pq$, and q^2, respectively)

laws governing polyploidy (e.g., all allotetraploids, when crossed with either ancestral species, produce few fertile offspring)

Immunology

Behring's law (the blood and serum of an immunized person, when transferred to another subject, will render the latter immune)

Island biogeography

the area law ($S = cA^z$, where S = the equilibrium number of species on an island and A = the island's area; constants c and z are specific to a given taxonomic group in a given island group—for example, Indonesian land birds)

Meteorology

the Buys-Ballot law (low pressure in the northern [southern] hemisphere is on the left [right] hand when facing downwind)

Mineralogy

Braggs' law of X-ray diffraction (X-rays of wavelength λ entering a crystal at an angle θ to the planes of atoms in the crystal, which are separated by distance d, are reflected only if $n\lambda = 2d \sin\theta$, where n is an integer)

Steno's law of the constancy of interfacial angles (measured at the same temperature, similar angles on crystals of the same substance remain constant regardless of the size or shape of the crystal)

Pathology

Courvoisier's law (when the common bile duct is obstructed by a stone, dilatation of the gall bladder is rare; when the duct is obstructed in some other way, dilatation is common)

Physics

Newton's laws of motion and gravitation

conservation laws for mass-energy, linear and angular momentum, and baryon number

the laws of thermodynamics

the laws of reflection and refraction

the Stefan-Boltzmann law (the radiation emitted by black bodies is such that its intensity per second per unit area of the body is proportional to the fourth power of the body's absolute temperature)

Ohm's law (the strength of an electric current varies directly as the electromotive force, and inversely as the resistance)

Physiology

Bowditch's law (any stimulus that will cause a contraction of the heart muscle will cause as powerful a pulsation as any greater stimulus)

Psychology

Thorndike's law of effect (a behavior that receives a pleasurable response will be performed more often; one that receives a painful response, less often)

Parody

Murphy's law

Of course, many claims customarily not explicitly designated as "laws" still function identically in scientific practice to claims formally titled "laws." In quantum physics, for instance, the laws intuitively include Schrodinger's equation, Pauli's exclusion principle, Heisenberg's uncertainty principle, and the axioms of quantum mechanics. Maxwell's equations subsume Ampere's, Faraday's, and Gauss's laws. There are Archimedes' and Bernouilli's principles, Debye's equation (relating the specific heat of a solid to its temperature), and Arrhenius's equation (relating the rate of a chemical reaction to the temperature). Mendel's laws are often called his "rules." The Lotka-Volterra logistic curve in population ecology functions much like the Hardy-Weinberg law in population genetics and the ideal gas laws in chemistry: each equation follows from various idealizing assumptions that often fail to hold, in that each expresses the effects of certain factors in the absence of certain others, and so each enables us to get a better grip on the factors at work in a given actual case by examining that case's departure from the equation. Furthermore, statements specifying the values of various physical constants—whether "fundamental" constants (e.g., the speed of light, the electron's mass) or "derived" constants (e.g., copper's electrical conductivity, its coefficient of thermal expansion)—are likewise laws. Apparently, there is no principled reason why some scientific discoveries are formally invested with the title "law" whereas others are not; the epithet "law" is honorific, its application and persistence often an accident of history.[27]

There are many distinctions that have sometimes been drawn among the laws. Some laws are termed causal, others as merely involving associations (albeit physically necessary ones). Some laws are deemed to be more theoretical, model-driven, or top-down, others more phenomenological, data-driven, or bottom-up. Some laws are termed fundamental, others derivative. Some laws are deterministic, others statistical. A law may concern a qualitative relation, a functional relation, or the value of a natural constant. It may concern a synchronic relation or a diachronic one. While these distinctions may be useful for many philosophical purposes, I focus on them only insofar as they can help to account for the laws' special roles in connection with counterfactuals, inductions, and explanations.

Appendix 2: Lawhood and nonlocal predicates

Many natural philosophers have defended the requirement that law-statements include neither indexicals nor demonstratives nor ineliminable references to particular times, places, events, objects, persons, nations, and so on. For example, James Clerk Maxwell (1952, p. 13) held that particular spatial or temporal coordinates cannot appear in law-statements because an event's spatiotemporal location cannot be causally efficacious; there is no logical bar to the same causes occurring at any spatiotemporal location, and the laws express the fact that the same causes produce the same effects. This requirement would prohibit proper names from appearing ineliminably in law-statements and would demand that no law require for its expression the use of a "local" predicate (that is, one whose meaning refers to particular times, places, and so on).

This requirement would account for the accidental character of many of the accidental generalizations listed in section 2.1. But it would not account for all of them. The example offered by Hans Reichenbach (1947, p. 368; 1954, pp. 10–11)—"All solid gold cubes are smaller than one cubic mile"—is intended to cover all gold cubes past, present, and future. As has often been remarked, it is as nonlocal as "All solid cubes of uranium-235 are smaller than one cubic mile," yet it lacks the physical necessity that the uranium generalization possesses.[28]

Although it is not the case that all true generalizations that essentially include no local predicates, proper names, indexicals, or demonstratives express laws, it might seem that no law-statement can essentially include local predicates and so on. I devote this appendix to evaluating the plausibility of this popular view. Ultimately, I conclude that no persuasive arguments for it can easily be made. If there is some such requirement on law-statements, we will discover it not by some argument that we are in a position to make now, but only after we have identified the laws' special scientific roles and determined whether these roles could in principle be played by claims that essentially involve proper names and the like. (I think they could.)

The requirement under discussion would account for the intuition that even if all diamonds formed in Australia are yellow, this is not a law. Surely the relevant law does not refer to the Australian provenance of these diamonds; the law might say instead that all diamonds formed in a certain magnetic environment are yellow, where this environment happens to obtain throughout Australia. However, such standard examples of laws as Galileo's law of free fall, which refers to Earth, and Kepler's laws of planetary motion, which refer to the solar system, would not qualify as laws under this requirement. The typical response is to exempt "derived laws" from the requirement (presuming that the sense in which Galileo's and Kepler's laws are "derived" can be explicated). That no proper names or local predicates are included ineliminably in statements of fundamental laws is one way that Reichenbach (1947, p. 264; 1954, pp. 10, 32–33), Hempel (1965, pp. 267–68), Rudolf Carnap (1966, pp. 211–12), and J. J. C. Smart (1968, pp. 59–62) have cashed out the intuition that natural laws are "universal" or "general" (roughly speaking: true anywhere and anywhen, with no unprincipled exceptions).

But it is not evident that all fundamental laws must be capable of being expressed without proper names, local predicates, and the like. To begin with, we might worry (with Smart 1968, p. 62) about whether this requirement can even be stated precisely. Apparently, it should preclude "All emeralds are green before the year 3000 and blue thereafter" from qualifying as a fundamental law, since this claim refers to a particular time. Of course, we could eliminate any *explicit* reference to the year 3000 by defining "*X* is grue" as "*X* is green when the time precedes the year 3000 and *X* is blue otherwise." (Note that "grue" contains the first two letters of "green" and the last two letters of "blue"; in the same playful spirit, we might also define "*X* is bleen" as "*X* is blue before the year 3000 and *X* is green otherwise." Nelson Goodman (1983, pp. 74–81) famously introduced predicates such as these.) What does the requirement dictate about whether "All emeralds are grue" could qualify as a fundamental law? It seems that there are two options, neither of which is attractive. On the one hand, the requirement could be that a statement of fundamental law must make no *explicit* reference to any particular time, place, and so on, in which case "All emeralds are grue" would satisfy this requirement. But then this requirement would impose no real constraint on the laws, since any *explicit* reference to a particular time, place, and so forth, could always be removed by the ruse of defining a predicate like "grue." On the other hand, the requirement could be that a statement of fundamental law must make no reference, *explicit or implicit*, to a time, place, and so on. Then it would seem that "All emeralds are grue" fails to satisfy the requirement, since the definition of "grue" refers explicitly to the year 3000. But claims about the definitions of predicates are notoriously fraught with difficulty. For instance, rather than defining "grue" in terms of "green," "blue," and the year 3000, and so portraying "grue" as a local predicate, why couldn't we define "green" in terms of "grue," "bleen," and the year 3000, thereby portraying "green" as a local predicate? If this maneuver is permitted, then the requirement will preclude "All emeralds are green" (or any other claim) from qualifying as a statement of fundamental law. In order for this requirement to be neither too permissive nor too restrictive, there must be some reason why a language could not use "grue" and "bleen" as the more basic terms, and define "green" and "blue" in terms of "grue" and "bleen." I find it hard to see what reason there could be (but this is a matter of considerable dispute, and I do not intend to discuss it further here).

Another reason to doubt the requirement that proper names and local predicates be eliminable from statements of fundamental laws is that this requirement appears inconsistent with the logical possibility of certain kinds of fundamental laws in biology, geology ("*Earth* science"), medicine, psychology, and other sciences more specialized than physics. Some philosophers have welcomed this consequence. For example, Reichenbach (1954, p. 32) notes that from this requirement, it follows that "on Earth" is inadmissible in statements of fundamental law. For this reason, Smart (1963, pp. 50–58; 1968, pp. 92–93, 106) and Karl Popper (1979, pp. 267–270) deny that there are any fundamental laws concerning particular biological species. For instance, consider "All robins' eggs are greenish-blue," which is one of Hempel's (1965, p. 267) examples of a law. A biological species, on one standard conception, is a group of individuals who are not reproductively isolated from one another but are reproductively isolated from members of other such groups (e.g., by geography or reproductive mode or timing). On this view, different species are evolutionarily uncoupled from one another; each species has its own isolated gene pool. By definition, a creature descended entirely from a lineage that is geographically isolated from robins—say, an extraterrestrial creature—cannot be a robin, even if it is so similar anatomically to a robin that in the absence of geographic barriers, it could breed freely with a robin and produce fertile offspring. That the robin is native to Earth, having evolved here in reproductive isolation from extraterrestrials, is accordingly thought to be part of the meaning of "robin"; this term refers implicitly to Earth. If local predicates are barred from statements of fundamental laws, then "All robins' eggs are greenish-blue" does not state a fundamental law.[29]

An alternative conception of biological species is that a species is picked out by its position in the evolutionary tree of life. That is, by definition an organism is a "robin" in virtue of its place in the genealogy of living things. In that case, extraterrestrial creatures again could not be robins, since their evolutionary origin would be distinct from that of robins. This view of biological species entails that the predicate "is a robin" is local and hence (on the requirement we are examining) inadmissible in statements of fundamental laws. Many of the philosophers who advocate this conception of biological species also maintain that a species is itself an individual object, a chunk of the genealogical nexus. Alexander Rosenberg (1987, pp. 195–97) and David Hull (1987, pp. 174–75), among others, maintain that since law-statements cannot refer to particular things and a species is a particular thing, it is logically impossible for there to be biological laws of the form "All organisms belonging to species *S* have property *T*." On this view, only the most general sciences discover laws of nature:

> [A]ll those special branches of biology, ecology, physiology, anatomy, behavioral biology, embryology, developmental biology, and the study of genetics are not to be expected to produce general laws that manifest the required universality, generality, and exceptionlessness. For each of them is devoted to the study of mechanisms restricted to one or more organism or population or species or higher taxa and therefore is devoted to the study of a particular object and not to a kind of phenomenon to be found elsewhere in the universe. [In physics the situation is different because physics deals with] natural kinds, whose instances can and do occur throughout the galaxy. Particular species do not recur, and they cannot be expected to be much like particular lines of descent either elsewhere on the planet or elsewhere in the universe. (Rosenberg 1985, p. 219)

But it is not obvious that specialized sciences do not aim to discover fundamental laws—fundamental, at least, to those sciences. Whether scientists regard some statement as expressing a law is exhibited not by what they call it (as I showed in appendix 1) but by what they do with it—how they use it in connection with counterfactuals, explanations, and inductions. Whether in these connections specialized sciences operate differently from physics is something that we will be in a position to know only after we have explicated the precise roles played by claims believed to state laws.

Of course, even if various specialized sciences treat certain claims involving local predicates as law-statements, an advocate of the requirement that we have been examining might hold that these claims are not believed to state *fundamental* laws. But why insist on this? If workers in these specialized sciences regard a claim involving local predicates as able to function as a law-statement, then why must they believe that there are law-statements involving no local predicates from which the given claim can be derived? Why should a claim's capacity to play the laws' scientific roles depend on its deriving from truths involving no local predicates?

The nonlocal-predicate requirement is dubious even in the physical sciences. It seems that P. A. M. Dirac (1938) wasn't making a *logical* error when he made his famous conjecture that the gravitational-force "constant" is inversely proportional to the time since the Big Bang. Yet this conjecture would require that a statement of the gravitational-force law include an implicit reference to some particular moment. The law-statement might, for instance, refer to the time elapsed since the Big Bang, which Dirac calls "a natural origin of time."[30] Of course, the Big Bang is a very special event, and so perhaps a principled exception to the nonlocal-predicate requirement could be made for it. On the other hand, presumably any event that was mentioned ineliminably by a statement of fundamental law would thereby qualify as very special.

It may be that no statement of fundamental law in the physical sciences refers ineliminably to any particulars. (Perhaps this is even a matter of natural law—a law governing laws.) But I see no good reason to believe that this fact holds simply in virtue of the *meaning* of "natural law." As Moritz Schlick (1949) wrote:

> Now again I certainly expect that all laws of nature will actually conform to Maxwell's criterion [i.e., "the absence of explicit space and time values" from statements of natural law] . . . none the less it remains theoretically possible that a future physics might have to introduce formulae which contain space and time in an explicit form, so that the same cause would never have the same effect, but the effect would also depend, in a definite way, e.g., on the date, and would be different tomorrow, or next month, or next year. (pp. 528–529)

As many have noted (e.g., Armstrong 1983, p. 26), the nonlocal-predicate requirement would permit an a priori argument to be made against Aristotelian physics, which held that certain law-statements refer to the center of the universe and that others refer to the path of the moon. We must be careful not to misconstrue a characteristic of laws according to the current research program in physics as built into the very concept of a law.

I find no reason why a claim's ineliminably including a local predicate or proper name would block its capacity to perform the laws' special functions in science. For example, generalizations involving local predicates seem entirely capable of being confirmed inductively; indeed, doing so seems to be one of the principal tasks of scientists who work in certain specialized disciplines. As Goodman (1947a, pp. 149–50) has noted, hypotheses with the local predicates "solar," "arctic," and "Sung" have surely been confirmed inductively. Admittedly, if we were told that all diamonds formed in Australia are yellow, we would not regard this regularity as very plausibly explained by the hypothesis "It is a law that all diamonds formed in Australia are yellow." We expect the explanation to involve not the diamonds' Australian provenance per se, but some nonlocal property they possess (such as their having formed in a certain magnetic environment). But I see no reason to believe that this expectation derives from the concept of a law rather than from our past experience.

Nevertheless, there remains considerable plausibility to the intuition that laws are distinguished by their "generality" or "universality" from such facts as the location of a mountain on Venus. Norman Swartz (1985) seems to be onto *something* when he says, "Our conception of physical law doesn't allow that one law should apply to the Northern Hemisphere of the Earth and another to the Southern Hemisphere of Venus; that one law should apply to the eighteenth century and another to the nineteenth" (p. 18). But I do not believe that this idea is built directly into the concept of a law. Roughly speaking, here is what I think is right about Swartz's intuition: from examining the electrical conductivity of a few pieces of copper from Earth's Northern Hemisphere (while pursuing the appropriate "inductive strategy"), we would be justly led to make accurate predictions regarding the electrical conductivity of pieces of copper from anywhere, even Venus's Southern Hemisphere. This inductive strategy makes more effective use of our evidence than does a narrower strategy that uses these observations to confirm predictions regarding the electrical conductivity of copper objects merely from Earth's Northern Hemisphere. Since the laws are associated with the best set of inductive strategies, there is no separate law covering the electrical conductivity of copper objects from Earth's Northern Hemisphere. This result does not depend on some express prohibition on local predicates in law-statements.[31]

Appendix 3: Lewis's reductive analysis as a useful contrast

Many reductive analyses of natural law have been proposed.[32] Necessitarian accounts call upon some irreducibly modal concept to perform the crucial work in explaining why our beliefs about the laws perform their characteristic functions in scientific practice. "Humean" (a.k.a. "regularity") accounts, on the other hand, hold laws to be contingent regularities that (unlike accidents) satisfy some further condition characterizable without using non-truth-functional sentential connectives or irreducibly modal, causal, normative, or nomic concepts.[33] Lewis, for example, suggests that the laws are the generalizations appearing in the deductive system of truths that possesses the best combination of simplicity and informativeness.[34]

Since I am not trying to reduce lawhood to ontologically prior facts, my account does not directly compete with these reductive accounts. (Lewis, e.g., purports to account for which counterfactual conditionals are correct as well as which regularities are laws by appealing only to the actual spacetime distribution of natural, intrinsic, local properties. I am prepared to take for granted the correctness of various counterfactuals in the course of explaining how laws are set apart from accidents.) However, these reductive accounts, like my approach, purport to explain why our beliefs about the laws play their characteristic roles in scientific practice. For this reason, it is occasionally useful to contrast my explanations to those proposed by these accounts—particularly Lewis's.

Lewis's approach nicely explains the lawhood of certain functional generalizations with uninstantiated values. For instance, suppose that in the entire history of the universe, there happen never to be objects of precisely 1.234 grams and 5.678 grams. An alternative to Newton's law of gravity that dictates a different value for the gravitational force between a pair of bodies with precisely those masses, but otherwise agrees with Newton's law, is not a law despite its accuracy to all of the gravitational forces ever present in the universe (supposing Newton's law so to be).[35] Newton's law is (arguably) simpler than any such alternative. Therefore, Lewis's account says that Newton's is the law.

Lewis's account must explain why believing that a given generalization appears in the best deductive system of truths is bound up with believing that it can perform certain functions in connection with explanations, inductions, and counterfactuals—for example, with believing that it would still have been true had some counterfactual circumstance obtained. Lewis says that whether the generalizations in the simplest, strongest deductive system of truths about the actual world are true in a given "possible world" greatly influences how "close" this possible world is to ours, and hence whether it is the "closest" possible world in which some counterfactual antecedent obtains.[36] And a counterfactual conditional "Had *p* obtained, then *q* would have obtained" is true exactly when *q* holds in the possible world in which *p* obtains that is closest to ours. But this response provokes the question, "Why, in measuring a possible world's closeness to our world, should we emphasize whether that world's laws are the same as the actual world's? Why should the fact that a given world includes all of the general truths belonging to the best deductive system in the actual world count more heavily, in determining that possible world's closeness to the actual world, than whether that possible world shares various other facts with the actual world?"[37]

We might reply by denying that a possible world is close *because* it preserves the facts about our world that derive from the axioms of the best deductive system of truths in our world—that is, by denying that those facts are more *important* than others in determining a possible world's closeness to ours. Rather, we might continue, a possible world preserves all of these axioms *because* it is close; a possible world that includes all of the facts about our world that derive from these axioms would necessarily have to be very similar to ours.

But this reply does not get us very far. Of course, if a law in the actual world is false in a given possible world, and this law is a theorem in the best deductive system in the actual world, then at least one of the truths in the actual world from which the law is derived in the system must be false in the given possible world, and in turn at least one of this truth's precursors must be false, and so on, all the way back to the axioms. But this need not lead to great differences between the actual world and the given possible world; the modifications that must be made to the system, in order to accommodate the initial deviation from the actual world, might be relatively minor. Of course, a sense of "similarity" among possible worlds might be employed in which these modifications never qualify as minor, but we would then require a non-question-begging reason for using such a metric.

Admittedly, the "best" deductive system must be strong. But the laws are fixed by which true deductive system is strongest *and simplest*. The reason that a law contributes more than an accidental regularity to the closeness of a possible world cannot be merely that it (so to speak) automatically brings more truths along with it. The reason must also be that the truths that a law brings along with it are more important to closeness, by some metric that has not yet been defended.[38]

John Earman (1984) suggests that the laws may simply fail to coincide with the generalizations in the system of truths having the best combination of simplicity and strength. He wonders whether Reichenbach's accidental generalization, "All gold cubes are smaller than one cubic mile," might constitute a counterexample:

> If this generalization is not to be counted as a law [by Lewis's approach], it is because it is not an axiom or theorem in the best (or each of the best) overall deductive systems for our world. Consider then what would happen if we were to add it as a new axiom. There would, by hypothesis, be a gain in strength. And, presumably, there would also be a loss in simplicity. The loss must, *pace* [Lewis], outweigh the gain. I will not say otherwise. But I do say that it is not compellingly obvious that the scales tip in this way while it is compelling that Reichenbach's generalization is not to be counted as a law. (p. 198)

It is likewise not obvious to me that there cannot be laws—perhaps "All human beings are mortal" (which is deemed a law by Ramsey 1931, p. 237) taken together with the other laws ascribing mortality to various kinds of creatures—having so little connection to the rest of the laws that the loss of simplicity from adding them outweighs the gain in strength from doing so. To pursue Earman's point a bit: By progressively strengthening Reichenbach's generalization—"All gold perfect solids . . . ," "All gold objects . . . ," "All objects consisting purely of one metallic element . . . "—we seem to be moving no closer to lawhood. Yet having added Reichenbach's generalization to the system of laws, we do not obviously decrease the system's simplicity at all by strengthening Reichenbach's generalization in this way.

I do not take any of these considerations to be a decisive argument against Lewis's account. (In section 3 of chapter 3 I present a more serious difficulty for any analysis like Lewis's.) In the upcoming chapters, I often find it useful to compare my approach to Lewis's.

• TWO

The Relation of Laws to Counterfactuals

Which counterfactual conditionals are correct seems to reflect whether a given fact is a law or an accident (as discussed in section 2.2 of chapter 1). After suggesting in section 1 of this chapter that a certain relation obtains between laws and counterfactuals, and that no accident joins the laws in bearing this relation to counterfactuals, in section 2 I work through various potential difficulties for this view. Many of these puzzles also raise issues of independent interest regarding counterfactuals.

Although I offer no account of what makes a counterfactual conditional correct (see chapter 1), I reveal some important constraints on any such account. In particular, I contend in section 3 that science treats a counterfactual's correctness as failing to supervene on the "non-nomic" facts. In other words, I contend that it is logically possible for different counterfactuals to be correct even when the non-nomic facts are no different. In view of the intimate relation between laws and counterfactuals, this nonsupervenience entails that the laws fail to supervene on the non-nomic facts. This result imposes a significant constraint on any reductive account of natural law. It also constrains any proposal regarding the root commitment associated with belief in a law (i.e., regarding the commitment that we undertake exactly when we believe in a law and that is responsible for the special roles played by beliefs about the laws).

Only after I am reasonably sure that I have correctly identified the laws' special relation to counterfactuals do I consider *why* this relation holds. In section 4, I suggest one part of the explanation that I later develop in chapters 4 and 5, along with some problems that afflict other proposed explanations.

For most of this chapter, I presuppose that laws, like accidental generalizations, involve exceptionless regularities—in other words, that $\blacksquare p \Rightarrow p$. Later in this book, I allow laws to be merely "reliable" and amend the results of this chapter accordingly. In section 2.2, I discuss a feature of counterfactual reasoning that supplies a key motivation for the view that the laws of a given possible world can have exceptions there. This view imposes another severe constraint on any proposal regarding the root commitment. I take this up more fully in chapter 6.

In chapter 3, I refine my account of the difference between laws and accidents in their relation to counterfactuals. This refined account explains why there is a kind of *necessity* associated with the laws but none with an accident, even if that accident is invariant under a broad range of counterfactual suppositions. I also show there just *how* p's physical necessity supervenes on the non-nomic facts *together with* the facts about which counterfactual conditionals (having non-nomic antecedents and consequents) are correct. But first things first: How, precisely, are laws set apart from accidents by their relation to counterfactuals?

1. Preservation under counterfactual antecedents

1.1. *Understanding counterfactuals*

A counterfactual conditional, such as "Had I gone shopping, then I would have spent a great deal of money," is a special kind of if-then statement. Its antecedent (as expressed in the indicative mood—"I went shopping") is false. Plainly, a counterfactual is not a material conditional; that its antecedent is false does not suffice to make a counterfactual correct. Rather, a counterfactual is a *subjunctive* conditional with a false antecedent. By a "subjunctive conditional," I mean a conditional such as "Were my car out of gas, then it would not go" (a present-tense example) and "Had Napoleon's hat been red, then its color would have been the same as one of the colors on the French flag" (a past-tense example). Let "$p > q$" represent the subjunctive conditional "Were it the case that p, then it would be the case that q" (where p and q are ordinary indicative sentences). If p is true, then "$p > q$" is correct exactly when q is true. If p is false, then "$p > q$" is a counterfactual.[1]

Counterfactuals are common in ordinary conversation. Ever since Roderick Chisholm (1946) and Nelson Goodman (1947b; 1983, pp. 3–27) forcefully exposed the difficulties involved in analyzing counterfactuals, philosophers have tried to specify the circumstances in which a counterfactual is correct. A counterfactual antecedent seems to demand of us, in Max Weber's (1949) words,

> the production—let us say it calmly—of "imaginative constructs" by the disregarding of one or more of those elements of "reality" which are actually present, and by the mental construction of a course of events

which is altered through modification of one or more conditions. . . . (p. 173)

Two principal means of cashing out this intuition have been proposed. Some philosophers have tried to develop accounts according to which a counterfactual $p > q$ is true (e.g., Chisholm, Goodman) or correct (e.g., J. L. Mackie) if and only if q is logically entailed by p together with a certain truth s [i.e., $(p \& s) \Rightarrow q$, letting "$\Rightarrow$" represent logical entailment]. Other philosophers, for example, Robert Stalnaker and David Lewis, have offered accounts that instead use talk of "possible worlds" to elaborate the intuition toward which Weber gestures. One simple way of doing so is to hold that $p > q$ is true if and only if q is true in the "p-world" (the possible world in which p is true) that is "closest" to the actual world (although obviously distinct from it, since p is actually false). If there is no single closest p-world, the criterion becomes that q is true in each p-world for which there is none closer to the actual world.

These two approaches, while not strictly incompatible, are suggestive in different ways. For example, the former approach lends itself to a "meta-linguistic" interpretation of counterfactuals. This may interpret $p > q$ as saying that the inference from $(p \& s)$ to q is valid. On this interpretation, counterfactuals have truth-values. Or a meta-linguistic view may follow Mackie in interpreting $p > q$ as an elliptical statement of the inference from $(p \& s)$ to q, so the counterfactual has no truth-value but is correct exactly when the inference is valid. The truth s is then said to "support" the counterfactual. In contrast, the possible-worlds approach has suggested to some philosophers that counterfactuals express objective facts about the relations among various "possible worlds" that are as real as the actual world. Though I do not adopt this view of possible worlds, I sometimes find talk of "possible worlds" to be helpful. Those readers who believe that "possible worlds" are real, and those who find it unhelpful to talk even metaphorically about the "closeness of possible worlds," should all be able to express my points in terms they find congenial.

Whatever our interpretation of counterfactuals, the crucial problem is to find a procedure that identifies what Weber calls the "course of events" that a given counterfactual antecedent directs us to consider. On a meta-linguistic view, the problem is to set out a procedure determining the truth s in each case. The procedure must lead to exactly those counterfactuals that we intuitively endorse (or nearly so), and there must be some principled reason why this procedure should succeed. Clearly, s cannot entail every truth. Otherwise (as noted by Goodman 1983), it would follow that the counterfactual "$p > q$" is correct for any q, since $\neg p$ is true (letting "$\neg$" represent negation) and any conclusion follows logically from $p \& \neg p$. We must be selective about which truths are entailed by s—which truths we preserve when contemplating the counterfactual antecedent.

For example, consider Goodman's classic example of a dry, well-made match that is surrounded by oxygen, never struck, and never lights. In contemplating what would have happened had the match been struck, we are typically supposed to preserve the fact that oxygen is present, the match is dry and well-made, and any well-made match lights if it is dry and struck in the presence of oxygen. We should therefore believe that had the match been struck, it would have lit. We are obviously not supposed to preserve the fact that the match never lights. If we did, and also preserved that oxygen is present, the match is well made, and any well-made match that does not light—despite having been struck in the presence of oxygen—is not dry, then we would accept the counterfactual, "Had the match been struck, it would not have been dry," which is incorrect.

The corresponding problem on the possible-worlds approach is to specify a procedure identifying the *p*-world(s) "closest" to the actual world. For example, we must find an independently motivated rule that deems some possible world in which the match is struck, dry, well made, and lights to be closer to the actual world (in which the match is not struck, dry, and well made and never lights) than any possible world in which the match is struck, not dry, and well made and never lights. Why is the fact that the match is dry, rather than that it never lights, preserved in the closest possible world in which the match is struck? The answer cannot be simply that the former is more important than the latter to a possible world's "proximity" to the actual world. This answer merely begs the question, "Why does *this* sense of 'closeness' govern counterfactuals?"

Preservation under counterfactual suppositions lies at the heart of the apparent difference between laws and accidents in their relations to counterfactual conditionals. As Chisholm and Goodman note, our intuitions about which counterfactuals are correct suggest that, roughly speaking, all of the laws are preserved under a counterfactual antecedent as long as it is logically possible for them all to be (i.e., as long as the counterfactual antecedent is itself logically consistent with all of the laws), whereas this principle cannot be extended to cover an accidental generalization as well; it is not the case that an accident is preserved as long as it and all of the laws logically possibly could all be. For example, consider again a dry book of well-made matches surrounded by oxygen. Suppose that as a matter of accidental fact, none of the matches in the book is ever struck and none ever lights. Now consider the counterfactual supposition that the first match from the book is struck. What would it take in order for the accident that none of the matches in the book ever lights to be preserved under this supposition—preserved, that is, along with the laws entailing that all dry, well-made matches, if surrounded by oxygen, light when struck? It would require that had the match been struck, then the match would not have been dry, well made, and surrounded by oxygen. But this is not so; the match would still have been dry, well made, and surrounded by oxygen. Hence, if the laws entailing "All dry, well-made matches, if surrounded by oxygen, light when struck" are preserved under this counterfactual supposition, then the accident "None of the matches in this book ever lights" cannot be preserved. Intuitively, the laws are preserved and the accident is not.

This chapter and the following are devoted to working out the ideas behind examples like this.

1.2. Laws "supporting" counterfactuals, and defining "preservation"

How a philosopher characterizes the laws' preservation depends on her conception of counterfactuals. Suppose she interprets counterfactuals as metalinguistic. Then she will maintain that when entertaining $p > q$, we are *always* permitted to regard the claims we believe to state laws as entailed by s—that is, as available to help justify q—whereas we are *sometimes* not permitted so to regard a claim that we believe expresses an accident. Chisholm (1946) offers an example:

> [C]onsider a small community where each of the lawyers happens to have three children. We may say: "(x) if x is a lawyer in . . . community in 1946, x has three children." But we should not want to say of Jones, whom we know not to be a lawyer there, that if he *were* to have practiced there he too would have had three children. The difficulty is that our universal conditionals about . . . the lawyers describe what are, in some sense, "accidents" or "coincidences." How are we to distinguish such "accidental" conditionals, of which examples are easily multiplied, from statements such as "all men are mortal," "all wolves are ferocious," etc., which describe "non-accidental" connexions? This is the crux of the whole problem. Our [procedure for identifying s] must exclude these "accidental" universal conditionals [because] they do not warrant the inference of certain contrary-to-fact conditionals. That is to say . . . in the case of the wolves we should not hesitate to infer "If *a* were a wolf, *a* would be ferocious." (p. 302)[2]

Talk of whether a claim "warrants the inference of certain contrary-to-fact conditionals," "is available to help support" a counterfactual, is entailed by s, and so on, sometimes raises issues that obscure the distinction between laws and accidents. For example, suppose that a bottle contains only arsenic during the short time that it exists, so that as a matter of accidental fact, all of the persons who drink from it are poisoned. Then, let us say, had Jones drunk from this bottle, she would have been poisoned. Does the accidental generalization "All of the persons who drink from this bottle are poisoned" help to support this counterfactual, or is this work really being done by "The bottle always contains arsenic and it is a law that arsenic is poisonous"? In other words, does s logically entail only the former claim (an accidental generalization), or does it logically entail the latter claim (which includes a law) as well? Is there a distinction between a truth that *supports* $p > q$ and a truth that would still have obtained had p obtained, entails q when conjoined with p, but is *not responsible for supporting* $p > q$? Presumably so, but do we really need to find a way to draw this distinction in order to understand the laws' special relation to counterfactuals?

I would prefer to avoid questions about which truth is responsible for supporting a given counterfactual by dispensing with all talk of "support" and instead expressing the laws' relation to counterfactuals simply in terms of whether a given truth is "preserved" under a certain counterfactual antecedent—that is, whether that truth would still have obtained had that counterfactual antecedent obtained. Let's say that some set of truths is "preserved under *p*" means that for any member *m* of the set, *m* would still have obtained had *p* obtained. In other words, some set Γ of truths is "preserved under *p*" exactly when for any $m \in \Gamma$, $p > m$. Regarding two sets (Γ and Δ) of sentences, that Γ is "preserved under Δ" means that for any $m \in \Gamma$ and for any $p \in \Delta$, $q \in \Delta$, $r \in \Delta$, and so on, all of the following are correct: *m* would still have obtained had *p* been the case; and had *q* been the case, then *m* would still have obtained had *p* been the case; and had *r* been the case, then had *q* been the case, *m* would still have obtained had *p* been the case; and so on—in other words, $p > m$, $q > (p > m)$, $r > (q > (p > m))$, and so on. We have here some nested subjunctive conditionals, such as $q > (p > m)$: "Were *q* the case, then were *p* the case, *m* would be the case." This conditional is correct exactly when $p > m$ is correct in the *q*-world closest to the actual world. I discuss nested subjunctive conditionals more fully in section 3. For now, I merely note that we do indeed use nested subjunctive conditionals. For instance, in the course of playing checkers, we might say: Had I moved my king to that square, then had you jumped it, I would have jumped you back.[3]

1.3. *Physical necessity, lawhood unconstrained by the non-nomic facts, and preservation*

My initial proposal, then, is

> Some fact *p* is a law (i.e., $\blacksquare p$) exactly when *p* is preserved under every subjunctive supposition that is logically consistent with the facts that are laws.

(Thus I set aside "counterlegals" such as "Had some emerald been red.")[4] This proposal is almost right, I think, but it must be refined in three respects: (1) Suppose that *p* and *q* are laws, and that *r* is a contingent truth that follows from (*p* & *q*). Must it be a law that *r*? Perhaps Jerry Fodor (1981, p. 40) is correct in suggesting that it is not a *law* that all objects that are emeralds or pendulums are green emeralds or pendulums having a period of $2\pi\sqrt{(l/g)}$.[5] Let's leave room for this possibility. But then if (as I have just proposed) there is a range of counterfactual suppositions under which all of the facts that are laws are preserved, then the laws may not be the *only* facts so preserved; any of their logical consequences is likewise so preserved even if it is not itself a law. Furthermore, if *p* and *q* are laws but (*p* & *q*) is not a law, then ($\neg p$ or $\neg q$) may be consistent with each of the facts that are laws. In that event, my proposal cannot be correct in holding that all of the facts that are laws (e.g., *p* and *q*) are preserved under the counterfactual supposition "Had either *p* been false or *q* been false." (2) A fact that is a law may fail to be preserved under a counterfactual antecedent that, although consistent with every fact that is a law, is inconsistent with some such fact's *lawhood*. For example, had it not been a *law* that every material object accelerating from rest remains at speeds less than 3×10^8 m/s (the speed of light), then perhaps it would not have been *true*—perhaps some object would have accelerated beyond that speed. (Had a law of nature not been standing in our way, then our particle accelerators would have broken that speed limit a long time ago!) Of course, if *p* is a law, then the supposition that *p* is not a law is inconsistent with the facts that are laws *as long as* "It is a law that *p*" entails "It is a law that it is a law that *p*." But this entailment may not hold; let's not assume it from the outset. (3) Likewise, a fact that is a law may fail to be preserved under a counterfactual antecedent that, although consistent with every fact that is a law, is inconsistent with the *nonlawhood* of some nonlaws. For example, had "All gold objects are smaller than one cubic mile" been a law, then perhaps the law of thermal expansion would have failed (since perhaps a gold object of slightly less than one cubic mile, upon being heated, would have been barred from expanding beyond one cubic mile). Of course, if *p* is a nonlaw, then the subjunctive supposition that *p* is a law is inconsistent with the facts that are laws as long as "*p* is a nonlaw" entails "It is a law that *p* is a nonlaw." But this entailment may not hold; again, let's not presuppose it.

To avoid these problems, let's define "*p* is physically necessary" to mean "*p* is entailed by the laws' lawhood and the nonlaws' nonlawhood" (remembering that some of the nonlaws are falsehoods). More precisely, "*p* is physically necessary in world W" means that *p* holds in every possible world with *exactly* W's laws. Recall that $\blacksquare p$ means "*p* is a law"; let $\Box p$ mean "*p* is physically necessary," so $\blacksquare p$ entails $\Box p$ but not vice versa. If $\Box p$ where *p* entails *q*, then $\Box q$; $\blacksquare p$ entails $\Box \blacksquare p$; $\neg\blacksquare p$ entails $\Box\neg\blacksquare p$; $\Box p$ entails *p*.[6] My proposal is now

> Some fact *p* follows from the laws exactly when *p* is preserved under every subjunctive supposition that is logically consistent with the facts that are physically necessary.

It may be fairly easy to tell whether science says that a given counterfactual supposition is consistent with the laws—with those claims *p* where $\blacksquare p$. But to say whether, according to science, a given supposition is consistent with the laws' *lawhood* (not to mention with the nonlaws' nonlawhood), we must make more substantive claims about what lawhood is. For instance, it appears to be a consequence of Lewis's account of law—according to which the laws are (roughly) the general truths belonging to the deductive system having the best combination of simplicity and informativeness (see appendix 3 of chapter 1)—that "There exists nothing in the entire history of the universe except a lone proton" is consistent with the actual laws' truth but inconsistent with the actual laws' lawhood and nonlaws' nonlawhood. A lone-proton world is so tremendously impoverished that "All copper objects are electrically conductive" does not belong to the best system, whereas "All material particles are protons" presumably does (I discuss this further in section 3 below). On other accounts of law, the counterfactual supposition of a lone proton *is* consistent with the facts that are actually physically necessary. So to apply the above principle regarding the laws' relation to counterfactuals, we must make some substantive claims about lawhood. Otherwise we would have no basis for regarding a given counterfactual supposition as consistent with the physical necessities.

Accordingly, I now tentatively advance some substantive claims about lawhood—the minimum that suffice for us to be able to apply the above principle to the sort of examples that we must examine in order to test this principle (as I do in section 2 below). Inevitably, there will be some proposed accounts of natural law with which my substantive claims about lawhood will fail to accord. (That's what makes them substantive!) Nevertheless, these substantive claims are prompted by some very basic intuitions about lawhood. I defend them a bit now, and more fully at various points in the course of this book.

Implicit in the concept of a natural law is the idea that the laws of nature *govern* the universe. For instance, a given emerald's color is *governed* by the law that all emeralds are green, comets *follow* the same laws of motion as the planets *obey*, and gravity was believed to act *in accordance with* (or *under*) Newton's inverse-square law. The laws governing various phenomena may themselves be governed by higher order laws. For instance, it is a law that all laws of motion are Lorentz-invariant. Eugene Wigner (1967, p. 43) regards such a principle of symmetry or invariance as "a superprinciple which is in a similar relation to the laws of nature as these are to the events." I agree with this idea (though I regard such a superprinciple as a law governing laws). In addition, I argue in chapter 7 that there are meta-laws specifying the sorts of laws in which natural kinds of various sorts must figure. (E.g., if the electron is a natural kind of material particle, then there must be a law of the form "All electrons have . . . rest mass.") So we have a hierarchy:

.
.
.
LAWS
that govern
LAWS
that govern
LAWS
.
.
.

At the base of this hierarchy are the facts about the actual world (*not* counterfactual conditionals) that are governed by laws but do not themselves govern anything (or describe what governs something). These are the "nonnomic facts"—for example, the facts about particles in motion that are governed by the laws *of motion*, as contrasted with the facts that are governed by laws *about laws*. That the hierarchy of laws governing laws governing . . . bottoms out seems to be Wigner's (1967) conception in his discussion of "three categories which play a fundamental role in all natural sciences: events, which are the raw materials for the second category, the laws of nature, and symmetry principles, for which I would like to support the thesis that the laws of nature form the raw material" (p. 38). Whereas the truths $\blacksquare p$, $\neg\Box q$, and $(r \supset \blacksquare r)$ are "nomic facts," the non-nomic facts include that all emeralds are green (a law) and that all gold objects are smaller than one cubic mile (an accident). The non-nomic facts, then, are the "bare" facts—the facts stripped of their status as laws or accidents. A "non-nomic claim" (a claim that, if true, expresses a non-nomic fact) does not concern the laws; it is not made true or false by which counterfactual conditionals are correct, and its truth-value does not depend in any obvious way on whether or not some fact is a law (or physically necessary).[7] Let U be a language for science containing exactly the non-nomic claims; identify a language with the set of its sentences. (So U does not include sentences with "$\blacksquare$" or "$\Box$.")

If it is a law that all emeralds are green ($\blacksquare p$), then although $\Box\,\Box p$ holds and entails $\Box p$, $\blacksquare p$ rather than $\Box\,\Box p$ is intuitively *responsible* for p's physical necessity. That is, a truth is physically necessary because it is entailed by the laws' lawhood and nonlaws' nonlawhood, and in p's case, what makes it physically necessary is $\blacksquare p$. Or suppose that $\blacksquare(\neg\blacksquare q \supset r)$ and $\neg\blacksquare q$ (where q and r are non-nomic). That r follows from these *shows* that $\Box r$—that r is no accident. But intuitively, that r so follows is not the *reason why* $\Box r$. If r is "Every emerald in my pocket is green," then r follows from a truth $\blacksquare p$ ("It is a law that all emeralds are green") where p is non-nomic, and $\blacksquare p$ is responsible for r's physical necessity. The law that $(\neg\blacksquare q \supset r)$ is not partly responsible for r's physical necessity because this law does not *govern* r; in the hierarchy, it is a law governing *laws*. To put the point generally: if some p is physically necessary, then it is certain facts about what laws govern p (together perhaps with certain facts at or below p's level in the hierarchy) that are responsible for p's physical necessity. In other words, if $\Box p$, then p must be entailed by facts of the form $\blacksquare q$ and $\neg\blacksquare r$ where q and r are facts at or below p's level in the hierarchy of laws and what they govern. [E.g., though q is non-nomic and thus below $(\neg\blacksquare r \supset q)$ in the hierarchy, $\blacksquare q$ could be responsible for $(\neg\blacksquare r \supset q)$'s physical necessity.] So if $\Box p$ where p is non-nomic, then the responsibility for p's physical necessity must fall entirely on the lawhood and nonlawhood of various non-nomic facts (i.e., on facts that govern p). So p must be entailed by facts of the form $\blacksquare s$ and $\neg\blacksquare t$ where s and t are non-nomic. This idea will shortly prove useful in deriving a principle regarding the laws' relation to counterfactuals—a principle that we can test against our intuitions and scientific practice.

The intuition that there are "non-nomic facts"—facts that laws govern but that govern nothing themselves—has sometimes been invoked in the service of the view that p's lawhood is ultimately reducible to, or at least supervenient upon, the non-nomic facts.[8] On this Humean view, lawhood adds nothing to the world beyond what the non-nomic facts already bring. I do not subscribe to this view. As I argue more fully in section 3 below, I believe that the distribution of $\blacksquare$s and $\neg\blacksquare$s over the non-nomic facts is not determined by these facts, any more than the rules of chess supervene on the actual moves made in a given chess game played according to the rules. The laws fail to supervene on the facts that they govern; two possible worlds can involve exactly the same non-nomic facts but differ in the laws governing them. Indeed, I think that once the non-nomic facts are fixed, the ■s and ¬■s could be distributed in just about *any* way among them.[9] (I tend to picture the ■s and ¬■s as having been sprinkled like powdered sugar over the doughy surface of the non-nomic facts.) In particular, once it is fixed which non-nomic claims possess ■ and which do not, the accidents may include any of those that are not entailed by but are consistent with those that possess ■.[10] If ¬■*q* and various facts of the form ■*p* logically entailed *r* (where *p*, *q*, and *r* are non-nomic and the facts ■*p* alone do not suffice to logically entail ■*q* or *r*), then ¬*r* together with those facts ■*p* would logically entail ■*q*, contrary to the notion that once the non-nomic facts are fixed, the ■s and ¬■s can be distributed among them in any fashion (e.g., *p* gets a ■ whereas *q* does not).[11] So my contention is that the only constraint that the facts of the form ■*p* and ¬■*q* (where $p,q \in$ U) impose on the facts in U (laws and accidents alike) is that the facts in U must be consistent with the *p*s where ■*p*.

The idea that, even after it is fixed which non-nomic claims possess ■ and ¬■, the accidental truths in U remain largely open seems implicit in the traditional contrast between the natural laws and the initial conditions. Having designated which non-nomic claims state laws and which do not, God, let's say, is free to set the non-nomic initial conditions in any manner consistent with the truth of the non-nomic claims that state laws. That two possible worlds could have exactly the same non-nomic facts, but differ in which non-nomic facts state laws, should not really be surprising, considering that the laws in a given world are tied up not just with what *in fact* comes to pass there, but also with what *would have* happened there had certain circumstances unrealized in that world instead come to pass there. Intuitively, two worlds can be identical in the former respect but different in the latter. This is especially evident when the two worlds are highly impoverished. For example, if a given possible world's history involves nothing but a single elementary particle moving uniformly forever, then presumably its laws could be exactly the same as the actual world's, but alternatively its laws could posit a gravitational force twice as strong as the actual one. This difference would make a difference to the counterfactuals holding in that world, but either set of laws is consistent with the non-nomic facts there (I discuss this example more fully in section 3 below).

Above I concluded that if □*p* for $p \in$ U, then there must be facts of the form ■*s* and ¬■*t* by which *p* is entailed, where $s,t \in$ U. I have just held that there are such facts only if *p* is entailed by facts *s* where ■*s* and $s \in$ U. So for a non-nomic fact to be physically necessary (i.e., to follow from the laws' lawhood and the nonlaws' nonlawhood), it must follow from the non-nomic facts that are laws.[12] That is, for a non-nomic fact to be an accident, it suffices that it fail to follow from the non-nomic facts that are laws; there is then no possibility of its following from the *lawhood* of those laws, or from any laws' lawhood and nonlaws' nonlawhood. To know whether some *p* in U is physically necessary, we do not need to know what *lawhood* is, only what the laws in U are.

Let Λ be the set of all facts *s* where ■*s* and *s* ∈ U, together with everything in U that they entail (so that Λ is "logically closed in U"). So we have found that Λ contains exactly the facts *p* in U where □*p*.[13] Thus, if a non-nomic subjunctive supposition *p* is consistent with Λ, then, ¬*p* (which is also non-nomic) is not in Λ, and so ¬*p* is not physically necessary, and so *p* is logically consistent with the facts that are physically necessary. Recall that this very constraint figured in my above proposal:

> A fact *p* follows from the laws exactly when *p* is preserved under every subjunctive supposition that is logically consistent with the facts that are physically necessary.

From this proposal, it follows in particular that for any non-nomic fact *p*, *p* ∈ Λ only if *p* is preserved under every subjunctive supposition in U that is logically consistent with the facts that are physically necessary. But we have just seen that if a subjunctive supposition in U is consistent with Λ, then it is consistent with the facts that are physically necessary. So,

> For any *p* in U, if *p* ∈ Λ, then *p* is preserved under every subjunctive supposition in U that is consistent with Λ.[14]

This consequence of my original principle and several substantive claims about lawhood is exactly what we need. It involves only those counterfactual antecedents and physical necessities *that belong to U*; unlike my original principle, no ■s appear in the counterfactual antecedents or physical necessities referred to by this new principle. So we can test this principle—we can see what it demands in various cases (as I do in section 2 below)—without having to make any substantive assumptions about what lawhood is; we already know how to tell whether a given supposition in U is logically consistent with some other claims Λ in U.

We can easily make this principle into a biconditional. If *p* ∈ U but *p* ∉ Λ, then *p* is either false or an accidental truth. If *p* is false, then under any truth *q* ∈ U, *p* is not preserved. If *p* is a non-nomic fact but ¬□*p*, then the counterfactual supposition ¬*p* is consistent with Λ (since otherwise *p* ∈ Λ), and obviously *p* is not preserved under this supposition. Thus, the above principle entails the principle

> For any *p* in U, *p* ∈ Λ exactly when *p* is preserved under every subjunctive supposition in U that is consistent with Λ.

Once we have drawn the distinctions that enable us to formulate principles like these, the apparently simple notion that "laws support counterfactuals" begins to dissolve into a variety of subtly different ideas. Any reductive analysis of natural law, in accounting for the laws' relation to counterfactuals, will have to make clear precisely what the relation is for which it purports to account. For instance, the above principle might be strengthened in various ways. Here is one possibility (another is discussed in the following section): Recalling the hierarchy of laws and what they govern, let U^+ contain just the claims that purport to describe either non-nomic facts or the laws governing those facts. In other words, U^+ is the set of claims that can be formed from $\blacksquare$, $\Box$, and the members of U, as long as any expression in the scope of $\blacksquare$ or $\Box$ belongs to U. [So $U \subset U^+$, "All laws of motion are Lorentz-invariant" $\in U^+$, "It is a law that all laws of motion are Lorentz-invariant" $\notin U^+$, and $(p \supset \neg \blacksquare q) \in U^+$ if $p,q \in U$.] If $p \in U^+$, $p \notin U$, and $\Box p$, then the responsibility for p's physical necessity intuitively rests with facts of the form $\blacksquare q$ and $\neg\blacksquare r$ (by which p is entailed) where q and r are claims at or below p's level in the hierarchy of laws and what they govern. That is, $q,r \in U^+$. Furthermore, I suggest that just as the distribution of $\blacksquare$s and $\neg\blacksquare$s over the non-nomic facts is largely unconstrained by those facts, likewise once the facts in U^+ are fixed, the $\blacksquare$s and $\neg\blacksquare$s could be distributed in just about *any* way among them—or, rather, among those outside U, since the facts in U^+ include the distribution of $\blacksquare$s and $\neg\blacksquare$s over the facts in U.[15] For example, if "All laws of motion are Lorentz-invariant" is true, it may be a "meta-law" or a coincidental regularity in the laws in U.[16] As in the above case, the only constraint that the facts of the form $\blacksquare p$ and $\neg\blacksquare q$ (where p,q are in U^+ but outside U) impose on the facts in U^+ but outside U is that $\blacksquare p$ entails p. We might then follow the course of the preceding reasoning. Let Λ^+ be the set of facts of the form $\blacksquare s$ and $\neg\blacksquare t$ where $s,t \in U$, together with all of the other claims in U^+ that hold in every possible world where those facts all hold. So Λ^+ is "logically closed in U^+" and contains exactly the facts p in U^+ where $\Box p$. Thus, if a subjunctive supposition p in U^+ is consistent with Λ^+, then p is consistent with the facts that are physically necessary, and we can derive

> $p \in \Lambda$ exactly when p is preserved under every subjunctive supposition in U^+ that is consistent with Λ^+.

We need eventually to determine whether this principle holds.[17] I now offer a different respect in which the preceding proposal might be strengthened.

1.4. Preservation of the laws

If Γ is a set of sentences, let $\mathrm{Con}(\Gamma,p)$ mean that p is logically consistent with every member of Γ. Consider, then, the forward direction of the principle given immediately above:

> Λ-*preservation*: Λ is preserved under $\{p \in U^+ \text{ such that } \mathrm{Con}(\Lambda^+,p)\}$.

This is the preservation merely of the *truth* of the laws in U, not of their status *as laws*. According to Λ-preservation, if it is a law that all emeralds are green and "There is an emerald under the carpet," although false, is consistent with Λ^+, then had there been an emerald under the carpet, it would have been green.

Principles like Λ-preservation are intended to be logical truths—to apply in any logically possible world, to *that world's* Λ. In other words, such principles are supposed to reflect *what it is* to be a law in a given possible world; it is not supposed to be a peculiarity of the actual world that such principles hold there. So Λ-preservation is applicable not only to the preservation of the *actual* laws' truth in the closest *p*-world to the *actual* world, but also to the preservation of the truth of *a given possible world's* laws in the closest *p*-world to *that possible world.* That is, if we contemplate some possible world, the counterfactual conditionals holding in that world are supposed to be governed by Λ-preservation, where "Λ" refers to the laws of that possible world. This possible world might itself have been called forth by a counterfactual antecedent; it might be the closest *r*-world. In that case, if Con(Λ^+,*p*) holds for the closest *r*-world's Λ^+ and ■*m* holds in the closest *r*-world (where $m \in$ U), then according to Λ-preservation, $r > (p > m)$.[18]

But what *are* the laws in the closest *r*-world if Con(Λ^+,*r*) holds for the actual world's Λ^+? Λ-preservation says nothing about this, since Λ-preservation concerns the preservation of the actual laws' *truth,* not their *lawhood.* We might consider a broader principle that covers the preservation of the laws' lawhood (and the nonlaws' nonlawhood):

Λ^+*-preservation*: Λ^+ is preserved under $\{p \in U^+$ such that Con(Λ^+,*p*)$\}$.

Notice that if $\neg p$ (where $p \in$ U), then ¬■*p*. So to preserve *p*'s nonlawhood, we do not need to preserve ¬*p*, only *p*'s status as either false or an accident. According to Λ^+-preservation, if it is a law that all emeralds are green and "There is an emerald under the carpet," although false, is consistent with Λ^+, then had there been an emerald under the carpet, then it would still have been a law that all emeralds are green. Obviously, Λ^+-preservation entails Λ-preservation.[19]

We can likewise apply Λ^+-preservation to the preservation of the lawhood of some *possible world's* laws. For example, suppose that the possible world is the closest *r*-world, such that $r \in U^+$ and Con(Λ^+,*r*) hold. Then according to Λ^+-preservation, the claims in U that are laws are exactly the same in the closest *r*-world as in the actual world. So Con(Λ^+,*p*) holds for the closest *r*-world's Λ^+ exactly when Con(Λ^+,*p*) holds for the actual world's Λ^+. Therefore, if $p,r \in U^+$ where Con(Λ^+,*r*) and Con(Λ^+,*p*) hold for the actual world's Λ^+ (and so for the closest *r*-world's Λ^+), then for any *m* in the actual world's Λ^+ (and so in the closest *r*-world's Λ^+), it follows from applying Λ^+-preservation *to the closest r*-world that $r > (p > m)$ holds in the actual world.

We might, then, entertain a "non-nested version" of Λ^+-preservation: for any $p \in U^+$ such that Con(Λ^+,*p*) and for any $m \in \Lambda^+$, it is the case that $p > m$. By applying this principle not only to the actual world, but also to various counterfactual worlds, we can generate all of Λ^+-preservation, even its implications regarding nested counterfactuals.[20]

Recall from chapter 1 that the correctness of counterfactual conditionals is extremely context sensitive; whether a fact is preserved under a given subjunctive supposition depends on our interests, which vary with the context.[21]

Principles such as Λ-preservation and Λ^+-preservation are intended to cover *all* contexts. In other words, Λ-preservation demands that *in any context*, $p > m$ is correct for any $m \in \Lambda$ and $p \in U^+$ where Con(Λ^+,p) holds. This invites many of the potential counterexamples to these principles that I discuss in section 2.

I argue in section 2 that Λ-preservation withstands these difficulties. I also argue that Λ^+-preservation holds—with a single exception, which I am not in a position to set out until the close of chapter 3 (and disregard until then). In section 3, I defend Λ^+-preservation insofar as it goes beyond Λ-preservation. Among the many philosophers who have been prompted by the intuitive connection between laws and counterfactuals to advance principles roughly similar to Λ-preservation and Λ^+-preservation are Jonathan Bennett (1984), John Carroll (1994, pp. 16–21, 59, 182–89), Chisholm (1946, 1955), Goodman (1947, 1983), Paul Horwich (1987), Frank Jackson (1977), Mackie (1962), John Pollock (1976), and P. F. Strawson (1952).[22] As I show in the following sections, Lewis rejects both of these principles, and David Armstrong denies Λ^+-preservation.

1.5. *Preservation of an accident along with the laws*

How shall we cash out the intuition that no accident joins the laws in bearing their special relation to counterfactuals? I have already shown that a nonnomic fact a, where $\neg\Box a$ (i.e., $a \notin \Lambda$), fails to be preserved under every counterfactual supposition that is consistent with Λ, since Con($\neg a$,Λ), and trivially, a fails to be preserved under $\neg a$. Now any extension of Λ-preservation to an accident *should* fail if Λ-preservation genuinely captures the laws' special relation to counterfactuals, but its failure should perhaps not be so trivial. It should fail because a's accidental character is betrayed by a's failure to be preserved under some counterfactual supposition. That a fails to be preserved under some p with which a is inconsistent is *not* attributable to a's lack of physical necessity.

Here is an alternative way in which Λ-preservation might be extended to some accident a:

> [$\Lambda \cup \{a\}$]-*preservation*: [$\Lambda \cup \{a\}$] is preserved under {$p \in U^+$ such that Con(Λ^+,p) and Con{a},p)}.

This principle fails—as (again) it should if it is to help capture the intuitive difference between laws and accidents in their relation to counterfactuals. Suppose that $\blacksquare l$ where $l \in U$. Let $p = (\neg l \text{ or } \neg a)$. Then, according to my remarks in section 1.3, Con(Λ^+,p) and Con({a},p) hold. So if Λ-preservation holds, then $p > l$, and so $p > \neg a$, and so [$\Lambda \cup \{a\}$]-preservation fails.

Of course, this p is inconsistent with (l & a), and so with [$\Lambda \cup \{a\}$]'s logical closure in U. We might consider a principle that requires Con([$\Lambda \cup \{a\}$]'s closure in U,p):

> [$\Lambda \cup \{a\}$]-*Preservation (first revision)*: [$\Lambda \cup \{a\}$] is preserved under {$p \in U^+$ such that Con(Λ^+,p) and Con([$\Lambda \cup \{a\}$]'s closure in U,p)}.

This principle fails—(again) just as it should. Let $p = (a \supset \Box a) = (\neg a \text{ or } \Box a)$. Then, according to section 1.3, Con(Λ^+,p) and Con([$\Lambda \cup \{a\}$]'s closure in U,p). So if Λ^+-preservation holds, then $p > \neg\Box a$, and so $p > \neg a$, and therefore [$\Lambda \cup \{a\}$]-preservation (first revision) must fail.

Obviously, for any b that logically follows from a, the counterfactual antecedent ($\neg b$ or $\Box a$) (i.e., $b \supset \Box a$) would work just as well in the above argument. This suggests a much more intuitive way to bring out the same point (for the accidental generalization a: "All *F*s are *G*")—by using the counterfactual antecedent "Had there been an x such that Fx and Px," where "Px" means $Gx \supset \Box a$. Consider Karl Popper's (1959, pp. 427–8) discussion of the moa, an extinct bird that lived only in New Zealand. (I mentioned this example in chapter 1.) Popper imagines "All moas die before age 50" to be an accidental generalization (let it be a); the moa had a hardy constitution, but New Zealand was heavily populated by a virus that kills moas young. Consider this counterfactual antecedent: "Had there been a moa, aged 49 years 364 days, living in the most promising circumstances—free of the virus, and not in the next day run over by a truck, eaten by a predator, starving, afflicted with pneumonia, tuberculosis, and so on" (I say "and so on" because at this point, I am deliberately expressing the counterfactual antecedent I have in mind in intuitive terms; I express it more explicitly below). The intuitive significance of this antecedent p is that what transpires in the closest p-world manifests the accidental character of "All moas die before age 50." To exhibit this generalization's lack of physical necessity, we select as our hypothetical "test moa" the ideal candidate for survival past age 50: the moa called up by the above counterfactual antecedent. This antecedent falls within the scope of [$\Lambda \cup \{a\}$]-preservation (first revision); the existence of such a moa is consistent with Λ^+ and with [$\Lambda \cup \{a\}$]'s closure in U, since it is *logically possible* for such a moa, breaking none of the actual laws, to die before its 50th birthday. Nevertheless, since Popper's generalization is merely accidental, the ideal test moa would have survived past age 50. So Popper's generalization is not preserved under this antecedent, in violation of [$\Lambda \cup \{a\}$]-preservation (first revision). Now intuitively, if *even* the ideal moa dies before age 50, then it must be physically necessary that all moas die before age 50. There must be—to put the point colorfully—some self-destruct mechanism, intrinsic to moas, that engages on the animal's 50th birthday. In other words, this antecedent, expressed explicitly, is "Had there been a moa possessing property P," where "Px" means $Gx \supset \Box a$.

Thus, [$\Lambda \cup \{a\}$]-preservation (first revision) fails *just where it should*; intuitively, a's accidental character is betrayed precisely by a's failure to be preserved under the above p. This strongly suggests that in contrasting the success of Λ-preservation (or Λ^+-preservation) with the failure of [$\Lambda \cup \{a\}$]-preservation (first revision), we have found a way to make explicit the difference that a truth's physical necessity makes to its preservation under counterfactual suppositions.

Notice that for any accident a, we now have a recipe for constructing a counterfactual antecedent where [$\Lambda \cup \{a\}$]-preservation (first revision) fails.

This is helpful. Of course, for a given accident a, the above p is not the *only* counterfactual antecedent for which $[\Lambda \cup \{a\}]$-preservation (first revision) breaks down. For instance, we typically regard "All of the marbles in the urn are black" as failing to be preserved under the counterfactual antecedent "Had we placed an additional marble in the urn." But perhaps there are some contexts in which we regard the urn generalization as preserved under this antecedent (as when our topic is the longstanding tradition of placing only black marbles into that urn). An advantage of discussing the counterfactual antecedent "Had there been an x such that Fx and Px," where "Px" means $Gx \supset \Box a$, is that for any accidental generalization in any context, this recipe yields a counterfactual antecedent under which $[\Lambda \cup \{a\}]$-preservation (first revision) breaks down.

Let's reassure ourselves of this with another example. All of the planets in the solar system revolve around the Sun in the same direction. Intuitively, this generalization's accidental character is manifested by its failure to be preserved under the counterfactual antecedent "Had there been a planet not produced by accretion out of the initial solar nebula, or by the breakup of already existing planets (either from tidal stresses or as the result of a collision), and so on." This counterfactual antecedent posits a planet having (I presume) no systematic reason to revolve around the Sun in the same direction as the actual planets do, unless it is physically necessary for all planets to orbit in the same direction. In some of the closest possible worlds in which this counterfactual supposition obtains, the planets do not all orbit in the same direction.

So we have found a counterfactual antecedent p under which $[\Lambda \cup \{a\}]$-preservation (first revision) must fail as long as Λ^+-preservation holds. (Although in chapter 3 (section 3.2) I ultimately find an exception to Λ^+-preservation, it will not affect the above argument.) Of course, this counterfactual antecedent p is inconsistent with $(a \ \& \ \neg\Box a)$, so although Con(Λ^+,p) and Con($[\Lambda \cup \{a\}]$'s closure in U,p), we also have $\neg$Con($[\Lambda^+ \cup \{a\}]$'s closure in U$^+$,p). We might consider a principle that requires Con($[\Lambda^+ \cup \{a\}]$'s closure in U$^+$,p):

> $[\Lambda \cup \{a\}]$-*preservation* (*second revision*): $[\Lambda \cup \{a\}]$ is preserved under $\{p \in$ U$^+$ such that Con($[\Lambda^+ \cup \{a\}]$'s closure in U$^+$,p)$\}$.

The special p for which $[\Lambda \cup \{a\}]$-preservation (first revision) is guaranteed to fail is intuitively a counterfactual antecedent in connection with which a's accidental character makes itself felt. Since this p falls outside the scope of $[\Lambda \cup \{a\}]$-preservation (second revision), there is some reason to think that this principle does not help to capture the intuitive difference between laws and accidents in their relation to counterfactuals. The ideal test moa and the ideal test planet are not relevant to $[\Lambda \cup \{a\}]$-preservation (second revision). On the other hand, in light of the way we arrived at it, this principle might be expected to help explicate the laws' relation to counterfactuals and, accordingly, to fail. I defer to chapter 3 any further treatment of this and other principles that might capture the laws' special relation to counterfactuals. We are already in quite deep enough; I must pause to test the above ideas more carefully.

2. Challenges to Λ-preservation

I now defend Λ-preservation against several types of prima facie counterexamples.[23] In each case, I show either that it is not a counterexample or how Λ-preservation should be refined to accommodate it.

At first sight, Λ-preservation may appear implausibly strong: it denies that there are any counterfactual circumstances in which the actual laws *could* still have held but *would not* had those circumstances come to pass. That is to say, it affirms that there are no circumstances, consistent with the laws, that would not have happened without the laws being different. It is a law that the half-life of iodine-131(I^{131}) is 8.1 days, so it is physically possible for there to be many atoms of I^{131} in the course of the universe's history and for each of them to decay upon turning 1 million years old. But since this is exceedingly unlikely under the actual laws, we might be tempted to hold that had this circumstance obtained, the laws would have been different, contrary to Λ-preservation.

But not so fast. Admittedly, if we are told that each of the many I^{131} atoms decays when it becomes 1 million years old, then (in the absence of any other relevant information) we strongly doubt that the laws set I^{131}'s half-life to 8.1 days. But that we *believe* this not to be a law, if we are *told* that every I^{131} atom decays at age 1 million years, does not entail that we believe that it *would not* still have been a law *had* every I^{131} atom decayed at age 1 million years. After all (following Lewis 1973, pp. 70–71, who is following Ernest Adams 1970, p. 90), if we are told that new evidence has definitively shown that Lee Harvey Oswald did not kill President Kennedy, then we conclude that someone else did. But despite our belief that if Oswald did not kill Kennedy then someone else must have done the deed, we do not believe that *had* Oswald not killed Kennedy, then someone else *would have*. (Here I presume that we are not "conspiracy theorists" concerning Kennedy's assassination.) Likewise, if every I^{131} atom decays at age 1 million years, then their decay must surely be governed by a different law, but it is not the case that had every I^{131} atom decayed upon turning 1 million years old, then the laws of nature would have been different. It is exceedingly unlikely for all of the many I^{131} atoms to decay upon turning 1 million years old. But had this coincidence occurred, the laws would have been no different. This example is like others involving unlikely coincidences that are permitted by the laws. For example, it is consistent with the laws for every elementary particle, 5 milliseconds after the Big Bang, to be moving at exactly the same speed, though this sort of "coordination" is very unlikely. Nevertheless, had this coincidence taken place, the laws of nature would have been just the same; only the universe's initial conditions would have been different.

I now turn to a different kind of apparent difficulty for Λ-preservation.

2.1. *Putative counterexamples involving counterfactual antecedents with implicit clauses*

2.1.1. Two examples. Suppose that two physicians, after work, are discussing their day. One says to the other: "The nurse rushed over to me and reported that the patient had been accidentally injected with the syringe marked 'A.' That syringe was intended for the lab; it was filled with arsenic—'A' for arsenic. So I hurried over to the patient's bedside, although I knew, of course, that there was nothing I could do. I waited for the inevitable. But the most remarkable thing happened: the patient did not die. So our dismay turned to excitement; we thought we had a reportable case on our hands, and prepared to write a stunning article for *The New England Journal of Medicine*. Then I checked the syringe. The label turned out to be 'H', not 'A'. So it contained no arsenic after all. Though the patient was out of danger, I must say that we were a bit disappointed. Had the syringe been filled with arsenic, then we would have discovered that it is not a natural law that such a large dose of arsenic is lethal."

Her assertion of this counterfactual $p > q$ seems correct in this particular context but contrary to respect for Λ-preservation: she believes $p > q$, $\text{Con}(\Lambda^+,p)$, but $\neg\text{Con}(\Lambda,p \,\&\, q)$. Respect for Λ-preservation seems to require that she instead accept "Had the syringe been filled with arsenic, it would still have been a law of nature that such a large dose of arsenic is lethal." Let's imagine the rest of the conversation. The second physician responds: "Don't feel too bad about just missing a landmark article; you didn't really come that close. Had the syringe actually contained arsenic, the patient would have died." It seems to me that the first physician would not disagree with this counterfactual. Somehow, it is consistent with the one that she asserted.

Consider a second example of the same phenomenon. Two foreign terrorists, disguised to be indistinguishable from the local population, board a public bus scheduled to make several stops along its hundred-mile route. They mix their luggage with that of the other passengers, all stowed in a compartment beneath the seats of the bus. Hidden in one of their bags is a time bomb, which explodes after the terrorists have departed the bus and left that bag behind. Many are killed. In the months that follow, the driver, who survives, sometimes cannot help feeling that he must be partly to blame for the calamity. But he also knows that these guilty feelings are unreasonable because he was in no way negligent; nothing even remotely suspicious took place, so nothing would have justified the driver's believing a hidden bomb to be unusually likely. He tries to ward off these guilty feelings by repeating to himself "I would have known there was a bomb aboard only if I had had ESP or something. Had I known, then I, a mere human being, would have had to have something like ESP or X-ray vision—something enabling me to see inside the luggage, for instance—because nothing the least bit suspicious went on." The driver believes it physically necessary that all human beings lack anything along the lines of ESP or X-ray vision. That he is correct in asserting the above counterfactual therefore appears to conflict with respect for Λ-preservation.

2.1.2. How to deal with these cases. I believe that these examples are compatible with respect for Λ-preservation. The context in which a counterfactual $p > q$ is uttered can influence its correctness in two ways. First, as I have already noted, the interests of the speaker and audience affect which *p*-worlds qualify as the closest. Second, the conversational environment in which the counterfactual is uttered may implicitly supply the antecedent with some additional content.

Lewis (1973) discusses the distinction I have just drawn between these two influences. He sees no work for it to do in accounting for our use of counterfactuals:

> It matters little how we divide up the influence of context between (1) resolution of the vagueness of comparative similarity, and (2) expansion of the explicit antecedent to the real, conversationally understood antecedent. My inclination, however, is to explain the influence of context entirely as the resolution influence. Better one sort of influence of context than two different sorts; we are stuck with the resolution influence whether or not we also admit the expansion influence; but given the resolution influence we do not need to admit the expansion influence as well. (pp. 66–67)

I disagree. There is separate work for these two influences to do and a principled way to distinguish their contributions. This result renders the arsenic and bus examples compatible with Λ-preservation. In the conversational context in which the counterfactual antecedent "Had the syringe been filled with arsenic" was entertained, a more complete statement of this antecedent turns out to be "Had the syringe used to inject the patient been filled with arsenic *and the patient lived*." Since *this* antecedent is believed to be inconsistent with Λ^+, Λ-preservation imposes no restriction on the conditionals with this antecedent that can be believed to be correct. Likewise, in the bus example, the counterfactual antecedent "Had I known" turns out to be shorthand for "Had I known there was a bomb aboard without my having known before the day's work [as I would have, had I been an accomplice] and without there having been anything even remotely suspicious for me to have observed [with actual human senses] that day [as there would have been, had one of the terrorists' suitcases accidentally opened in plain view]." This antecedent is also believed inconsistent with Λ^+. Hence, these examples do not constitute exceptions to Λ-preservation.

Of course, this defense of Λ-preservation requires some principled means of showing that these additional clauses were implicit in these counterfactual antecedents all along, and were not merely contributed by the criterion for the closest *p*-world.[24] When the bus driver says "Had I known, I would have had to have X-ray vision or something, because there was nothing remotely suspicious," is " . . . there was nothing remotely suspicious" retained in the closest possible world in which the counterfactual antecedent obtains because it is implicitly part of that antecedent, or solely because of the criterion of closeness? If the latter, then Λ-preservation must be rejected. Likewise, when the physician says "Had the syringe contained arsenic, we would have discovered that it is not a natural law that such a large dose of arsenic is lethal, because the patient lived," the fact that the patient lived is retained in the closest possible world in which the counterfactual antecedent holds. But how can we tell whether "the patient lived" is implicitly part of the counterfactual antecedent or dependent entirely upon the criterion of closeness?

Suppose someone asserts "$p > q$, because r" where r is a truth about the actual world that the person who asserts $p > q$ takes also to hold in the possible world(s) to which we are directed by the counterfactual supposition. We want to know whether r is added to the explicit antecedent by the conversational context, or whether r is preserved under this antecedent solely because of the criterion of closeness. Then we should immediately ask the person who asserted the counterfactual the question, "Does $p > r$ hold?" If r was originally implicit in the antecedent of the counterfactual $p > q$ that she asserted, then by asking "$p > r$?" we remove r from p because this is the charitable interpretation of the question; if r is retained implicitly in p, then $p > r$ is *trivially* true. Once r is no longer possibly implicit in p, the correctness of $(p > r)$ requires that the closeness criterion be sufficient for retaining r in the closest p-world. So if she replies to our question by *not* accepting $p > r$, then r was not originally retained by virtue of the closeness criterion, so r must have been implicit in p; had r originally been contributed by the closeness criterion, then r would not have been lost by this operation.[25] On the other hand, if she replies to our question by affirming $p > r$, then r must have initially been supplied by the closeness criterion, though it may *also* have been implicit in p.

Let us apply this procedure to the bus and arsenic examples. After the bus driver asserts "Had I known, I would have had to have X-ray vision or something, because there was nothing even remotely suspicious," we ask him, "Do you believe that had you known, there would still have been nothing remotely suspicious?" Surely he replies, "No. Had I known, I would have had to have been tipped off by something suspicious. That was the point I was trying to make." That there was nothing suspicious must then have been implicit in the antecedent of the counterfactual that the driver initially asserted. Likewise, after the physician asserts "Had the syringe been filled with arsenic, we would have discovered that it is not a natural law that such a large dose of arsenic is lethal, since the patient lived," we ask her, "Do you believe that had the syringe been filled with arsenic, the patient would still have lived?" Surely she replies, "No. Had the syringe been filled with arsenic, the patient would have died." (Recall that as I told the story, the second physician replies by asserting this counterfactual, which the first physician accepts.) That the patient lived must then have been implicit in the antecedent of the counterfactual that the first physician initially asserted.

By distinguishing the clauses implicit in the counterfactual antecedent from the facts retained by the closeness criterion, I have tried to explain why someone might assert "$p > q$, because r" and then deny $p > r$. Contrary to

Lewis, the distinction does some work in accounting for the way we use counterfactuals. My means of identifying clauses implicit in a counterfactual antecedent does not eliminate the contribution of the closeness criterion. For example, in a typical context in which I assert "Had the match been struck, it would have lit, because the match was dry and oxygen was present," I also accept "Had the match been struck, it would still have been dry and oxygen would still have been present." That the match was dry and that oxygen was present must then have been supplied by the closeness criterion.

2.1.3. Other applications. This approach likewise explains away a putative counterexample to Λ-preservation suggested to Bennett by Peter van Inwagen: If I reached Jupiter within the next ten seconds, that would be a miracle because Earth is more than ten light-seconds from Jupiter. Bennett (1984, pp. 84–85) claims that this counterfactual's antecedent is shorthand for something like "If I reached Jupiter within the next ten seconds from my present position, which is more than ten light-seconds from Jupiter." Bennett's claim is supported by my procedure, which requires that we ask the person who asserted the original counterfactual, "Do you believe that if you reached Jupiter within the next ten seconds, you would now be at your present position, which is more than ten light-seconds from Jupiter?" It seems to me that she should answer, "No. My point was that in order to reach Jupiter within the next ten seconds, I would have to be closer to Jupiter now than I actually am, that is, closer than ten light-seconds."[26]

Goodman (1983, p. 15) holds that if a truth r is entailed by s (the special truth s such that $p > q$ holds exactly when q is logically entailed by p together with s—recall section 1.1 of this chapter), then r is "cotenable" with p, that is, $\neg(p > \neg r)$. The match example I just gave bears this out nicely, as Goodman notes: it is not the case that had the match been struck, it would not have been dry. But "the patient lived" helps to support "Had the syringe contained arsenic, then it would not have been a natural law that such a large dose of arsenic is invariably lethal" even though $p > \neg r$: Had the syringe contained arsenic, the patient would not still have lived. This apparent counterexample to Goodman's cotenability requirement can be explained away. I have argued that the antecedent of the counterfactual "Had the syringe contained arsenic, we would have discovered that it is not a natural law . . . " is not really the same as the antecedent in the question "Had the syringe contained arsenic, would the patient still have lived?" that we find ourselves asking when applying Goodman's cotenability requirement to this example.

2.2. Putative counterexamples involving backward-directed counterfactuals

2.2.1. The problem. Another kind of putative counterexample to Λ-preservation involves "backward-directed counterfactuals." These are counterfactuals $p > q$ where q is false and concerns a period of time prior to that with which p is concerned. For example, a backward-directed counterfactual might say that had the present been different in some way, then the past would have been different from what it was. Respect for Λ-preservation requires that we believe some backward-directed counterfactuals, since we believe certain claims of the form "All *F*s are preceded by *G*s" to express physical necessities. For example, we believe "All claps of thunder are preceded by flashes of lightning" to be physically necessary, so respect for Λ-preservation demands that we assent to "Had a thunderclap just occurred, a lightning bolt would have preceded it."

Lewis (1986, pp. 32–66, 159–213) has been prominent in maintaining that the typical backward-directed counterfactual required by Λ-preservation is false. Lewis says that when we are presented with a backward-directed counterfactual, charity occasionally moves us to employ a "nonstandard" criterion for identifying the closest *p*-world and that, under this criterion, a backward-directed counterfactual stands a better chance of being true. But according to what Lewis calls the "standard" measure of closeness, the events prior to *p* in the closest *p*-world are exactly the same as the events in the actual world up until the moment with which *p* is concerned, at which time the two worlds obviously diverge. Actually, this is not quite correct; Lewis says that the divergence sometimes occurs a bit before the moment with which *p* is concerned. But this in no way changes the fact that according to Lewis, the closest *p*-world and the actual world coincide in all respects at one moment but diverge thereafter. It follows that some event in the closest *p*-world violates the laws of the actual world if those laws are deterministic, that is, consistent with only one past and one future given the world's state at a certain moment.[27]

Contrary to Λ-preservation, then, Lewis holds that for some counterfactual antecedents *p*, the laws in the actual world are broken in the closest *p*-world, even though it is logically possible for all of the claims in the actual world's Λ^+ to remain true in the closest *p*-world. This permits Lewis (1986b, p. 56) to maintain that typical backward-directed counterfactuals are not true (under the "standard" closeness metric). The closest *p*-world conforms to the actual laws except for a moment (or perhaps more than one) at which there occurs a "miracle," thereby enabling *p* to obtain although it follows events that *p* could not follow according to the actual laws. In the closest *p*-world, this miracle (a violation of the actual laws, though not of the laws of the closest *p*-world) occurs at the moment to which *p* refers (in order to maximize the period of time during which the events of this possible world are the same as those in the actual world) unless too great a miracle would then be required to accommodate *p*. That is, it may be that a "small" miraculous departure from what actually happened, if this miracle occurs early enough, is sufficient for the actual laws to bring about *p* later, whereas if the miracle is deferred until the latest moment, then a "large" miracle, consisting of many and varied small miracles, is needed. In that case, the *p*-world with the small miracle is closer, Lewis says, although it matches the actual world for a shorter period. For this reason (as noted by Bennett 1984, p. 77), Lewis's "standard" criterion for identifying the closest *p*-world endorses the backward-directed counterfactual "If Stevenson had been president of the United States in February 1953, he would have been president in January 1953" because the only small miracle that could have produced *p* is Stevenson's having won the election held in November 1952. Here *q* describes a moment during the "transition period" (Lewis 1986b, p. 39) between the miracle and *p*.

Lewis deliberately designs his standard criterion for identifying the closest *p*-world so that it deems false all backward-directed counterfactuals except for those similar to this one about Stevenson. Since my concern is with whether Lewis is correct in abandoning Λ-preservation, I can leave aside the further details of his criterion and turn instead to whether the backward-directed counterfactuals that Lewis rejects are false, and so constitute counterexamples to Λ-preservation.

2.2.2. Are backward-directed counterfactuals odd? In this passage, Lewis (1986b) explains one of his motives for denying the truth of many backward-directed counterfactuals that Λ-preservation demands:

> Under determinism *any* divergence, soon or late, requires some violation of the actual laws. If the laws were held sacred [as Λ-preservation demands], there would be no way to get rid of [the fact that ¬*p*] without changing all of the past; and nothing guarantees that the change could be kept negligible except in the recent past. That would mean that if the present were ever so slightly different, then all of the past would have been different—which is absurd. So the laws are not sacred. (p. 171)

Likewise, Michael Slote (1978) says that "even assuming determinism, there is something odd about [holding] counterfactuals like 'if I had been depressed yesterday, the whole previous history of the universe would have been different' [to be] true" (p. 12). Slote elaborates an account of counterfactuals that, although quite different from Lewis's, also rejects these backward-directed counterfactuals.

Of course, it is misleading to say that, according to Λ-preservation, determinism requires that, had the most trivial present circumstance been different, then "all of the past" or "the whole previous history of the universe would have been different," as if every fact about the past would have been different. All that follows is that each past moment would have been different in some respect—perhaps trivial, perhaps not. If it strikes you as remarkable that a trivial counterfactual supposition could propagate through all past times, then I believe it should strike you as equally remarkable that it could propagate through all future times (yet neither Lewis nor Slote denies that it would). I put it to you that determinism portrays the cosmos as a remarkable place.

The counterfactuals noted by Lewis and Slote admittedly *do* sound weird, but they are not false (much less "absurd"). They seem odd because our standard reason for contemplating a counterfactual antecedent such as "Had I been depressed yesterday" is not to consider whether *every* past moment would then have been different in *some* respect or other, but instead to consider some *particular* moment and to examine whether it would then have been different in some *particular* way. That is so whether the counterfactual is forward directed ("Had I been depressed sometime yesterday, would I then have cried all day or phoned you?") or backward directed ("Had I been depressed sometime yesterday, would someone have first had to remind me of something sad?")

2.2.3. Other motives for denying backward-directed counterfactuals. A second consideration that may have encouraged Lewis to maintain that the laws are not to be held "sacred" is his conception of natural law. Recall from chapter 1 that, according to Lewis, laws differ from accidental generalizations in belonging to the deductively closed set of truths having the best combination of simplicity and informativeness. Suppose that this account is correct in finding only this fairly superficial difference between laws and accidents. Then it would sound unreasonable to insist that the laws be preserved at all costs under a counterfactual antecedent; it would seem much more natural to join Lewis (1973) in saying, "Laws are very important, but great masses of particular fact count for something too" (p. 75). That is, Lewis's view of lawhood suggests that laws are not so fundamentally different in kind from accidents that one cannot countenance a few departures from the laws in return for greater retention of accidental facts. Lewis (1986b) apparently thinks that to forbid this sort of trade-off would be narrow-minded:

> We should not take it for granted that a world that conforms perfectly to our actual laws is *ipso facto* closer to actuality than any world where those laws are violated in any way at all. It depends on the nature and extent of the violation, on the place of the violated laws in the total system of laws of nature, and on the countervailing similarities and differences in other respects. Likewise, similarities or differences of particular fact may be more or less weighty, depending on their nature and extent. Comprehensive and exact similarities of particular fact throughout large spatio-temporal regions seem to have special weight. It may be worth a small miracle to prolong or expand a region of perfect match. (p. 164)

Of course, this reason for doubting Λ-preservation is only as good as Lewis's account of lawhood. I argued in chapter 1 that, under Lewis's account, it is difficult to explain why laws carry any kind of special weight in determining what *p*-worlds are closest to the actual world. (Although Lewis rejects Λ-preservation in favor of assigning to laws a more modest role in determining the closest *p*-world, Lewis's standard criterion for closeness still assigns greater weight to the preservation of a law than to the preservation of an accident, for it assigns "miracles" a special significance.) In section 3 below (as I anticipated in section 1.3 of this chapter), I argue against Lewis's account of natural law—in particular, against his view that the laws in a given possible world supervene on the non-nomic facts there. Rather than taking Lewis's account of lawhood as a reason to reject Λ-preservation, I recommend rejecting Lewis's account of lawhood in favor of one that is more sympathetic to Λ-preservation.

Lewis (in the passage I just quoted) doubts "that a world that conforms perfectly to our actual laws is *ipso facto* closer to actuality than any world where those laws are violated in any way at all." I agree, if one uses an intuitive sense of "closer to actuality" that has nothing to do with counterfactual conditionals. For example, a world in which the gravitational constant differs from its actual value by a trillionth of a trillionth of a percent (too small to make any appreciable difference) would intuitively be closer to the actual world than a world in which the actual laws hold but there was so much matter shortly after the Big Bang as to produce a Big Crunch before stars and life had time to form. But to dispute Λ-preservation, Lewis must use the sense of "closer to actuality" that informs our judgements regarding the correctness of counterfactual conditionals. Lewis has agreed (1986b, pp. 42–3) that this is not the sense of "closer to actuality" that we would ordinarily use in judging a possible world's "similarity" to the actual world. Otherwise, as Kit Fine (1975, p. 452) noted, we would have to deny such intuitively correct counterfactuals as "Had Nixon pressed the button, there would have been a nuclear winter" because a world with a nuclear winter is intuitively quite different from the actual world; hence, a world in which Nixon presses the button, but a small miracle interrupts the signal, is intuitively closer to the actual world than a world in which a nuclear winter ensues in accordance with the laws. That Λ-preservation conflicts with our intuitive sense of the similarity of possible worlds is no argument against it.

Falling well beyond the scope of this book are two of Lewis's (1986b) other motives for rejecting the backward-directed counterfactuals that Λ-preservation demands. One motive derives from his analysis of causation (pp. 35–38, pp. 170–171); another arises from his account of our sense that the future is open whereas the past has been fixed (pp. 36–38). Discussion of these issues would take us too far afield, so I turn to what seems like Lewis's principal reason for deeming such backward-directed counterfactuals false under the standard criterion for the closeness of possible worlds. He expresses it thus:

> Today I am typing words on a page. Suppose today were different. Suppose I were typing different words. . . . Would yesterday also be different? If so, how? . . . I do not think there is anything you can say about how yesterday would be that will seem clearly and uncontroversially true . . . Seldom, if ever, can we find a clearly true counterfactual about how the past would be different if the present were somehow different. (p. 32)

But surely, we can think of many plausible backward-directed counterfactuals (other than Stevenson-like examples, which Lewis countenances), such as

> Had there been a thunderclap, it would have been preceded by a lightning bolt.
>
> Had this pendulum's period been half of its actual value, its cord would (have to) have been cut shorter by a factor of one-quarter.[28]

Had the die fallen with the side with six dots upmost, it would (have to) have been thrown differently.[29]

If President Kennedy had survived the shooting, he would have had to have been in a bullet-proof car.[30]

If Stirling Moss had won, he would have had to have used wet-weather tires from the start of the race.[31]

Had the match been alight, it would (have to) have first been struck.[32]

Had the match not been alight, although dry and in the presence of oxygen, it would (have to) not have been struck.[33]

Had I been depressed sometime yesterday, someone would have to have first reminded me of something terribly sad.

In the face of these examples, why does Lewis insist that backward-directed counterfactuals pose special difficulties?

2.2.4. The threat of backtracking. Lewis's concern is that if we permit backward-directed counterfactuals to go through, then we will become committed to incorrect counterfactuals that are not backward directed. Lewis (1986b, p. 33) recalls an example suggested by P. B. Downing (1959, p. 125). Suppose (as Bennett 1984, p. 70, tells the story) that Darcy and Elizabeth quarrel, and that Darcy, being a proud man, then refrains from asking Elizabeth for a certain favor. It seems correct to say that had Darcy asked for this favor, then Elizabeth, holding a grudge, would not have granted his request. But suppose that we also assent to the backward-directed counterfactual "Had Darcy asked for this favor, then there would (have to) have been no prior quarrel." We surely believe that if there had been no quarrel, then Darcy would have asked for his favor and Elizabeth, unfailingly generous except when cross, would have granted his request. Both intuitively and by the kind of transitivity that counterfactuals respect (Lewis 1973, pp. 32–35), it follows that had Darcy asked Elizabeth for a favor, then she would have granted his request, contrary to the above claim.

Jackson (1977, p. 9) offers another example. Suppose that I am standing by a window far above the street and that there is no net between me and the ground. It seems correct to say that had I jumped out of the window, I would have been gravely injured. But my cautious nature suggests the backward-directed counterfactual "Had I jumped, I would have arranged earlier for a net to be in position." Had I arranged for a net, I would have jumped and been unhurt. So had I jumped, I would have been unhurt.

Lewis (1986b, pp. 34–35) maintains that we avoid these inconsistencies because in each example, we reject the "backtracking" counterfactual (i.e., the counterfactual that ultimately follows from a backward-directed counterfactual). Lewis says that while these backward-directed counterfactuals might be true under a nonstandard criterion of closeness that we might charitably employ when our interlocutor utters such a counterfactual, they are false under the standard criterion.

I agree with Lewis that the two pairs of counterfactuals,

Had Darcy asked for a favor, Elizabeth would not have granted his request

Had Darcy asked for a favor, Elizabeth would have granted his request

Had I jumped, I would have been gravely injured

Had I jumped, I would have been unhurt

are similar to such pairs as W. V. O. Quine's

If Caesar had been in command in Korea, he would have used the atomic bomb

If Caesar had been in command in Korea, he would have used catapults (1960, p. 222)

and Goodman's

If New York City were in Georgia, New York City would be in the South

If New York City were in Georgia, Georgia would not be entirely in the South (1983, p. 15).

Which member of the pair is correct depends on which features of the actual world are preserved in the closest possible world in which the counterfactual antecedent holds. Different features should be held fixed in different conversational contexts; when we have different interests, different possible worlds qualify as the closest in which, say, Caesar was in command in Korea. Where I disagree with Lewis is in his calling the contexts in which the past is held fixed "standard" and the others "nonstandard." I see no reason to privilege some context as the "standard" one for entertaining the counterfactuals concerning the favor and the jump, just as we privilege no context as the "standard" one for entertaining the counterfactuals concerning Caesar and Georgia. No inconsistency arises as long as there is no context in which both counterfactuals of a pair are correct. To my mind, there is nothing substandard about a context in which it is correct to say "Had I jumped, I would have arranged for and so there would have been a net." Indeed, there are equally respectable contexts in which one would correctly say "Had I jumped, the window would (have to) have been much closer to the ground."

I agree with Bennett's point (1984, p. 71) in presenting this example. Whether a given physically unnecessary fact in the actual world is retained in the closest *p*-world depends on the context in which the counterfactual is posed. This is responsible for the alternative counterfactuals in the jump and favor cases. To deem all backward-directed counterfactuals false as standardly interpreted (except for Stevenson-like cases), and on this basis to deny Λ-preservation, is a misdirected overreaction to the threat of inconsistency that arises in these cases, since the same threat arises in cases (such as the Caesar and Georgia examples) that do not involve backward-directed counterfactuals.

2.2.5. "Would have" versus "would have to have" counterfactuals. Lewis (1986b) remarks:

> Back-tracking counterfactuals, used in a context that favors their truth, are marked by a syntactic peculiarity. They are the ones in which the usual subjunctive conditional constructions are readily replaced by more complicated constructions: "If it were that . . . then it would have to be that . . . " or the like. A suitable context may make it acceptable to say "If Jim asked Jack for help today, there would have been no quarrel yesterday," but it would be more natural to say " . . . there would have to have been no quarrel yesterday." (pp. 34–35)

Many others have noticed the same "syntactic peculiarity," apparently without finding it especially puzzling or suspicious.[34] Without argument, they all treat "Had the match been alight, then it would have to have first been struck" as equivalent to " . . . then it would have first been struck." But why, then, does the latter sound so much less "natural"? (Notice that on my list of plausible backward-directed counterfactuals, I often gave the more "natural" would have to have.") If we can convince ourselves of "Had the match been alight, it would have first been struck" principally by thinking of it as equivalent to "Had the match been alight, it would *have to* have first been struck," then we could undermine our confidence in the backward-directed counterfactuals that Λ-preservation requires by finding reason to doubt that they are equivalent to these other, more appealing counterfactuals.[35] In other words, if "would have to haves" and the corresponding "would haves" are not equivalent, then Λ-preservation is threatened, because friends of Λ-preservation use the intuitive plausibility of backward-directed "would have to haves" to argue for the corresponding "would haves," which sound unnatural but which Λ-preservation requires.

Indeed, it is evident that "Had I lit the fuse, I would *have to* have used a match" is not a different way of expressing "Had I lit the fuse, I would have used a match," since "would have to have" can be false when "would have" is true. Suppose that whenever I must light a fuse, I automatically reach for a match; my habit is to use matches to light my fuses. Accordingly, had I lit the fuse, I would have used a match. But suppose that by using something other than a match—say, an acetylene torch—I could have succeeded in lighting the fuse. Then it is not the case that had I lit the fuse, I would have to have used a match. Suppose I forget my matches and so do not light the fuse. Imagine the rebuke I receive from my supervisor: "Why didn't you light the fuse? I know that you happened not to have any matches with you, and I know—indeed, I am now even more confident than I was before—that had you lit the fuse, you would have used a match. But it isn't as if you would *have to* have used a match, and you did have an acetylene torch. You should have improvised!"

So "Had *p* been the case, then *q* would have to have been the case" is not synonymous with "Had *p* been the case, then *q* would have been the case." However, I believe that "would have to have" entails the corresponding "would have." So we can use the intuitively plausible backward-directed would-have-to-haves to motivate the corresponding unnatural would-haves, and thereby defend Λ-Preservation.

If we are willing to say that "Had *p* been the case, then *q* would have been the case" means that *q* holds in all of the closest *p*-worlds, then I propose that "Had *p* been the case, then *q* would have to have been the case" (whether forward or backward directed) means that *q* holds in all of the *p*-worlds that are either the closest or nearly the closest. Consequently, the "would have to have" counterfactual entails the corresponding "would have" counterfactual (but not vice versa). For example, in the closest worlds in which I light the fuse, I retain my habit of invariably using a match, whereas I do not retain this habit in some worlds in which I light the fuse that are nearly but not quite among the closest worlds in which I light the fuse. To say "Had I lit the fuse, I would *have to* have used a match" is to say " . . . I would *necessarily* (or *inevitably*) have used a match," " . . . I would have been *forced* (or *obliged*) to use a match," and so on. This counterfactual is true only if I would have used a match had I lit the fuse even if certain other circumstances had been different in some contextually allowable respects. That is what makes my using a match "inevitable"; there is no avoiding it by shifting to any of the other circumstances in the admissible range. The context in which the counterfactual is uttered influences not only which possible worlds qualify as closest, but also which are nearly the closest. But I suggest that as long as Con(Λ^+,*p*), the actual laws hold in all *p*-worlds that are nearly the closest.

The context dependence of the range of possible worlds that qualify as *nearly* the closest is manifested in this conversation between physicians A and B:

> Dr. A: I heard that your patient died.
>
> Dr. B: Yes. Had I aggressively treated the illness, I would have to have used drug *X*. It isn't merely that I would have used *X*, considering my success with it in similar, though less advanced cases. I would have had to use *X*, since this is the only drug licensed for the patient's illness. As you know, *X*'s administration would have caused nasty side effects, greatly diminishing the patient's quality of life, and because the illness in this case was so advanced, *X*'s administration would have prolonged the patient's life only a little, if at all. So the patient and I agreed that I would not aggressively treat the illness, but simply try to make her more comfortable.
>
> Dr. A: But you could have used drug *Y* instead, which widely known but preliminary unpublished reports suggest would have been effective without side effects.
>
> Dr. B: But drug *Y* is not yet legal.
>
> Dr. A: Well, perhaps you could have smuggled some in. I have heard it done. I don't think that you would *have* to have used drug *X*, although I agree that you would have.
>
> B: Well, of course, if you wish to take that kind of option seriously, then admittedly, I wouldn't *have* to have used drug *X*. But it would have been wrong to have used *Y*. . . .

Dr. B has not exactly changed his mind about whether he would have to have used drug *X*. Rather, in the course of the conversation, there is a change in the criterion being used for a world that is nearly the closest in which Dr. B aggressively treats the patient's illness. We can imagine similar exchanges revolving around such would-have-to-haves as "Had there been no butter in the house, I would have to have used margarine" and "Had I invited Sally, I would have to have invited her husband." (Note once again that Dr. A can with consistency assert "Had Dr. B aggressively treated the illness, he would have used drug *X*" and deny " . . . he would have to have used drug *X*.")

So why is it that when the backward-directed "Had *p* been the case, then *q* would have been the case" is correct, then "Had *p* been the case, then *q* would have to have been the case" often sounds more natural? And why is this anomaly peculiar to backward-directed counterfactuals, if the foregoing analysis applies both to forward- and to backward-directed would-have-to-haves? I suggest that the ways in which we standardly use backward-directed counterfactuals are different from the paradigmatic uses of forward-directed counterfactuals and require would-have-to-haves. Therefore, a backward-directed would-have often sounds "unnatural." On the other hand, the roles that forward-directed counterfactuals ordinarily play can be performed by would-haves and cannot be played by would-have-to-haves.

To defend this view properly, I would have to survey and catalog the ways in which backward-directed and forward-directed counterfactuals are commonly used. I do not attempt this here. But let's look at one of these standard uses to see whether the line I am suggesting seems plausible.

We use both forward-directed and backward-directed counterfactuals when we discuss whether a given agent chose the best course of action. We try to determine whether any alternative course of action that was open to the agent would have led to better results than the course that she actually chose to take. We might identify an alternative course of action that we believe to have been available to the agent, and investigate what its results would have been. This involves a forward-directed counterfactual—"Had the agent done *p*, then *q* would have resulted"—where the course of action at issue is the agent's doing *p* along with whatever else she would have done had she done *p*. To ascertain that this alternative would have been less effective than the agent's actual course of action was, it suffices to know what would *in fact* have happened in the event that the agent had done *p*. We do not need to know whether this would *have to* have resulted from her doing *p*.

The further information conveyed by the would-have-to-have counterfactual may even be misleading with regard to whether the agent acted for the best. That is because the would-have-to-have conditional may depend not only on what would have happened had the agent done *p* (i.e., on the closest possible world in which the agent does *p*), but also on what would have happened had the agent done *p* in slightly different circumstances (i.e., on the possible worlds in which the agent does *p* that are nearly but not quite the closest). Whether the agent should have done *p* in the circumstances in which she *actually* found herself does not depend on what *p*'s results would have been in *other* circumstances. For example, suppose that I am driving my car and start to make a left turn. Suddenly, a pedestrian darts across my intended path. I slam on the brakes and barely avoid colliding with her. Did I do the right thing in braking? Yes—had I not braked, I would have struck her. But suppose it is not the case that had I not braked, I would *inevitably* have struck her. Admittedly, had I failed to brake as I did, I (or she) would *have to* have done *something* or else I would have struck her. But in some of the possible worlds where I don't brake that are nearly (but not quite) the closest, she sees me turning just in time to jump out of the way. So I was right to brake; had I not done so, I would have struck the pedestrian, since she would not have seen me in time. But it is not the case that had I not braked, I would *have to* have struck her; there are other ways that the accident could have been avoided.

In this example, "would have to have" is misleading with regard to whether my conduct was the best I could have chosen. In other examples, "would have to have" reveals that even if the agent had done *p* in the course of a slightly different program of action—different from what she would have done, had she done *p*—then she would still not have done as well as she actually did. For example, suppose I am picking up a friend at the airport and there are two ways to get there—via the bridge or the tunnel. I choose the tunnel and am late in picking up my friend. But, I argue, I chose the best option; had I gone by way of the bridge, I would have been even later. My friend suggests that had I taken the bridge and left a little earlier than I did, I would have arrived on time. No, I explain: had I taken the bridge, I would *inevitably* have arrived even later than I did, had I left a little earlier or not, because several lanes of the bridge were closed for scheduled repairs today, and traffic across the bridge has consequently been very slow all day. Had I taken the bridge and arrived on time, there would have to have been no repair work on the bridge today. In this example, "would have to have" is not irrelevant to evaluating my choice. It reveals that none of the courses of action in a certain range—all of which involve my taking the bridge—would have succeeded. Nevertheless, consider any *particular* course of action in that range, such as my taking the bridge and doing whatever else I would have done had I taken the bridge, such as leaving no earlier than I actually did. To know whether this particular course of action would have been better than the course I actually pursued, it suffices to know the forward-directed would-have conditional ("Had I taken the bridge, I would still have been late—even later than I actually was").

We also use *backward-directed* counterfactuals when discussing whether an agent chose the best option. There may be a course of action that we believe would have worked out better than the course that the agent actually selected. We must determine whether this promising alternative was actually available to the agent—whether she could somehow have managed to perform it. In particular, we must determine whether certain past events for which the agent is not responsible closed off this alternative. To know this, it does *not* suffice to know that, had the agent performed this action, the past *would have* been different in a certain respect. The agent could still perhaps have managed to perform the action; perhaps the past would not *have to* have been different in that respect. That is, although in the *closest* possible world in which the agent performs the action, the past is different in a respect that was actually beyond her control, there may still be a possible world in which she performs the given action that is *nearly* the closest such world and where the past is not different in such a respect. In that event, the action was available to her. To know that the course of action was not a live option, knowledge of the backward-directed would-have is insufficient; we must know the relevant would-have-to-have. That is why Dr. A needed to consider whether the government would *have to* have legalized drug *Y* in order for Dr. B to have used it, and likewise why my supervisor emphasizes that it is not the case that I would *have to* have used a match had I lit the fuse.

To know whether a given course of action is open to the agent, it is not enough to know whether past events for which the agent is not responsible *would* have been different had she performed the action; we must know whether they would *have to* have been different. I believe that many backward-directed counterfactuals are most easily imagined functioning in such contexts. Since these contexts require "would have to have," many backward-directed counterfactuals seem more natural as would-have-to-haves than as would-haves. The same is not true of forward-directed counterfactuals, since for their characteristic uses, "would have" is appropriate.

2.2.6. The challenge posed by nonbacktracking contexts. I return now to Darcy and Elizabeth. I have argued that in different contexts, we take different approaches to the counterfactual antecedent "Had Darcy asked Elizabeth for a favor." In some contexts, we are right to "backtrack"; we should allow the closest possible world in which Darcy requests a favor from Elizabeth to differ from the actual world in the events prior to Darcy's making his request. For example, in a context in which we are illustrating Darcy's pride, we should say that Darcy would have asked Elizabeth for a favor only if there had been no prior quarrel between them, in which case Elizabeth would have granted his request. In other contexts, it is inappropriate for us to backtrack; in the closest possible world in which Darcy requests a favor from Elizabeth, the actual events prior to Darcy's request are preserved. For instance, if our concern is to explain why Elizabeth did not perform a certain favor for Darcy, we might assert that Darcy did not ask Elizabeth for the favor, and it is no wonder—for he knew that if he had, she would not have granted it, in light of yesterday's quarrel.

Now the problem is that if the actual laws are deterministic and their truth is preserved in the closest possible world, and if the period in that possible world before Darcy makes his request is just as it is in the actual world (as occurs in nonbacktracking contexts), then the rest of the possible world's history must also be the same as the actual world's.[36] But this precludes Darcy's asking Elizabeth for a favor in that world—the closest possible world in which Darcy asks Elizabeth for a favor! So the actual laws cannot be preserved in that possible world. In other words, that Darcy asks Elizabeth for a favor is logically inconsistent with the actual laws together with the actual past, since the actual past omits the causal antecedents of Darcy's action that the laws require. Since, in a nonbacktracking context, we are preserving the actual past under the counterfactual supposition that Darcy asks Elizabeth for a favor, we must not be preserving the actual laws. Here we have an argument against Λ-preservation.[37]

What are we to make of this argument? In certain contexts, when we entertain a counterfactual supposition, we do regard the actual course of events as preserved prior to the moment with which that supposition is concerned. We do not worry about whether the laws require the counterfactual supposition to have certain causal antecedents, and those antecedents themselves to have certain causal antecedents, and so on. If we had to take into account all these causal antecedents, stretching back forever, then we could hardly be justified in asserting various counterfactual conditionals that intuitively we have every right to assert. As Bennett (1974) puts it, we "need not envisage a history for the antecedent-situation" (p. 391); we need not consider how it managed to be brought about. We proceed as if the counterfactual supposition manages to obtain without any causal antecedents at all, in violation of the actual laws—as if through a miracle, as Lewis says.[38] Nevertheless, it remains highly counterintuitive to assert that had Darcy asked Elizabeth for a favor, then the natural laws would have been different. Remember, the laws of nature would have been no different had I not been born, or had no asteroid crashed into Earth to wipe out the dinosaurs, or had the Sun never formed. So why should the laws have been different had Darcy asked Elizabeth for a favor? *Pride and Prejudice* is fiction, but it is not *science* fiction.[39]

We might try to reconcile these intuitions by again taking context into account. (I endorse only part of the following view, but it serves as a bridge to the view that I adopt at the close of this section.) When we assert $p > q$, certain features of the closest *p*-world are not our concern. For instance, I might say, "Had Abraham Lincoln watched the U.S. presidential campaign debates between Bill Clinton and Bob Dole in 1996, then he would surely have compared them unfavorably to his own debates with Stephen Douglas in 1858."[40] Of course, in the relevant possible world, Lincoln must somehow be reincarnated (with his old memories intact), and so either Lincoln is not a human being in that world or the actual laws are violated there. But how Lincoln could have been in a position to compare the two debates is not our concern; we are interested only in the content of that comparison. Likewise, when we say (in a nonbacktracking context) that Elizabeth would not have granted Darcy's request, we are not concerned with the causal antecedents of Darcy's request. So although Darcy makes his request miraculously, we do not assert that had Darcy requested a favor from Elizabeth, then that request would not have possessed the causal antecedents required by the actual laws of nature. That counterfactual conditional is irrelevant. If later the context changes, shifting our concern to the causal antecedents of Darcy's request, then we allow backtracking. We call to mind a different possible world, where no quarrel takes place and Darcy's request possesses the causal antecedents demanded by the actual laws. So in neither backtracking nor nonbacktracking contexts do we assert that a miracle would have occurred—that the laws would have been different—had Darcy made his request, even where the closest possible world includes a miracle. Any miracle in that world takes place "offstage," out of sight of the counterfactuals that we entertain in that context. If we try to bring the miracle "onstage," we inevitably alter the context to permit backtracking and so the possible world with the miracle no longer counts as the closest possible world in which Darcy requests a favor from Elizabeth.

I think that this view is correct in holding that, in backtracking and nonbacktracking contexts alike, we fail to assert that, had Darcy requested a favor from Elizabeth, the laws of nature would have been different. But considering that, in a nonbacktracking context, a miracle takes place in the closest possible world in which Darcy requests a favor from Elizabeth, why are we not obliged to say in that context that the laws of nature would have been different? The explanation I just floated was that, in this context, we are not interested in the causal antecedents of Darcy's request, and so we do not assert counterfactuals about them, just as in the Lincoln example certain aspcts of the closest possible world lie beyond our concerns. And if we turn our attention to the causal antecedents of Darcy's request, then a *different* possible world becomes the closest—a world where no miracle precedes Darcy's request. But this explanation has two defects. First, even if we are not interested in the causal antecedents of Darcy's request, we *are* interested in the natural laws in the closest possible world. We use those laws in figuring out what happens in that world. Since an offstage miracle is still a miracle, we should apparently say in a non-backtracking context that the laws of nature would have been different had Darcy requested a favor from Elizabeth. The second problem with this explanation is that it supposcs that, if we turn our attention from whether Elizabeth would have granted Darcy's request to whether the laws would have been different had Darcy made his request, then a different possible world becomes the closest. But this shift does not seem to happen.[41]

Consider an example in which such a shift does occur. Suppose that we initially assert that Elizabeth would not have granted Darcy's request, but later, in a different context, we allow backtracking and assert that she would have granted Darcy's request. We did not contradict ourselves. Rather, the context shifted, and accordingly, there was a shift in which possible world qualified as the closest. We set aside the possible world that we were originally considering (in which Darcy quarrels with Elizabeth) and called to mind a different possible world (in which no quarrel occurs). Now, I see no reason to think that the same sort of shift occurs when we start by asserting (in a nonbacktracking context) that Elizabeth would not have granted Darcy's request had he made one, and sometime later assert that the laws of nature would have been no different had Darcy made his request. In asserting a counterfactual about the laws, we seem to be contemplating *the very same possible world* that we had brought to mind when we asserted that Elizabeth would not have granted Darcy's request. Whereas we would not say "Had

Darcy requested a favor from Elizabeth, she would not have granted his request *and* she would have granted it," since this would run together two distinct possible worlds, we *are* willing to say that had Darcy requested a favor from Elizabeth, the laws of nature would have been no different *and* (partly as a consequence, perhaps!) she would not have granted Darcy's request, in view of their earlier quarrel. The natural laws we are talking about must then be the laws in the possible world where Elizabeth does not grant Darcy's request. Despite the "miracle" in that world (i.e., the violation of the actual world's laws), we say that the laws in that world are the same as the actual laws.

So I find myself coming to the conclusion that the actual laws *can be violated* in a possible world *and still remain laws there*, as long as every violation occurs "offstage," outside the range of events with which we are concerned in contemplating this world. The laws in a possible world depend only on those aspects of the world with which we are concerned, and in connection with *those* aspects, the laws are *not* violated.

I realize that many will find this conclusion unpalatable—too bizarre to merit serious consideration. But the nature of counterfactuals in nonbacktracking contexts suggests that *p* can express a physical necessity in a given possible world even if it is false there, as long as it is "reliable" there—that is, close enough for our purposes to being true in that world.[42] As I discuss in chapter 6, a violation may be beyond our concern either because it is negligibly small, considering the degree of inaccuracy that our purposes in the given context can tolerate, or because it involves facts in which we are uninterested in this context.[43]

To accommodate $\neg(\blacksquare p \Rightarrow p)$, we must redefine "$\Lambda$-preservation." We must say, regarding two sets (Γ and Δ) of sentences, that Γ is "preserved under Δ" if and only if for any $m \in \Gamma$ and for any $p \in \Delta$, $q \in \Delta$, $r \in \Delta$, and so on, all of the following are *reliable* (i.e., close enough, for the relevant purposes, to being correct): $p > m$, $p > (q > m)$, $p > (q > (r > m))$, and so on. We must also redefine "Con(Γ,*p*)" as holding exactly when the claims to which we would be entitled by reasoning from *p*, in accordance with the inference rules corresponding to the members of Γ, could all be reliable, considering what reliability requires in this context.[44] In nonbacktracking contexts, our purpose in conjuring up the world in which Darcy asks Elizabeth for a favor is to predict something about what would have happened *after* Darcy's request. So the laws of that world can be reliable there even if they are violated *prior to* Darcy's request. The actual laws' reliability is preserved under this counterfactual antecedent, in accordance with Λ-preservation. The counterfactual "Had Darcy asked Elizabeth for a favor, then the actual laws would still have been true" is (in the relevant context) close enough to being correct.

Of course, the notion that the laws of a given possible world can be violated there, as long as these violations are not of any concern to us, runs contrary to the traditional view of natural laws as associated with exceptionless uniformities. In chapter 6, I give an independent argument for my heterodox view: that it is needed in order to account for the ways in which the concept of a natural law figures in our reasoning *about the actual world.* This later argument, let me emphasize, does not concern *counterfactual* reasoning at all, and so should relieve us of the suspicion that in arguing against $\blacksquare p \Rightarrow p$ in the present section, I have been misled by certain strange features of counterfactuals. After offering this argument in chapter 6, I resume my discussion of nonbacktracking contexts.

2.3. *Putative counterexamples involving metaphorical uses of counterfactuals*

If a counterfactual is true or false, or merely correct or incorrect in some sense, then there is presumably a difference between interpreting it literally and interpreting it figuratively. I once encountered a charming children's book by Sarah Perry entitled *If* (Children's Library Press, 1995). It contains illustrations accompanying such imaginative counterfactuals as "If dogs were mountains, their noses would form a jagged river bank while their bodies turned into wooded slopes," "If toes were teeth, they would wiggle between a little girl's lips when she smiles," and likewise for "If fish were leaves . . . " and "If butterflies were clothes. . . . " These counterfactuals are surely not intended literally.

We sometimes assent to a counterfactual that would conflict with Λ-preservation if it were interpreted literally, but we are supposed to interpret it metaphorically. Let me offer an example in the spirit of the remark: "When he ran away, he was so scared that had he been a dog, he would have had his tail between his legs." Suppose that it is a natural law that all robins have red breasts. Here is the example: "He is so contrary, contrary just for the sake of being contrary, that if he had been a robin, he would have had a blue breast." I regard this counterfactual as a figure of speech. (After all, who is to say that a blue breast is really more contrary than a green breast?)

2.4. *Putative counterexamples involving contexts in which beliefs about laws are set aside*

Consider a weather glass that is guaranteed by natural law to predict accurately whether the next day's weather will be fair or foul, as long as the glass is in working order.[45] Suppose that it is fair today and the weather glass was in working order yesterday. Then we would ordinarily accept the counterfactual, "If the weather glass had read 'foul' and been in working order yesterday, the weather would have been foul today." In accepting this counterfactual, we respect Λ-preservation; we regard the laws responsible for the glass's accuracy as preserved under an antecedent p where $\text{Con}(\Lambda^+,p)$. But sometimes we accept the counterfactual, "If the weather glass had read 'foul' and been in working order yesterday, the weather glass would (contrary to law) have been inaccurate" since it would still have been fair today.

For example, a vendor of weather glasses, who knows that the accuracy of her products is secured by natural law, nevertheless finds herself asserting this counterfactual in the course of this conversation with a potential customer:

> Customer (pointing to a glass): Is this weather glass reliable?[46]
>
> Vendor: Yes. For instance, it read "fair" yesterday, and you can plainly see that it is fair today.
>
> Customer: But maybe it read "fair" yesterday because it was broken so that it would always read fair. Then its "accuracy" yesterday would not constitute good evidence of its accuracy when it is in working order. Even a broken clock is right twice a day.
>
> Vendor: It was in proper order yesterday.
>
> Customer: Good. But does the fact that it read "fair" yesterday and was in working order, and that today's weather is fair, confirm (to some degree) that whenever it is in working order, its prediction is accurate? For the test to supply us with good evidence, it must be that the hypothesis could have failed the test—that the test could have revealed the weather glass to have been inaccurate yesterday. For instance, you cannot confirm "All ravens are black" by asking your robot to bring only black objects to you and noticing that one of them is a raven. By this procedure, you could not have discovered a counterexample, a raven that is not black.
>
> Vendor: Yes, I too have studied a little philosophy of science. But had the weather glass read "foul" yesterday and been in working order, it would have been inaccurate, since today is fair. So yesterday's "fair" reading, made while the weather glass was in working order, confirms the instrument's accuracy whenever it is in working order.

It seems, then, that the vendor regards Λ-preservation as violated, since she accepts $p > q$ ("Had it read 'foul' yesterday and been in working order, it would have been inaccurate"), $\text{Con}(\Lambda^+, p)$, and $\neg\text{Con}(\Lambda, p \;\&\; q)$.[47] In other words, the vendor is willing to sacrifice the actual laws (which suffice to ensure the glass's reliability) for the sake of preserving today's weather, a physically unnecessary fact.

We can reconcile this example with Λ-preservation by noting that the vendor utters this counterfactual while trying to convince the customer of the very laws whose preservation is at issue. The vendor presents yesterday's weather-glass reading and today's weather as evidence for those laws. Therefore, although the vendor herself already believes in those laws, she cannot presume them when entertaining the counterfactual; from the customer's viewpoint, she would then be begging the question. After all, had the vendor been willing to invoke her knowledge of laws that are unknown to the customer, she could have answered the customer's initial question ("Is the weather glass reliable?") simply by citing the relevant laws. Instead, she offers evidence for these laws.

In short, the vendor is entertaining the counterfactuals on behalf of the customer. Since the customer does not initially agree with the vendor that certain claims express laws, respect for Λ-preservation does not oblige the vendor to treat those claims as truths that are preserved in the manner of laws. The vendor herself already believes ¬Con(Λ,*p* & *q*), but the customer does not (at least initially).[48]

2.5. *Putative counterexamples involving ambiguous counterfactual antecedents*

Van Inwagen (1979) notes that "All human beings are mammals" expresses a law. Therefore, "If Cleopatra's asp had been human, it would have been a mammal" appears to be required by Λ-preservation. Yet he demurs:

> It seems absolutely inconceivable that an asp should be or should have been a human being. And who can say *what* would have been the case if something absolutely inconceivable had happened? One *could* say . . . that just anything would be the case if the inconceivable were to happen, and thus that [this counterfactual] is true. But some philosophers might not be satisfied with an account of counterfactual conditionals that confers truth upon "If Cleopatra's asp had been human, it would have been immaterial." And some philosophers might even argue that at least some instances of "Cleopatra's asp is human > *p*" (or, worse, "The number 5 is human > *p*") are *meaningless*. (p. 443)

Adam Morton (1973) raises a similar difficulty in an article with the memorable title "If I Were a Dry Well-Made Match":

> [C]onsider the law "All tailless offspring of second generation tailless Manx cats are feeble kittens." It surely does not support the counterfactual "If this pebble were a tailless offspring of second generation tailless Manx cats it would be a feeble kitten." For it could as easily (that is to say, not at all easily) come about that this pebble was the offspring of second generation tailless Manx if the copulation of two second generation Manx resulted in a pebble as it could if this pebble were a cat. (p. 323)

These examples all rest on the ambiguity of their counterfactual antecedents. By supposing that this pebble had been a tailless offspring of second-generation tailless Manx cats, is one supposing that the given object, which happens actually to be a pebble, is *instead* the offspring of cats, or is one supposing that a particular object is a pebble and *also* the offspring of cats? In other words, is "this pebble" in the counterfactual antecedent being used merely to identify the object in question (by describing how it happens to be in the actual world), or is it intended to describe the object in the closest possible world in which the counterfactual antecedent obtains? Perhaps the object in question is *essentially* a pebble. In that case, the object's pebblehood must be preserved in the closest logically possible world in which the counterfactual antecedent obtains, since for that object to exist in that world at all, it must exist as a pebble there. If (to switch examples) the counterfactual antecedent "Had Cleopatra's asp been human" directs us to a possible world in which a certain animal, Cleopatra's "pet," is instead a human being, then we readily assent to the counterfactual "If Cleopatra's asp had been human, it would have been a mammal," as Λ-preservation demands. If, on the other hand, the counterfactual antecedent "Had Cleopatra's asp been human" directs us to consider the closest possible world in which there is a human asp—perhaps because the species of Cleopatra's asp is essential to it—then the antecedent is inconsistent with Λ^+, and so Λ-preservation does not require "Had Cleopatra's asp been human, it would have been a mammal." In either event, then, Λ-preservation is upheld.

The same applies to "Had I been a dry, well-made match." Assume it is physically necessary that all dry, well-made matches light when struck. If it is conceptually or metaphysically impossible for me to have been a match, then this counterfactual antecedent falls outside the scope of Λ-preservation, and so Λ-preservation does not demand "Had I been a dry, well-made match and been struck, then I would have lit." If I am not prohibited by my essence from being a dry, well-made match, then Λ-preservation presumably requires this counterfactual, and that seems correct. I take no position on whether I have an essence or, if I do, whether it is incompatible with my being a match.

Likewise, perhaps the antecedent of the Manx counterfactual ("If this pebble were a tailless offspring of second generation tailless Manx cats, then it would be a feeble kitten") does not ask us to consider the closest possible world in which this object is simultaneously a pebble and a tailless offspring of second generation tailless Manx cats. In that case, the conditional holds, in accordance with Λ-Preservation. On the other hand, perhaps the antecedent asks us to consider the closest possible world in which this object is a pebble and also a tailless offspring of second generation tailless Manx cats; perhaps the object's pebblehood is one if its essential properties. Then the antecedent is inconsistent with Λ^+. Either way, Λ-preservation is upheld.

Some counterfactual antecedents are clear about their demands. When I say "Had this copper wire been made of rubber, it would have been an electrical insulator," you understand me to be interested in a possible world in which that wire, which actually happens to be copper, is instead made of rubber. You know that I am using the fact that the wire is copper merely to identify it, and since it is law-contravening for a rubber object also to be copper, the wire is not copper in the closest possible world in which it is rubber. In other cases, it is unclear whether a certain definite description is merely identifying the object in question. We are perhaps tempted to think, with van Inwagen, that "Had Cleopatra's asp been human" expressly demands a possible world where there is a human asp. But then the antecedent is law-contravening, and so outside the scope of Λ-preservation.

2.6. *Putative counterexamples involving our beliefs about the actual world*

Had an emerald been the same color as stop signs in the actual world, then there would have been a red emerald. Under this counterfactual antecedent, then, Λ is not preserved (since it is a law that all emeralds are green). This poses a difficulty for Λ-preservation since "An emerald is the same color as the stop signs in the actual world" appears consistent with Λ^+. It seems logically possible for all actual stop signs to be green; they merely happen not to be.

However, this is too quick. If the stop signs in the actual world are red, then in all logically possible worlds, the stop signs in the actual world are red. (Perhaps it is misleading to say "in" all logically possible worlds, but that's precisely the point.) Therefore, if the stop signs in the actual world are red, then there is no logically possible world in which the stop signs in the actual world are green. (We must be very careful to understand talk of the "actual world" rigidly.) Hence, if in the actual world all of the stop signs are red and it is a law that all emeralds are green, then there is no logically possible world in which Λ^+ holds and there is an emerald of the same color as the stop signs in the actual world. Therefore, the counterfactual antecedent we were examining ("Had an emerald been the same color as the stop signs in the actual world") falls outside the scope of Λ-preservation.

A counterfactual antecedent can refer to the actual world even if it does not explicitly include the word "actual." The context, a tone of voice, or the word order can suggest that the antecedent implicitly includes a rigid reference to the actual state of things. For instance, consider the difference between these two counterfactual antecedents: "Had emeralds been the color of stop signs" versus "Had stop signs been the color of emeralds." The former (in a certain tone of voice) directs us to imagine how the world would have been had emeralds been the color of actual stop signs. The latter tells us to imagine how the world would have been had stop signs been the color of actual emeralds.

2.7. *Putative counterexamples involving no "connection" between antecedent and consequent*

It has sometimes been suggested that even if *q* obtains in the closest *p*-world, the counterfactual $p > q$ may be incorrect; to secure the counterfactual, there must be some further connection between *p* and *q*. For instance, consider the counterfactual "Had I been wearing a green shirt, then all emeralds would have been green." If this counterfactual's correctness requires not merely that all emeralds would still have been green, but also that some causal connection exists between the emeralds and my shirt, then this counterfactual is incorrect, even though all emeralds are green in the closest possible world in which I am wearing a green shirt. Since Λ-preservation requires that this counterfactual be correct, Λ-preservation would then be violated.

One way to address this problem would be to amend the definition of "preservation" so that *m* is "preserved under *p*" exactly when *m* holds in the closest *p*-world. This would leave room for *m* to be preserved under *p* even if $\neg(p > m)$ because *m* is not properly connected to *p*. But I don't think that this step is necessary; I am not persuaded that Λ-preservation fails when "*m* is preserved under *p*" is defined as $p > m$. Consider an example. Suppose that, as an accidental matter of fact, all of the coins I receive in change today are pennies. Consider counterfactual *c*:

> Had a certain lion in Africa succeeded in catching a certain zebra, then all of the coins I received in change today would have been pennies.

Although *c*'s consequent *q* obtains in all of the closest possible worlds in which *c*'s antecedent *p* holds, there is no "connection" between *p* and *q*. Indeed, that is why *c* is correct; it follows from *d*:

> Even if the lion had caught the zebra, all of the coins I received in change today would still have been pennies.

But Fine (1975, p. 453), for instance, deems false any counterfactual like *c*, that is, with a "true and unconnected consequent." However, Fine accepts *d*. He joins Chisholm (1946, p. 301) in regarding *d* as the negation of "If the lion had caught the zebra, some of the coins I received in change would not have been pennies," and hence, as the negation of an incorrect conditional. In this case, according to Fine, $p > q$ (i.e., *c*) and $p > \neg q$ are both incorrect, as when we deny both "Had I flipped this fair coin, it would have landed heads" and "Had I flipped this fair coin, it would not have landed heads" (since a heads-world and a tails-world tie for closest, and $p > q$ holds only if *all* of the maximally close *p*-worlds are *q*-worlds).

Fine's interpretations of *c* and *d* run contrary to my intuitions, according to which both *c* and *d* are correct. Fine's motivation is presumably that *d* sounds considerably more natural than *c*. I agree, but I believe that I can account for this impression without holding *c* to be incorrect. It seems to me that *c* and *d* make the same correct counterfactual claim, but *d* additionally makes a true claim about the actual world: that all of the coins I receive in change today are pennies.[49] Therefore, to utter *c* rather than *d* is to suggest that we do not know that actually all of the coins are pennies, since if we did, we would likely have uttered *d* instead.[50] For this reason, *c* is misleading when its consequent is known to be true.

Sometimes a conversational implicature of asserting *c* is that the events involving the lion and zebra are causally related to the change I receive today. It is then misleading to assert *c*. Analogous remarks apply to "Had I been wearing a green shirt, then all emeralds would have been green."

It seems to me, then, that we need not reject Λ-preservation in order to account for our preference for *d* over *c*.

2.8. Conclusion

The preceding section examined several kinds of potential counterexamples to Λ-preservation and found that each either fails to undermine this principle or suggests some refinement of it. This is hardly the final word, of course; we can continue to investigate any apparent difficulties for Λ-preservation, to develop a typology of problem cases, and to elaborate the resources available to Λ-preservation for dealing with them. But I shall leave matters as they are for now.

3. Λ^+-preservation

3.1. *Preservation of the laws' lawhood and nested counterfactuals*

Having examined whether the actual laws' *truth* is preserved under counterfactual antecedents *p* such that Con(Λ^+,*p*), I now turn to whether their *lawhood* (along with the rest of Λ^+) is also preserved there.[51] In many examples, it seems manifestly to be. Had I failed to brush my teeth this morning, the natural laws would have been no different. Even if no human being in 1993 had believed that $E = mc^2$, the natural laws would have been exactly the same—it would still have been a law that $E = mc^2$.

What is the source of these intuitions? As I suggested briefly in section 1.3 above, they are closely tied to the distinction between the natural laws and the initial conditions. We imagine that the universe is initially set up in some condition, and the natural laws then take over to direct its course through time. The universe's unfolding in accordance with the laws generates a self-consistent history as long as the initial conditions are consistent with Λ. To put the point colorfully: God might be pictured as first designating which claims in U express laws, and as then grasping the knobs that set the initial conditions in U; God can twiddle them in any manner within the range consistent with the Λ that God has already designated.[52] When we contemplate how the world would have been had certain conditions *p* (in U) obtained, we are in our own minds "playing God" in that we are hypothetically twiddling the knobs to set *p* as the universe's condition at a given time. We do not thereby reset the laws unless *p* is inconsistent with Λ.

Scientists do sometimes "play God" in this way. For example, cosmologists investigate how the universe would have developed had the Big Bang possessed certain characteristics—had it been more or less energetic, or had the initial distribution of matter been more or less homogeneous. They do this by running computer simulations of the universe's development under various hypothetical initial conditions.[53] Intuitively, the hypothetical initial conditions are what the scientists feed into the computer program on a particular run, whereas the scientists have built the actual laws of nature directly into the program, where they function as the natural laws of the possible universes being simulated. Likewise, in his charming popularization *What If the Moon Didn't Exist? Voyages to Earths That Might Have Been* (Harper Collins, 1993) Neil F. Comins (professor of astronomy and physics at the University of Maine) considers what Earth would have been like had the Moon never existed, or had Earth been more or less massive, or been more or less tilted on its axis, and so on. He writes:

> [B]y making small changes in the collapsing, swirling cloudlet that became our solar system, it is entirely plausible that the planets, moons, and other matter here would all be different than they are. . . . [C]hanges in the collapsing cloudlet would have led to the earth having a different mass or spinning on its axis at a different angle as it orbits the sun. The earth's global properties (such as its mass, chemical composition, and rotation rate) could all have been different, since the laws of physics and chemistry would have allowed matter to be distributed differently during the formation of our solar system. The worlds in this book do not exist. We will develop and explore them by extrapolating from conditions on the present earth and by using plausible astronomical and geological conditions or theories. (pp. xiv–xv)

Of course, we might understand Comins as contending merely that the actual laws would still have been *true*, but might not still have been *laws*, had the solar system's initial conditions been different. But it seems much more natural to interpret these remarks as contending that the laws would have been no different.

According to the preceding section, the laws remain *true* in the closest world in which I failed to brush my teeth this morning; I am now concerned with whether they remain *laws* in that world. For this reason, it is important to examine the *counterfactual conditionals* obtaining in that world, since they reflect the laws of that world. So rather than asking "Had I failed to brush my teeth this morning, would the natural laws (e.g., that all emeralds are green) have been any different?" and noting that our intuitions conform to Λ^+-preservation, we could have approached the same issue by asking "Had I failed to brush my teeth this morning, then had there been emeralds on that table, would they all have been green?" Intuitively, the answer is "Yes." This question concerns a nested counterfactual $p > (q > r)$ that falls within the scope of Λ-preservation applied to the actual world.[54] The counterfactual conditionals that obtain in the closest p-world, where p (in U^+) is consistent with the actual world's Λ^+, tell us something about the closest p-world's Λ^+, since by Λ-preservation, the closest p-world's Λ must be preserved in the q-world closest to the closest p-world (as long as $q \in U^+$ and q is consistent with the closest p-world's Λ^+). The correctness of "Had I failed to brush my teeth this morning, then had there been emeralds on that table, they would all have been green" is consistent with the actual world's Λ^+ remaining the Λ^+ in the closest possible world in which I failed to brush my teeth this morning.

Comins offers us some nested counterfactuals that conform to Λ-preservation. He refers to Earth as it would have been, had the moon not existed, as "Solon." He then writes (p. 7) that, like the young Earth, young Solon would have rotated once every six hours, thereby dragging the water in the high tide away from the point directly below the Sun. But if Solon had not rotated, then high tide would have been located directly between the Sun and the center of the planet. In other words, had Solon existed (i.e., had Earth possessed no moon), then had Solon not rotated, high tide would have been located directly between the Sun and Solon's center.

3.2. *The laws in the closest lone-proton world*

The actual laws' lawhood is evidently preserved under a counterfactual antecedent positing my failure to brush my teeth this morning, or my failure to be born, or the failure of an asteroid to have hit Earth and wiped out the dinosaurs, or the failure of the Sun to have coalesced; had any of these come to pass, the laws of nature would have been just the same. But the preservation of the laws' lawhood might seem considerably less evident under a counterfactual antecedent positing an even greater departure from the actual world. For instance, consider how things would have been had there been nothing in the entire history of the universe except a single proton moving uniformly at 5 meters per second. I presume that in this possible world, all of the claims expressing laws in the actual world are still *true*.[55] But do they all still express *laws*? For instance, consider the actual laws that govern the interaction of protons and electrons, or the law specifying that copper is electrically conductive. These are uninstantiated and therefore *trivially* true in the closest possible world containing nothing but a lone proton. But are they *laws* in that world?[56]

That the actual laws remain the laws in the closest lone-proton world is prompted by the intuitions suggesting that had I failed to brush my teeth this morning, the laws of nature would still have been the same. When we contemplate the closest lone-proton world, we imagine a world where there *happens to be* only a single proton; we imagine taking the actual world and setting its initial conditions so that a lone proton is the result generated by the actual laws. Cosmologists might run their computer simulation for these rather boring initial conditions—perhaps as a test of their program.[57] Consider this question: "Had there been nothing but a lonely proton, then had there been an electron at the same distance from that proton as the electron in an actual hydrogen atom lies from that atom's proton, then what would their mutual electrostatic attraction have been?" Intuitively, it would have been the same as it is in an actual hydrogen atom, because Coulomb's law of electrostatics would still have been in force. We do not treat this question by throwing up our hands, as we do at the question, "Had there been a kind of elementary particle that repels electrons by a new kind of force, how would that particle have interacted with protons?"

Why might it nevertheless strike some of us as counterintuitive to say that had there been nothing forever but a lonely proton, then it would still have been a natural law that all copper objects are electrically conductive? Of course, this law would be uninstantiated in a lone-proton world. But why should this be an impediment to its being a law there? It has often been remarked that there are plenty of uninstantiated laws in the actual world. Newton's first law of motion (a body acted upon by no forces feels no acceleration) is uninstantiated because there happens never to be only a single material particle, so every actual particle feels the gravitational influences of others. Various laws specify the reaction of one chemical substance with another under certain conditions that samples of these substances happen never to experience. There are also functional laws with uninstantiated values of certain quantities (as discussed in appendix 3 of chapter 1). Admittedly, none of these examples is entirely convincing: the functional laws have some instantiated values, the reaction laws arguably derive from more fundamental laws that are instantiated, and Newton's first law is instantiated if it concerns a body acted upon by no *net* force (and might just as well be regarded as concerning an uninstantiated value of the quantities functionally related by Newton's second law). It seems now, though, that the examples are getting in the way of the larger question. Even if as a matter of fact all natural laws are actually instantiated, we should allow the logical possibility of uninstantiated laws—even many of them. Some intuitive resistance to this idea seems to derive from the contrast that this would create between the richness of the laws in the closest lone-proton world and the poverty of the goings-on there. But the problem cannot be simply that the laws of that world would then far "outreach" what happens there, since we should expect there to be laws concerning many circumstances that in fact never occur. After all, laws support counterfactual conditionals.

Why might the copper law's being uninstantiated in the closest lone-proton world seem far worse than the case of Newton's first law, various reaction laws, and functional laws in the actual world? The copper law concerns a *kind* of thing that is unrepresented in the closest lone-proton world. Copper is a chemical element, a natural kind of a certain sort. It might be thought that there cannot be a law concerning the members of a given natural kind if there are in fact no instances of that kind. But that is not right: in the actual world, it may well be a law that all atoms of the element having atomic number 200 have valence +2, even if, as it happens, no atoms of the element with atomic number 200 ever exist. However, I take it that this law follows from other, "more basic" laws about electrons and protons, and there are electrons and protons in the actual world, even if there are no atoms of the element with atomic number 200. In the closest lone-proton world, by contrast, there are no electrons to compose an atom of the element with atomic number 200, so the existence of even these more basic laws is in dispute.

It might be argued, then, that if there can be a copper law in a world inhabited only by a single proton, then the natural laws are so little constrained by the non-nomic facts that nothing prevents some of the actual laws from concerning uninstantiated kinds of particles—and not just kinds that fill out symmetry groups involving *instantiated* kinds of particles (in the way that the element with atomic number 200 fills a preexisting place in the Periodic Table), but kinds entirely unconnected to instantiated kinds. If a world containing nothing but a lone proton can have as one of its laws that all electrons are negatively charged, then what prevents the actual world from including myriad laws such as "All hypermesons are exactly two and one-half times as massive as the electron," where the hypermeson is intuitively not one of the real kinds of subatomic particles at all? In other words, the laws of nature are supposed to reflect the natural kinds (as I discuss further in chapter 7), but if there could be laws such as "All copper objects . . . " in a world containing nothing but a proton, there could be laws such as "All hypermesons . . . " in a world with the same non-nomic facts as the actual world. But, I presume, there are no hypermeson laws in the actual world; hypermesons are not physically possible.[58] Then what makes copper physically possible in the closest lone-proton world? If the laws in the actual world fail to extend very far beyond the instantiated kinds (they extend as far as element 200, but not to hypermesons), then why do the laws in the closest lone-proton world extend so far beyond the kinds instantiated there?

I think there is a good answer to this question. The reason why the laws in the closest lone-proton world extend to so many kinds of things that are unrepresented there is that this possible world is picked out by its relation to the actual world. In the closest lone-proton world, the electron is a natural kind of particle because we call forth this world by impoverishing the actual world—where the electron is a natural kind of particle. The supposition that there is nothing except a lone proton diminishes the population of the world but not the *kinds* of things that are there. (The cosmologist running her computer simulation of how things would have been, had there been nothing but a single proton, must twiddle the knob setting the electron parameter to zero, but the parameter itself does not disappear from the program—does not disappear from the list of initial conditions that must be set—and there never was a parameter to be set regarding hypermesons.) Copper remains a natural chemical kind in the closest lone-proton world because it is a natural chemical kind in the actual world and the closest lone-proton world is arrived at by beginning with the actual world and then severely depopulating it. The same argument cannot be used to show that the *actual* world may contain as yet undiscovered (and presumably undiscoverable) laws concerning hypermesons since the actual world is not posited as the closest possible world to some other world. We refer to the actual world in bringing some counterfactual world under our gaze, but we do not arrive at the actual world by hypothetically impoverishing, augmenting, or otherwise altering some *other* world. The actual world is picked out from among the possible worlds not by virtue of its standing in some relation to another possible world, but simply by its being actual.

Because the closest lone-proton world is derived by starting with the actual world, it remains a law there that all copper objects are electrically conductive. We can test this argument by using a *different* means to pick out a lone-proton world: by designing it "from scratch"—fashioning it "out of whole cloth," as in a science-fiction story. We might, for example, imagine a universe with protons, electrons, and so on, but specify the comparative strengths of the electromagnetic, gravitational, and nuclear forces to be . . . , so copper is not electrically conductive. Now take this universe and set the populations of the various kinds of particles so that there is nothing but a single proton moving uniformly forever. *This* lone-proton world is not the lone-proton world closest to the actual world. We have not reached this lone-proton world through a counterfactual conditional; we have constructed it from scratch. And the laws of nature in this universe are different from those in the closest lone-proton world.

Notice that these two possible worlds agree in their non-nomic facts but differ in their laws (and, accordingly, in the counterfactual conditionals correct therein). The natural laws fail to supervene on the truths in U. Moreover, this example suggests that if we call up a lone-proton world "from scratch," then we may stipulate any of the non-nomic facts there to be its laws. I introduced these ideas in section 1.3, and now discuss them further.

3.3. *Nonsupervenience*

Suppose there had been nothing but a single proton moving uniformly forever. Then had there been an electron at 5.3×10^{-11} meters from the proton (their average separation in an actual hydrogen atom), their electrostatic attraction would have been 8.2×10^{-8} newtons (as dictated by Coulomb's law). But we can imagine a lone-proton world where an alternative counterfactual holds: had there been an electron at 5.3×10^{-11} meters from the proton, their electrostatic attraction would have been 3.14159×10^{-8} newtons. So I am not contending that in *every* lone-proton world, Coulomb's "law" is a law—merely that it is a law in the *closest* lone-proton world. There are many possible worlds identical in all of their non-nomic facts but differing in their nomic facts (and so in the counterfactual conditionals that are correct there).

Now perhaps I am mistaken in thinking that there is a logically possible world containing a genuine *proton* where a proton–electron interaction, had it occurred, would have deviated from Coulomb's law. If part of the proton's *essence* is to conform to Coulomb's law, then an entity deviating from Coulomb's law is no proton. Perhaps to be a proton, an object must (by virtue of what protonhood *is*) have a positive electric charge of one unit, which conceptually implicates the electrostatic force law, which dictates how the proton would interact with other particles having various characteristics, including those particles that would constitute a copper object, and so the supposition of a lone proton conceptually demands that all copper objects be electrically conductive.[59]

I cannot examine this view here with all of the care it deserves; it raises issues about the content of theoretical terms that do not fall within the scope of this book. But while prudence perhaps dictates that I leave it at that, too much of this sort of prudence can ruin an otherwise useful discussion. In this spirit, let me aver that I find implausible the strong version of this view—so strong as to hold that, in any context, every actual law is essential to the properties and kinds that figure in it. For a particle to have a positive electric charge of one unit would still make sense if the electrostatic force law were different—say, if it were an inverse cube force, or if the force constant were twice as large. As John Earman (1984, p. 211) remarks, physicists who think about the possibility that magnetic monopoles exist, such that Maxwell's equations have to be modified, do not mean something different from the rest of us by the term "electromagnetic field." Or consider scientists in the 1920s who referred to certain subatomic particles as "protons" but were mistaken about the laws governing them. In some conversational contexts, I would say that these scientists were using the same "proton" concept we use now, and so were correct in their belief that protons exist, but were making non-conceptual mistakes in holding various beliefs about the laws governing protons (e.g., in believing that all protons at all times have definite positions). In other contexts, I would say that these scientists were using a different "proton" concept from ours and that there are no "protons" in their sense, since actual particles differ in important respects from the particles they posited. Protonhood in a given context has to do with whatever features of protons are most important there, and presumably these may involve some, all, or none of the laws concerning protons.

There is another reason that I resist deeming the laws concerning protons to be like logical or conceptual truths: it seems to me to dodge the issues that make the concept of natural law so philosophically intriguing in the first place. The challenge posed by this concept is precisely that the laws are necessary, but not as necessary as the logical and conceptual truths. I am working toward elaborating a sense in which physical necessity constitutes a "grade" of necessity inferior to logical and conceptual necessity, but superior to no necessity at all, where an accident lacks any "grade" of necessity even if it is preserved under a broad range of counterfactual circumstances. I take up these issues more fully in chapter 3. Only after failing to find a way to elaborate the notion of an intermediate grade of necessity would I be tempted to conclude that the natural laws must belong with the logical and conceptual truths.

Accordingly, I shall continue to suggest that two worlds containing exactly the same non-nomic facts can disagree in their natural laws. Moreover, as suggested in section 1.3, the non-nomic facts impose very minimal constraints on the natural laws, except insofar as they make certain claims false and so inadmissible as laws. Intuitively, when we are conjuring up a possible world from scratch and have specified the non-nomic facts there, we can deem just about any combination of them to be its laws, and thereby stipulate the correctness of some counterfactuals there.[60]

These considerations impose severe constraints on any philosophical account of law. For example, they run contrary to many "regularity" accounts, according to which the non-nomic facts are responsible for determining which regularities in the non-nomic facts are laws.[61] For instance, as discussed in appendix 3 of chapter 1, Lewis says that the laws are the generalizations in the deductive system of truths that possesses the best combination of simplicity and informativeness about the non-nomic facts. Since a lone-proton world is so much simpler than the actual world, there are deductive systems of truths in that world that, in their combination of simplicity and informativeness regarding the non-nomic facts there, surpass the deductive system that is best for the actual world. Therefore, according to Lewis's account, some of the actual laws are not laws in the closest lone-proton world (or in any other lone-proton world) and vice versa. Plausibly, "Every particle at every moment is a proton" and "At every moment, there exists exactly one particle" belong to the best deductive system for any lone-proton world. I see no reason why supplementing these generalizations with "All copper objects are electrically conductive" would contribute any information at all regarding the non-nomic facts in this world—and certainly not enough to outweigh the loss of simplicity that would result. (For that matter, I do not see why this generalization would be a better addition than "All copper objects are electrical insulators.") Lewis's account accommodates uninstantiated laws (such as Newton's first law of motion) on the grounds that they follow from instantiated laws (such as $F = ma$) that greatly simplify the system. But there is no such ground for including "All copper objects are electrically conductive"; the actual laws from which it follows will nearly all themselves be uninstantiated in this world.[62]

4. Why Λ-preservation?

Ultimately, my account of why Λ-preservation holds appeals to the concept of "inductive" confirmation to be introduced in chapters 4 and 5. For now, I merely set out the proper context in which to view this proposal by briefly discussing some other explanations for Λ-preservation that have been offered.[63]

I begin with the explanation offered by Michael Tooley, who (with Fred Dretske and Armstrong) has elaborated the view that the law expressed by "All *F*s are *G*" is the fact that the universals *F*ness and *G*ness stand in the relation of "nomic necessitation." Tooley (1987) offers this account of why a principle like Λ-preservation holds:

> This view of the truth conditions of nomological statements explains the relation between counterfactuals and different types of generalizations. Suppose that it is a law that (x)(Px → Qx). This will be so if the relevant universals stand in the appropriate relation. Let us now ask what would be the case if some object, *b*, which presently lacks property P, were to have that property. In particular, is it reasonable to assume that it would still be a law that (x)(Px → Qx)? It seems reasonable to reply that the supposition about the particular object, *b*, does not give one any reason for concluding that properties P and Q no longer stand in the relation of nomic necessitation. If this is right, then one can justifiably conjoin the supposition that *b* has property P with the proposition that the nomological relation in question holds between properties P and Q, from which it will follow that *b* has property Q. And this is why one is justified in asserting the counterfactual: If *b* were to have property P, it would also have property Q. (p. 138)

It is difficult to improve upon James Woodward's (1992) reply to Tooley's remark:

> It is hard to see how this explanation differs from simply stipulating that relations between universals are of such a character as to support counterfactuals, while relationships between particulars are not. Why is it "reasonable" to suppose that if the properties P and Q stand in the relationship of nomological necessitation, they would continue to do so if *b* were to be P, but not reasonable to make the corresponding supposition if the relationship in question is a relationship among particulars? Tooley does not derive this result from some independent theory of what it is for two universals to stand in a contingent and irreducible relationship. Instead such relationships between universals are simply invested with, or assumed to have, the standard counterfactual supporting properties of laws, while relationships involving particulars are not. (p. 191)

Admittedly, it seems intuitively reasonable for universals and their relations to be generally unaffected by the behavior of particulars. But of course, as long as the relations in question ("nomological necessitation") are logically contingent, they must be capable of being undermined by *certain* logically possible facts regarding particulars. Precisely which facts are these? Does the general metaphysical intuition that relations among universals are relatively unaffected by the behavior of particulars suggest Λ^{+}-preservation? Or merely Λ-preservation? Or that a law's truth is preserved under any counterfactual antecedent *p* (in U?) with which that law's truth is consistent? Or that a law's lawhood is preserved under any such *p* with which its lawhood is consistent?

Recall from section 1 that once I drew a few distinctions, the apparently simple notion that "laws support counterfactuals" began to dissolve into a variety of subtly different ideas. Any reductive analysis of natural law, in accounting for the laws' relation to counterfactuals, will have to make clear precisely what the relation is for which it purports to account. It is not evident how talk of universals standing in relations of nomological necessitation manages to be selective in suggesting some of these ideas rather than others.

Moreover, I have shown that accidental generalizations can also sometimes sustain counterfactuals. But apparently, this supplies no reason to regard accidental generalizations as relations among universals. Why then, on Tooley's view, can accidental generalizations support counterfactuals? Why will that explanation not also work for laws, obviating the need for any appeal to "nomic necessitation"? What specifically is it about the difference between laws and accidents in their relation to counterfactuals that makes appeal to a relation among universals necessary in the laws' case but not in the accidents'? It is hard to see how this level of detail can be funded solely by a general metaphysical intuition regarding universals and particulars. The details must either be expressly built into the notion of "nomic necessitation," or a much more substantial account of this relation must be offered.[64]

As noted in chapter 1, our belief that a given hypothesis may state a physically necessary truth is intuitively related to our willingness to regard instances of that hypothesis as confirming it "inductively." In chapters 4 and

5, I argue that the special relation between our beliefs about the laws and our beliefs about counterfactual conditionals is explained by the special, "inductive" way in which we regard evidence as having confirmed a physical necessity. In other words, I argue that our respect for Λ-preservation is accounted for by the root commitment we undertake in believing ■*h*, which involves our regarding various facts as inductive evidence for *h* (even if *h* was not originally discovered by being confirmed inductively).[65] I also argue that we regard [Λ ∪ {*a*}]-preservation (first revision) as violated—we regard no accident as joining the laws in bearing their special relation to counterfactuals—because we cannot regard a claim that we believe to express an accident as having been confirmed inductively.

Other philosophers (e.g., Goodman 1983, p. 20; Strawson 1952, pp. 199–200) have also noted an analogy between the fact that laws are somehow distinctively used to make predictions regarding *actual* unexamined cases and the fact that laws are somehow distinctively used to describe *counterfactual* (and perforce unexamined) cases. Mackie (1961) has notably tried to use this analogy to explain why we respect something like Λ-preservation. He supposes that we believe "All of the coins in my pocket are shiny" to be an accidental generalization, and, holding that we can inductively confirm only claims that we believe to be laws if true, he supposes that our justification for believing this generalization must be our having checked every coin in my pocket:

> [A]lthough it is true that all the coins in my pocket are shiny, I know this only by some process of complete enumeration, by having checked the coins actually in my pocket one by one. That being so, as soon as I add the (false) supposition that some *other* coin is in my pocket, I completely undermine the reason that I had for adhering to the universal, for this, being *another* coin, which is not actually in my pocket, has not been checked by this process of complete enumeration. Once I add the supposition that corresponds to the antecedent of a *counterfactual*, then, I have no reason for adhering to this universal, and therefore I cannot use this supposition and this universal as joint premises in an argument. Consequently I cannot assert the counterfactual conditional which is a condensation of an argument with these as joint premises. That is why the true proposition, "All the coins in my pocket are shiny," known merely by complete enumeration, does not sustain the counterfactual "If this (other) coin had been in my pocket it would have been shiny."
>
> On the other hand, if we know that whenever potassium is exposed to air it bursts into flame, our reasons for believing this, whatever they may be in detail, are not undermined in any corresponding way by the supposition that this bit of potassium (which has not actually been exposed to air) has been exposed to air. The ultimately inductive evidence is related in exactly the same way to the universal proposition whether we add the (false) supposition or not. Hence this causal universal, supported in the way in which scientific propositions are supported, and

> this (false) supposition, can be used as joint premises in an argument, and consequently we can assert the counterfactual conditional which is a condensation of this argument. That is why the causal law, known in the usual inductive way, does support the counterfactual "If this bit of potassium had been exposed to air it would have burst into flame." (pp. 284–285; see also 1962, pp. 71–73; 1973, pp. 114–119)

Clearly, this idea does not depend on Mackie's view (mentioned in section 1.1) that a counterfactual conditional is the "condensation" of an argument and so is neither true nor false; Mackie's idea can be adapted, for instance, to a possible-worlds construal of counterfactuals. Mackie's key point is that our justification for believing "All *F*s are *G*" generally survives our supposing counterfactually that *c* is *F* only if we believe "All *F*s are *G*" to be a law.

This explanation presumes that how our belief in a given claim is justified has some influence on whether we are justified in using that claim in conjunction with a given counterfactual supposition. (Tooley makes the same presumption.) *Why might this be?* By what right does Mackie say "therefore" in this sentence: "Once I add the supposition that corresponds to the antecedent of a *counterfactual*, then, I have no reason for adhering to this universal [the accidental generalization], and therefore I cannot use this supposition and this universal as joint premises in an argument," for which, Mackie believes, the counterfactual is shorthand?

Mackie appears to be saying that we are justified in asserting $p > q$ exactly when we justly believe *q* if we come to believe *p* and adjust our other beliefs so as to make room for this new belief. This interpretation is suggested by Mackie's repeated reference to *add*ing the counterfactual antecedent to our stock of beliefs and noting whether our justification for believing the generalization survives. Suppose, then, that we believe that "All *F*s are *G*" does not follow from the laws. Mackie holds that our justification for believing that all *F*s are *G* then depends on our believing justly that we have checked every *F*.[66] Hence, if we suddenly come to believe justly that there is an unchecked *F*, we are in general no longer justified in believing "All *F*s are *G*." So if we believe justly that "All *F*'s are *G*" is an accident, then we cannot in general believe justly that had *c* been *F*, then *c* would have been *G*. In contrast, we must believe this counterfactual if we believe justly that "All *F*s are *G*" expresses a law, since the generalization has then received inductive support. If we come to believe that there is an additional unchecked *F* beyond those we already knew about, we regard it as having been confirmed (along with every other unchecked *F*) to be *G*.

But the view that we are justified in believing $p > q$ exactly when we are justified in believing *q* if we adopt *p* justly and made the appropriate revisions in our other beliefs—the "Ramsey test" (Ramsey 1931, p. 247)—faces familiar difficulties. I mentioned these originally at the start of section 2, in connection with the iodine-131 example. We don't accept "Had Oswald not killed Kennedy, then someone else would have," but if we are told that Oswald did not in fact kill Kennedy, we conclude that someone else must have rather than that Kennedy wasn't assassinated after all. Richmond Thomason (see Lewis 1986b, p. 155) has likewise noted that the Ramsey test cannot account for the remark: "My wife Sally is very clever. Had she been unfaithful to me, I would (still) have believed that she was faithful, since her deception would have been so clever." If he modifies his stock of beliefs to include that Sally is unfaithful to him, surely he then believes this to be one of his beliefs.

Ultimately, I believe I take more seriously than Mackie the analogy between projecting *h* onto actual unexamined cases and projecting it onto counterfactual cases. To explain why we respect Λ-preservation and regard no accident as joining the laws in their special relation to counterfactuals, I do not discuss whether our justification for believing *h* survives our believing *c* to be *F*. Rather, I shall understand *h*'s *inductive* confirmation (the sort of confirmation unavailable to hypotheses that we believe physically unnecessary) as involving confirmation of *h*'s holding in actual cases *and* confirmation of its holding in various counterfactual cases. On this view, evidence bears upon various counterfactuals for the same reason as it bears upon various claims about actual unexamined cases. So whether or not we conclude that the hypothesis being confirmed is physically necessary, we become justified in believing various counterfactuals in the course of becoming justified in believing that actual unexamined cases accord with the hypothesis. There is no need for some *special, new* kind of reason for *going beyond* our beliefs about the actual world and undertaking various beliefs about counterfactual cases (see also Woodward 1992, pp. 187–90).

In chapters 4 and 5 I argue that, when we confirm inductively a hypothesis *h* that we believe may be physically necessary, the range of counterfactual cases that we confirm to accord with *h* corresponds to the range of counterfactual cases in the scope of Λ-preservation. Furthermore, the range of counterfactual cases that we must confirm to accord with a hypothesis *a*, in order to confirm *a* inductively while believing that *a* lacks physical necessity, corresponds to the range of counterfactual cases under which [Λ ∪ {*a*}]-preservation (first revision) requires that *a* be preserved. It then turns out that we cannot confirm a hypothesis "inductively" if we believe that it lacks physical necessity. For the same reason, we treat no accident as bearing the same relation to counterfactuals as do the laws.

• THREE

Why Are the Laws of Nature So Important to Science (I)?

1. A "Theoretician's Dilemma" for laws

I take for granted that science is interested in discovering the non-nomic facts. But for $p \in U$, why should science care about discovering whether or not *p* is a law insofar as this goes beyond discovering whether or not *p* is true? By discovering a regularity to be a matter of natural law, we learn nothing more about the non-nomic facts than we already knew simply from discovering the regularity itself. Why, then, should science be concerned with discovering whether some regularity is accidental or a matter of natural law? Of course, if science's goal is *stipulated* to be the discovery of *all* facts, including nomic facts, then science must be interested in discovering the laws. But this answer merely begs the question; it cannot persuade anyone who did not already accept that the laws matter to science.

The question I have just asked is like Carl Hempel's celebrated "Theoretician's Dilemma" (Hempel 1965, pp. 173–226): Why should science care about the reality of the unobservable entities posited by a theory when exactly the same empirical predictions follow from the theory's empirical adequacy as from its truth? A theory's truth, over and above its empirical adequacy, is unobservable, just like a truth's physical necessity. Hempel's question cannot be answered merely by *stipulating* the goal of science to be the discovery of all facts, observable and unobservable alike.

If the distribution of $\blacksquare$s and $\neg\blacksquare$s over the truths in U is fixed entirely by those truths, then in ascertaining the lawhood (or nonlawhood) of some truth *p* in U, we may learn something more about the non-nomic facts than we already knew simply from discovering *p*'s truth. However, as I argued in chapter 2 (sections 1.3 and 3), the only constraint that the facts of the form $\blacksquare p$ and $\neg\blacksquare q$ (where $p,q \in$ U) impose on the facts in U is that the facts in U must be consistent with the *p*s where $\blacksquare p$.

If we know that it is a law that all *F*s are *G*, then we know that there is no point in trying to produce an *F* that is not *G*; we know that any attempt to do so fails. (E.g., we know that anyone who attempts to accelerate a material particle from rest to beyond the speed of light fails, even if she has superconducting magnets and other advanced technology.) But to know this, it suffices to know merely that all *F*s *are G*; we do not need to know that they *must* be *G*—that it is a law.[1]

Still, by discovering that it is a law rather than an accident that all *F*s are *G*, we learn not merely that all *actual* efforts fail to produce an *F* that is not *G*, but also that all efforts *would have* failed even if scientists had tried harder or more often. This suggests that scientific interest in identifying the laws *can* be accounted for, once it is granted that science is concerned with non-nomic facts in a *broad sense* (broader than U): not only with what actually happens as far as facts in U are concerned, but also with what would have happened as far as non-nomic matters are concerned. In other words, the non-nomic facts "in a broad sense" include the facts about which subjunctive conditionals having antecedents and consequents in U are correct. Let the non-nomic facts in a broad sense be termed the facts in U*, which is the set of claims that can be formed from ">" and the members of U. I shall presume science to be concerned with the correctness of counterfactual conditionals in U* just as it is concerned with the truth of claims in U. Science can ascertain the correctness of some of those counterfactual conditionals by discovering, of some fact in U, whether it is physically necessary or accidental.

Before elaborating this point, I must make one remark concerning this strategy for addressing the nomic version of the Theoretician's Dilemma. We should feel no regret at this strategy's having to invoke counterfactual conditionals. As I showed in section 3 of chapter 2, nomic facts fail to supervene on the truths in U. So when contemplating an appeal to counterfactuals, we should not allow ourselves to be inhibited by any nostalgic remnant of an ambition to reduce the facts in Λ^+ to non-nomic facts. Indeed, my aim is not to reduce nomic facts even to the facts in U*. (Again, I do not know whether the laws obtain in virtue of the correctness of various counterfactual conditionals or whether, conversely, various counterfactual conditionals hold in virtue of the laws.) Rather, my aim is to answer this chapter's title question—to determine whether and why beliefs about the laws, insofar as they go beyond beliefs about non-nomic matters (actual and counterfactual), play an important role in scientific reasoning.

Of course, many of the philosophers who originally investigated the Theoretician's Dilemma were not prepared to grant that science is interested in the correctness of counterfactual conditionals—or even that counterfactual conditionals are meaningful. I am not so squeamish (though I do not presume counterfactuals to have truth-values, only that they are correct or incorrect in some sense). I suggest in chapters 4 and 5 that observations confirm various counterfactuals to be correct in the course of confirming various predictions about actual facts. As I show in this chapter, the Theoretician's Dilemma regarding nomic facts assumes an especially interesting form when we are prepared to countenance *all* non-nomic facts, here in the broad sense of facts in U*, as having immediate scientific relevance, and we then consider what the scientific significance of facts like $\Box p$ and $\blacksquare p$ might be.

I now return to the thought that I began three paragraphs above, that we can account for the scientific relevance of $\Box r$ (insofar as it goes beyond truths in U, such as r) by noting how its discovery enables us to discover the correctness of various *counterfactual* conditionals with non-nomic antecedents and consequents. In section 1.3 of chapter 2, I derived

Λ is preserved under any $p \in$ U where Con(Λ,p).

This principle explains how science can discover that some counterfactual conditional is correct by discovering that a given non-nomic fact is physically necessary. For example, by discovering that it is a law that all copper objects are electrically conductive, we may ascertain (if we believe it physically possible for me to hold a copper object in my hand) that were I holding a copper object in my hand, then I would be holding an electrical conductor. So having already discovered that r (where $r \in$ U), science remains *interested* in discovering that $\Box r$.

However, this account does not explain why it is so *important* in science to discover that $\Box r$ over and above discovering r (where $r \in$ U). Admittedly, the members of Λ are the *only* non-nomic claims that are preserved under every $p \in$ U that is consistent with Λ. [This is easily shown: if $q \notin \Lambda$ because q is false, then for any truth $p \in$ U, p is consistent with Λ but $\neg(p > q)$. And if $q \notin \Lambda$ and q is a truth in U, then $\neg q$ is consistent with Λ (since otherwise q follows from members of Λ, so $q \in \Lambda$, contrary to our stipulation), so if we make $p = \neg q$, we again have $\neg(p > q)$ where p is consistent with Λ.] So to explain the special scientific *importance* of identifying Λ's membership, we must explain why it is especially important to science that some truth in U is preserved under every $p \in$ U that is consistent with Λ. But why is preservation under *this* particular set of subjunctive antecedents in U especially notable—why is it more important than preservation under some *other* set of subjunctive antecedents in U? If we cannot say why, then we cannot in this way explain why identifying the natural laws is more important to science than identifying the truths that are preserved under some other, arbitrarily chosen set of subjunctive antecedents. (After all, no special title is bestowed on those truths that are preserved under every subjunctive antecedent logically consistent with, say, "George Washington was the first president of the United States"; such preservation lends a truth no special importance.) On the other hand, if we say that preservation under every $p \in$ U that is consistent with Λ is especially important because science especially cares about counterfactual suppositions that are physically possible, then we have just traversed a very tight circle: the laws are important because they are preserved under a certain range of counterfactual suppositions, where this range is important because it is delimited by the laws and the laws are important.[2]

We could solve this problem only if we had some means of distinguishing the laws from the accidents, by virtue of their relation to counterfactual conditionals, that does not itself appeal to the distinction between laws and accidents. This is what I now propose to give.

In section 2 below, I present a respect in which Λ^+ is distinguished from all other sets of claims: it nontrivially possesses "stability." Likewise, Λ is the unique set that nontrivially possesses "non-nomic stability."[3] I do not use the concept of a natural law in specifying what it is for a set to be "stable." The laws' unique stability will show us just how physical necessity supervenes on the truths in U*. That is, I will show how the physical necessities can be read from the non-nomic truths (in the broad sense that includes the truths about which counterfactual conditionals, having antecedents and consequents in U, are correct). Stability also allows us to find a kernel of truth in the intuition that the laws collectively possess a range of invariance that is *maximally* broad—in particular, broader than the range of invariance of any set containing an accidental truth. Section 2.2 of chapter 1 discussed how a law sometimes fails to be preserved under a counterfactual antecedent under which some accident is preserved. So to cash out the intuition that a law's range of invariance (a.k.a. resilience, robustness, autonomy, constancy, permanence, etc.) is broader than any accident's, I cannot say that an accident's range is a proper subset of a law's.

That stability is a kind of maximal invariance suggests that there is a close relation between stability and necessity. In particular, below I use the laws' nontrivial non-nomic stability to explain what it is for the laws to possess a kind of necessity "between" logical or conceptual necessity and no necessity at all. I also explain why there is no "grade" of necessity—perhaps inferior to physical necessity in the way that physical necessity is inferior to logical necessity—that an accident possesses by virtue of the fact that it would still have obtained under some wide range of counterfactual suppositions. I also discuss whether the laws could be organized into *multiple* grades of necessity between logical or conceptual necessity and no necessity at all. (I am then able to explain a novel sort of exception to Λ^+-preservation.)

In chapter 5, I resume my explanation of why the laws are so important to science by explaining why science must take a special interest in discovering which set nontrivially possesses non-nomic stability. Of course, I would be begging the question if I explained the scientific significance of nontrivial non-nomic stability by appealing to the scientific importance of identifying the laws. Rather, I appeal to science's interest in *predicting* truths in U*—getting to know them in advance of having directly observed them to obtain. (In the case of truths in U* that concern what *would have* happened under various *counterfactual* circumstances, their observation is out of the question; any knowledge of them must be a "prediction.") How can we successfully make such predictions in a case where (as I later explain) our prior opinions fail to supply us with a good reason to regard any observation as confirming any such prediction? To discover how is to discover which set of claims in U nontrivially possesses non-nomic stability—or so I argue in chapter 5. To seek the best way to arrive at such predictions "starting from scratch" is to seek the natural laws.

I do not mean to suggest that this is necessarily the only reason that the truths of the form $\Box r$ or $\neg\Box r$ (where $r \in$ U), insofar as they go beyond the truths in U, are important for science to discover. Whereas I shall take for granted science's interest in discovering the correctness of various counterfactual conditionals with non-nomic antecedents and consequents, another philosopher might just as well begin by presuming science to be interested not merely in predicting but in explaining why the non-nomic facts obtain. But any account of the laws' value to science risks traversing a very tight circle in the same manner as did our initial appeal to counterfactuals. For instance, if we say that a non-nomic fact's lawhood has special scientific importance because science is interested in explaining the non-nomic facts and the laws have a special explanatory power, then we must say *why* the laws have this power. If we say that a scientific explanation essentially involves subsumption under the laws, then in attributing the laws' importance to their explanatory power, we have really done little more than *stipulate* the laws' importance; we have not explained why science should especially care about subsuming non-nomic facts under laws, just as my above proposal failed to explain why science should especially care about counterfactual conditionals in U* with antecedents consistent with Λ. To explain what it is about laws that makes this *particular* range of counterfactual suppositions so special or makes subsumption under *them* so explanatory, we need to specify a relation to the non-nomic facts (in the broad sense) that is borne uniquely by the laws and whose scientific importance can be appreciated without presupposing that we already (for some reason) take special interest in the laws. Such a property has, to my knowledge, never before been identified. That is what I now propose to do.

2. Stability

2.1. *Defining stability*

According to Λ^+-preservation, the set Λ^+ is preserved under every p in U^+ that is consistent with Λ^+. Moreover, recall that for any accidental truth $r \in$ U, no set of truths (in U^+) containing r is preserved under every such p. But as I have just shown, this fact cannot help to explain the scientific importance of Λ^+ without begging the question, since Λ^+ itself is used to fix the range of p's under which preservation is at issue. This seems like stacking the deck—gerrymandering the relevant range of subjunctive antecedents precisely to suit the laws. Suppose we had instead begun by considering a range of counterfactual antecedents in U^+ that includes some that are inconsistent with Λ^+ but none that are inconsistent with a given accident a. Then perhaps a is preserved under all of *these* counterfactual suppositions, whereas a given law has a narrower range of invariance among them.

So let us avoid using the laws to fix the relevant range of subjunctive antecedents. Instead, let us allow *the set whose preservation is at issue* to determine the relevant range of subjunctive antecedents. Let us say that a set Γ is "stable" exactly when

(i) $\Gamma \subseteq U^+$;
(ii) if $q \in \Gamma$, then q is true;
(iii) if $q \in U^+$ and $\Gamma \Rightarrow q$, then $q \in \Gamma$; and
(iv) Γ is preserved under $\{p \in U^+$ such that $\mathrm{Con}(\Gamma,p)\}$.[4]

That is, a set of truths logically closed in U^+ is "stable" if and only if it is preserved whenever it *could* be preserved, that is, under any subjunctive antecedent (in U^+) that is consistent with each of its members. Plainly, Γ's stability is one way of understanding the "maximum degree" of invariance that Γ could possess.

Notice that Λ^+-preservation just means that Λ^+ is stable. Two obviously stable sets are the set of logical truths in U^+ and the set of all truths in U^+. These two are the sets (apart from the null set) that possess stability *trivially*. The set of logical truths in U^+ is stable because for any p and any logical truth m, *trivially* m is logically entailed by p, and so automatically $p > m$ is correct, at least where p is consistent with the logical truths. [The same argument applies to nested counterfactuals $p > (q > m)$, $p > (q > (r > m))$, etc.] The set of all truths in U^+ is preserved under each subjunctive antecedent consistent with each of its members because $p > m$, $p > (q > m)$, $p > (q > (r > m))$, and so on, are *trivially* true if m, p, q, r, and so on, are true, and no false supposition in U^+ is consistent with each of this set's members.

Now I present the argument that the only stable set besides these two—in other words, the only *nontrivially* stable set—is Λ^+. Ultimately I do not endorse this argument *in full*, but I think that it is *nearly* right. (I make the relevant qualifications in section 3.2 below and in section 4 of chapter 7.) This argument points us toward recognizing a property—nontrivial *non-nomic* stability—that picks out the physical necessities in U.

2.2. *Which sets are nontrivially stable?*

Let us begin with a lemma. Suppose that both Γ and Γ′ are stable and neither is a subset of the other. Take any

$$r \in \Gamma,\ r \notin \Gamma',\ r' \in \Gamma',\ r' \notin \Gamma.$$

Then ($\neg r$ or $\neg r'$) is consistent with Γ, since otherwise, some members of Γ must logically entail $\neg(\neg r$ or $\neg r')$, that is, $(r \,\&\, r')$, but then [from (iii) in the definition of stability, recalling that Γ is stable] $r' \in \Gamma$, contrary to our initial assumption. Likewise, ($\neg r$ or $\neg r'$) is consistent with Γ′. Therefore, by the stability of Γ and Γ′, respectively, it follows that

$(\neg r \text{ or } \neg r') > r, (\neg r \text{ or } \neg r') > r'.$

But these are mutually inconsistent, since the former implies

$(\neg r \text{ or } \neg r') > \neg r'.$[5]

We have, then, a *reductio*. We have shown that if there are two distinct, stable sets, then one must be a proper subset of the other. Since Λ^+ is stable, it follows that if there is any other nontrivially stable set Γ, then either $\Lambda^+ \subset \Gamma$ or $\Gamma \subset \Lambda^+$. By eliminating these two possibilities, I show below that Λ^+ is the only nontrivially stable set.

First consider the case where $\Lambda^+ \subset \Gamma$; let Γ satisfy requirements (i), (ii), and (iii) in the definition of stability. (I.e., Γ is a set of truths, logically closed in U^+.) So Γ contains some truth $a \in U^+$ such that $a \notin \Lambda^+$—in other words, some accident a. Some ps in U^+ where Con(Λ^+,p) are such that $\neg$Con(Γ,p). Intuitively, then, the range of subjunctive antecedents under which Γ must be preserved, in order to be stable, is narrower than the corresponding range for Λ^+ to be stable. However, the relevant ps must direct us to possible worlds in a narrower range—namely, to possible worlds in which not only Λ^+ but also a obtains.

To see why Γ fails to be stable, consider this example: let Γ be the logical closure in U^+ of "All of the matches now in this book remain forever unlit" (a) together with the members of Λ^+. Suppose that all of the matches in the book are dry and well made, oxygen is present, and so on, but as it happens, none of the matches is ever struck. Consider the counterfactual antecedent p: "Had one of them been struck." Now Con(Γ,p) holds, since the laws of nature plus the fact that one of these matches is struck do not logically entail that one of them lights; for the match to light from being struck, oxygen must also be present, the match must be dry and well made, and so on—which is not contained in Γ. Since Con(Γ,p), Γ is stable only if all of the matches would still have remained unlit even if one of them had been struck. But when standard conditions prevail, then (in an ordinary conversational context, it is correct to say that) had one of the matches been struck, oxygen would still have been present, the matches would still have been dry and well made, and so the match would have lit. Hence, Γ is not stable.

We could have shown Γ's instability in a different way. Let b be some accidental truth that is unrelated to a, such as "All gold cubes are smaller than one cubic mile." Consider ($\neg a$ or $\neg b$) as a counterfactual antecedent: "Had either one of the matches in the book been lit or there been a gold cube exceeding one cubic mile." Now Con(Γ,($\neg a$ or $\neg b$)). But in a great many conversational contexts, we would be correct in denying $(\neg a \text{ or } \neg b) > a$, though this counterfactual is required by Γ's stability—since we would be correct in denying $(\neg a \text{ or } \neg b) > \neg b$ ("Had one of the matches in the book been lit or there been a gold cube exceeding one cubic mile, then there would have been a gold cube exceeding one cubic mile"). Indeed, in a great many contexts, I daresay we would be correct in denying $(\neg a \text{ or } \neg b) > \neg b$ *and* correct in denying $(\neg a \text{ or } \neg b) > \neg a$.[6]

This argument can be generalized to show the instability of any set Γ where $\Lambda^+ \subset \Gamma$ and (to preclude Γ's *trivial* stability) b is a truth in U^+ where $b \notin \Gamma$. (It follows that b is accidental.) If Γ satisfies requirements (i), (ii), and (iii) in the definition of stability, then there is an accidental truth a such that $p \in \Gamma$ just in case p is entailed by a and some members of Λ^+. Now Con(Γ, ($\neg a$ or $\neg b$)), since otherwise some members of Γ logically entail $\neg$($\neg a$ or $\neg b$), that is, (a & b), and so [by (iii) in the definition of stability] $b \in \Gamma$, contrary to our supposition. So since $a \in \Gamma$, Γ's stability requires that ($\neg a$ or $\neg b$) > a, and hence ($\neg a$ or $\neg b$) > $\neg b$. It requires, in other words, that under this counterfactual antecedent, b is always sacrificed for the sake of preserving a. But if there are some contexts in which this counterfactual is correct, there are surely others in which (at least for some such b) we would be correct in asserting instead ($\neg a$ or $\neg b$) > $\neg a$ & b. In addition, there may well be contexts in which neither of these counterfactuals can be correctly asserted. Since there are contexts in which a is not preserved under a counterfactual antecedent p where Con(Γ,p), Γ is unstable. In short, Γ is stable only if a's preservation is more important than b's in *every* conversational context, for *any* $b \notin \Gamma$—which is highly implausible.[7]

A similar argument suggests the instability of any set Γ satisfying requirements (i), (ii), and (iii) in the definition of stability, where $\Gamma \subset \Lambda^+$ and (to preclude Γ's *trivial* stability) Γ contains some logically contingent truth. Consider a counterfactual antecedent that is consistent with every member of Γ, though inconsistent with some member of Λ^+—so Λ^+'s stability does not require Γ's preservation under this counterfactual antecedent. Then Γ is not always preserved under this counterfactual antecedent.

Consider an example. Suppose Γ is Λ—the logical closure in U of the claims $p \in$ U such that $\blacksquare p$. Now here is a counterfactual antecedent of the sort I just mentioned: "Had it not been a *law* that every material particle accelerated from rest remains at a velocity beneath the speed of light." That no material particle is ever accelerated from rest to beyond the speed of light can logically possibly be true even if it is not a *law*. As I discussed in section 3 of chapter 2 (in connection with lawhood's failure to supervene on the nonnomic facts), our counterfactual antecedent is consistent with every member of Λ. So the following counterfactual conditional must be correct in order for Λ to be stable: "Had it not been a law that every material particle accelerated from rest remains at a velocity beneath the speed of light, it would still have been true." This counterfactual conditional is false, since there might well have been a particle accelerator (natural or artificial) capable of accelerating particles from rest to beyond the speed of light had there been no law of nature standing in its way. Hence, Λ is not stable.[8]

Again, we can generalize this argument to apply to any $\Gamma \subset \Lambda^+$ (where Γ is logically closed in U^+) except the set of logical truths in U^+. Let $a \in \Lambda^+$ where $a \notin \Gamma$. Let $b \in \Gamma$ where b is logically contingent. Now Con(Γ,($\neg a$ or $\neg b$)), since otherwise some members of Γ entail $\neg$($\neg a$ or $\neg b$), that is, (a & b), and so (by Γ's logical closure) $a \in \Gamma$, contrary to our supposition. So since $b \in \Gamma$, Γ's stability requires that ($\neg a$ or $\neg b$) > b, so that ($\neg a$ or $\neg b$) > $\neg a$—in other words, requires that under this counterfactual antecedent, *a* is always sacrificed for the sake of preserving *b*. But it is implausible that *b*'s preservation takes precedence over *a*'s in every context, for any such *b*. For example, suppose *a* is that Hooke's law is a law and *b* is that Snell's law is a law. It is not at all plausible that, in every context, it is correct to assert that had Hooke's not been a law or Snell's not been a law, then Snell's would still have been a law and Hooke's would not. In many contexts, both this counterfactual and "Hooke's would still have been a law and Snell's would not" are rightly denied.[9]

So, apparently, nontrivial stability is to be found neither in a proper subset of Λ^+ nor in a superset of Λ^+. The only nontrivially stable set seems to be Λ^+. The laws are not invoked to designate the range of counterfactual antecedents under which Γ must be preserved in order for Γ to qualify as stable. That range is instead picked out entirely by Γ.

2.3. *Non-nomic stability, maximal invariance, and physical necessity*

Stability's scientific significance is ultimately best appreciated by means of "non-nomic stability." Recall from section 1.3 of chapter 2 that Λ is preserved under any $p \in$ U where Con(Λ,*p*). In other words, Λ possesses *non-nomic* stability, where a set Γ is non-nomically stable exactly when

(i) $\Gamma \subseteq$ U;
(ii) if $q \in \Gamma$, then q is true;
(iii) if $q \in$ U and $\Gamma \Rightarrow q$, then $q \in \Gamma$; and
(iv) Γ is preserved under $\{p \in$ U such that Con(Γ,*p*)$\}$.[10]

Roughly, a set of truths logically closed in U possesses non-nomic stability if and only if all of the set's members are preserved whenever—*non-nomically speaking*—they could all be preserved, that is, under any *non-nomic* subjunctive antecedent consistent with each member of this set.[11]

The set of all truths in U and the set of logical truths in U (along with the null set) possess *trivial* non-nomic stability. That Λ is the only set that *nontrivially* possesses non-nomic stability[12] follows from an argument similar to one I just gave according to which Λ^+ is the only nontrivially stable set. Briefly stated, by the same sort of reasoning as above, it follows that if there are two sets possessing non-nomic stability, then one must be a proper subset of the other. Suppose, then, that we remove some member(s) from Λ and the resulting set remains logically closed in U. Then the resulting set's members will all be consistent with some physically impossible $p \in$ U, namely, something contrary to what we removed. But, as long as the resulting set goes beyond the logical truths in U, it need not be the case that the members of the resulting set would all still have held had *p* obtained. For instance, had Hooke's law been false—deprived not merely of its lawhood, but of its truth—would Snell's law still have been true? No guarantees (I qualify this in section 3.1). On the other hand, suppose we take some accidental truth *a*, add it to Λ, and then render the result logically closed in U. The non-nomic stability of this set would require that a's preservation always take precedence over the preservation of any accidental truth b outside the set. (There is such a b as long as the resulting set is not the set of all truths in U.) In other words, since b is outside the set, $(\neg a$ or $\neg b)$ is consistent with each member of the resulting set, so the resulting set's non-nomic stability would require that $(\neg a$ or $\neg b) > (a \,\&\, \neg b)$ be correct in every context—which I have shown to be highly implausible. Once again, if a is "All of the matches in this book remain forever unlit," then we would have to say, in every context, that had one of the matches in this book been lit or there been a gold cube exceeding one cubic mile, then all of the matches in the book would still have been unlit but there would have been a gold cube exceeding one cubic mile.

Instead of considering whether a's preservation always takes precedence over every b's preservation, I might have argued in a different way that $[\Lambda \cup \{a\}]$'s logical closure in U lacks non-nomic stability.[13] We know from Λ^+-preservation that accident a is not preserved under the counterfactual antecedent $p = (a \supset \Box a)$; I showed in section 1 of chapter 2 that this counterfactual antecedent posits the "ideal test moa" for betraying the accidental character of Karl Popper's "All moas die before age 50." Now Con($[\Lambda \cup \{a\}]$'s closure in U,p). Hence, $[\Lambda \cup \{a\}]$'s closure in U lacks stability. Since $p \notin$ U, this argument fails to show that $[\Lambda \cup \{a\}]$'s closure in U lacks *non-nomic* stability. Still, in this light its non-nomic stability seems highly doubtful. Perhaps a counterfactual antecedent outside of U must be used by any *general recipe*, applicable to any accident a, for constructing a counterfactual antecedent p where Con($[\Lambda \cup \{a\}]$'s closure in U,p) and under which a fails to be preserved. But on a case-by-case basis, we could presumably always proceed within U. For example, whereas the "ideal test moa" is posited by "Had there been a moa, aged 49 years 364 days, free of the virus, and not in the next day run over by a truck, eaten by a predator, starving, afflicted with pneumonia, tuberculosis, *and so on*," there is also a *non-nomic* counterfactual antecedent under which the moa generalization is not preserved: "Had there been a moa, aged 49 years 364 days, free of the virus, and not in the next day run over by a truck, eaten by a predator, starving, or afflicted with pneumonia or tuberculosis (*period*)." It seems highly implausible that in any context, we should say that such a moa would have died within the day of some illness unmentioned by the antecedent. This suggests that although there is no general recipe for designing such a counterfactual antecedent, there is such an antecedent for any accidental generalization.

I have argued, then, that for any $p \in$ U, $\Box p$ holds exactly when p belongs to a set that nontrivially possesses non-nomic stability. Obviously, it follows that the truths in U* suffice to determine whether or not p in U states a physical necessity; p's physical necessity supervenes on the non-nomic truths in the broad sense that includes the correctnesses of counterfactual conditionals having non-nomic claims as their antecedents and consequents.[14]

We have here a distinction between the laws and the accidents that is not drawn by referring to the laws; the range of non-nomic counterfactual antecedents under which the members of some set must be preserved, for that set to possess non-nomic stability, is not specified by appealing expressly to the laws. Rather, it is fixed by the set's own members. Of course, in order to answer the title question, we still have to understand why science should especially care about identifying the members of the only set that nontrivially possesses non-nomic stability. I cannot resume this argument until after I have discussed "inductive" confirmation in chapter 4, so I pick up this thread in chapter 5.

Non-nomic stability is a kind of *maximal* invariance (resilience, robustness, autonomy, constancy, permanence . . .) under hypothetical suppositions; a non-nomically stable set would still have obtained under any $p \in$ U under which it *could* still have obtained. So in virtue of Λ's non-nomic stability, the laws (in U) *collectively* possess a kind of *maximal* invariance—greater invariance (in U) than any set (in U) containing some accidental truth (except trivially the set of all truths in U). There is then *something* to the intuition that a law has greater invariance than an accident.

Intuitively, the laws possess a grade of *necessity* "between" logical necessity and no necessity at all. This intuition can be captured by positing that there is a grade of necessity corresponding to a set in U if and only if that set is non-nomically stable—is *maximally* invariant in the sense I have just elaborated. In other words, for *any* set Γ of truths in U, there is *some* range Δ of subjunctive antecedents in U such that Γ is preserved under Δ. But this is not enough for there to be a type of *necessity* associated with membership in Γ. For that, Γ must be preserved under every subjunctive antecedent in U under which it *could* logically possibly be preserved. Only then does Γ possess a kind of "must"-ness.

Corresponding to each of the two trivial cases of non-nomic stability, there is a grade of necessity. Because the logical truths would still have held under any non-nomic circumstance with which they are all consistent, the logical truths possess "logical" necessity. At the other trivial extreme sits the set of all truths in U, corresponding to which is the zeroth grade of necessity: no necessity at all. Consider the range of subjunctive antecedents in U with which the logical closure in U of the laws in U is logically consistent. This range is obviously narrower than the corresponding range for the logical truths. Hence, the corresponding grade of necessity ("physical" necessity) is "weaker" than that possessed by the logical truths; it is closer to the zeroth grade of necessity. The range of subjunctive antecedents in U that are logically consistent with the set of all truths in U is the narrowest—no broader than that set itself.

The laws, then, differ from the accidents not merely in degree but in kind. It is not the case that Λ's range of invariance in U is broader than that of every other set of truths in U that is logically closed in U and contains some logically contingent truths. But Λ's range of invariance in U is as broad as it could be.

2.4. *Why define stability in terms of nested counterfactuals?*

Stability was defined in section 2.1 above in terms of preservation, and Γ is preserved under Δ exactly when for any $m \in \Gamma$ and any $p \in \Delta$, $q \in \Delta$, $r \in \Delta$, and so on, all of the following are correct: $p > m$, $p > (q > m)$, $p > (q > (r > m))$, and so on. So Λ's non-nomic stability requires that various *nested* counterfactuals be correct. However, none of these nested counterfactuals had to be consulted in the arguments I have just given to show that any set containing an accidental truth lacks nontrivial non-nomic stability. Apparently, for any such set, even some of the *non-nested* counterfactuals required for its non-nomic stability fail to be correct. This raises the question: Why define Γ's non-nomic stability in terms of the *nested* counterfactuals? Why not require only that for any $m \in \Gamma$ and any $p \in U$ where Con(Γ,*p*), the subjunctive conditional $(p > m)$ is correct (where Γ is a set of non-nomic truths that is logically closed in U)?

The answer is that (i) it is *logically possible* for a set containing an accidental truth to have nontrivial non-nomic stability in this *non-nested* sense, though this would require an exquisitely delicate coordination among the facts in U*, and (ii) it is intuitively plausible to identify kinds of necessity with sets possessing nontrivial non-nomic stability only if "non-nomic stability" is defined in the *nested* sense. Let me explain.

Suppose that *p* and *q* are each in U, false, and consistent with Λ. By Λ's non-nomic stability in the *non-nested* sense, $(q > m)$ for every $m \in \Lambda$. But suppose that for some $m \in \Lambda$, $\neg(p > (q > m))$. In other words, suppose that Λ lacks non-nomic stability in the *nested* sense; it is not the case that the counterfactual conditionals that are responsible for Λ's non-nomic stability in the non-nested sense would all *themselves* still have held had *p* obtained (though $p > m$, for every $m \in \Lambda$). Then it is mere coincidence that Λ possesses non-nomic stability in the non-nested sense, since it is an accident that $\neg p$ obtains and it is not the case that Λ would still have possessed this stability had *p* obtained instead. In other words, although Λ's members *would* all still have obtained under every counterfactual antecedent in U under which they *could* all still have obtained, this invariance is itself merely accidental—it is not the case that it would still have obtained under each of these counterfactual antecedents. Intuitively, Λ then fails to correspond to a kind of necessity. Therefore, we do not countenance this as logically possible. On the contrary, we believe not only that had we tried to violate the natural laws (e.g., to accelerate a material particle from rest to beyond the speed of light), we would have failed, but also that if the non-nomic circumstances had been different in some manner that is consistent with Λ (e.g., if we now had access to twenty-third century particle-acceleration technology), then it would still have been the case that had we tried to accelerate a material particle from rest to beyond the speed of light, we would have failed. Any set (hold on now!) whose nontrivial non-nomic stability in the non-nested sense is *accidental*—in the sense I have just described—must contain an accidental truth.

Intuitively, it would take extraordinary coordination for all of the counterfactual conditionals needed to make Γ nontrivially non-nomically stable in the non-nested sense to be correct even though Γ contains an accidental truth. Nevertheless, this seems to be a logical possibility. Therefore, to identify the special relation that holds in any possible world between the laws there and the counterfactual conditionals that are correct there, we need to identify *p*'s physical necessity (for $p \in U$) with *p*'s belonging to a set possessing nontrivial non-nomic stability in the nested sense.

3. Counterfactuals and the root commitment

3.1. *Multiple grades of physical necessity*

What is the relation between the root commitment that we undertake in believing ■*p* and our belief that *p* belongs to the unique set that nontrivially possesses non-nomic stability? Does this latter belief follow from that root commitment? If so, this would impose a considerable constraint on what that root commitment could be. However, it is not obvious that the root commitment should require this belief. Perhaps the root commitment requires us to believe that *p* belongs to *some* set that nontrivially possesses non-nomic stability, but not that there is only one such set. How much is built into the root commitment?

Let me back up a bit. We tend to presume that in the *actual* world, there is a set that nontrivially possesses non-nomic stability. But is it *logically necessary* that there be such a set? I see no reason why. In a possible world lacking such a set, there are no laws of nature; there is no sense of necessity "between" logical necessity and no necessity at all. The *concept* of a natural law should not suffice to guarantee that there are, in fact, natural laws. It is up to science to discover whether there are.

By the same token, although we tend to presume that there is a unique set that nontrivially possesses non-nomic stability, it seems at least *logically possible* for there to be *more than one* such set. We can design "from scratch" a possible world where this is so by stipulating some of the non-nomic claims and some of the counterfactual conditionals holding in that world. Consider the following claims:

- *a* = All particles are X-ons or Y-ons, pointlike, and never created or destroyed.
- *b* = All X-ons have one unit of positive electric charge and one unit of mass.
- *c* = All Y-ons have one-half unit of negative electric charge and one unit of mass.
- *d* = Coulomb's "law" governs the force between two electric charges, and applies not merely to electrostatic cases but also to dynamic ones (i.e., describes instantaneous rather than retarded action-at-a-distance).
- *e* = Newton's "laws" of motion obtain.

ıths in the world being constructed—and suffice
ditions) to determine all events in that world. (If
de, then we can enrich the above set. All that is
n generalizations regarding the relation between
ralizations specifying the forces present, and gener-
arges with respect to each of these forces that are
pecies of particle.) Now stipulate that the logical
vially possesses non-nomic stability. Stipulate fur-
re in U of *a*, *d*, and *e* nontrivially possesses non-
ws of motion and force laws would all still have
acteristic charges with respect to these forces been
nstead possessed two-thirds of a unit negative elec-
ier that the logical closure in U of *a* and *e* nontrivi-
stability. (So the laws of motion would still have
the particles' charges been different—had, say, the
or diminished with the cube of the distance.) We
three nontrivially non-nomically stable sets, and
ical necessity.

ıat for any two sets that nontrivially possess non-
oe a proper subset of the other. So since a possible
ıere, any other such set must either be a proper
ıe of that world's accidental truths along with all
n U). Which of these is logically possible? If there
he smaller set be the possible world's Λ and the
that world's accidental truths? Could the larger set
ınd the smaller set contain a more select group of
the concepts of a "natural law" and a "physical
ply to such a world, because they presuppose that
nontrivially possesses non-nomic stability—that
ecessity "between" logical necessity and no neces-

at arbitrary how (or whether) to extend the con-
"physical necessity" to a possible world with more
ially possesses non-nomic stability. But since the
ı their nontrivial non-nomic stability, and since (I
ientific importance of the physical necessities de-
n their nontrivial non-nomic stability, I urge that
le world as having multiple grades of physical
ween" logical necessity and no necessity at all.[15]
possible worlds where more than one set nontrivi-
tability. But though my arguments in the preced-
e intuitively expect there to be a unique grade of
tual world, there might in fact turn out to be more
will discover there to be a small subset of laws
recedence, in any context, over the preservation of
eories of an inflationary multiverse, where fermion

masses and coupling constants are tunable parameters, seem to posit multiple grades of physical necessity.) I regard my arguments in the preceding section as showing not that multiple grades of physical necessity are impossible, but rather that our intuitive presumption of a unique grade of necessity between logical necessity and no necessity at all lines up with the intuition (brought out by that argument) that there is a unique nontrivially non-nomically stable set. What is important for my purposes is the *connection* between physical necessity and stability, not whether Λ is the *unique* nontrivially non-nomically stable set. It is up to science to discover whether or not it is.[16]

I conclude that the root commitment associated with belief in ■*p* is responsible for this biconditional: □*p* (for *p* in U) if and only if *p* belongs to *some* set that nontrivially possesses non-nomic stability. This leaves room for many "grades" of physical necessity but prohibits an accident's belonging to any nontrivially non-nomically stable set.[17]

3.2. *Preservation of physical necessity and its lack*

Recall that I have defined "non-nomic stability" in a *nested* sense. So Λ's non-nomic stability demands that for any $p, q, r \ldots \in$ U such that Con(Λ,*p*), Con(Λ,*q*), Con(Λ,*r*) . . . , all of the following are correct: $p > m$, $p > (q > m)$, $p > (q > (r > m))$, and so on, for any $m \in \Lambda$. In other words, in the closest *p*-world, not only does *m* obtain, but so do $q > m$, $q > (r > m)$, and so on. Hence, in the closest *p*-world, the actual world's Λ has non-nomic stability. Therefore (as long as this non-nomic stability is nontrivial), the actual Λ's members all possess a grade of physical necessity in the closest *p*-world: for any $m \in \Lambda$, □*m* would still have held had *p* obtained. (Here we have part of Λ^{+}'s stability—i.e., part of Λ^{+}-preservation—that goes beyond Λ's non-nomic stability.)[18]

Of course, I have just derived the preservation only of the *physical necessity* of the laws in U, not of their *lawhood*. Although I have left room for a distinction between laws and physically necessary nonlaws, I have not yet officially drawn this distinction. Once I have done so (in chapter 7), I posit a root commitment that is sensitive to this distinction and entails the preservation of the laws' *lawhood*, not merely their physical necessity.

Now Λ^{+}'s stability (i.e., Λ^{+}-preservation) requires also that the lack of physical necessity of the physically unnecessary claims (and the nonlawhood of the nonlaws) be preserved under all counterfactual antecedents in U that are consistent with Λ.[19] Suppose we knew there to be no greater number of nontrivially non-nomically stable sets in the closest *p*-world than in the actual world. Then we could conclude that no claim actually lacking physical necessity belongs to a nontrivially non-nomically stable set in the closest *p*-world, and hence that no such claim possesses physical necessity there. However, I see no reason why it could not happen that in the closest *p*-world, there are more nontrivially non-nomically stable sets than in the actual world. So I see no contradiction in (say) an accidental generalization in the actual world belonging to such a set in the closest *p*-world. In that case, we would have an exception to Λ^{+}'s stability (Λ^{+}-preservation).

I argued in chapter 2 that Λ^{+}-preservation is intuitively plausible. The logical possibility of the above sort of exception does not, I think, stand in any tension with those intuitions. I think we intuitively tend to expect of any world that it contain a unique grade of necessity "between" logical necessity and no necessity at all. If we implicitly take a counterfactual supposition $p \in$ U where Con(Λ^{+},*p*) as invoking such a possible world, then we must believe that in the closest *p*-world, the physical necessities in U are exactly the actual physical necessities, as Λ^{+}-preservation demands.[20]

3.3. Why we have not yet found the root commitment

The root commitment associated with belief in ■*p* must explain why a claim in U expresses a physical necessity in a given possible world if and only if, in that world, the claim belongs to a set that nontrivially possesses non-nomic stability. By proposing a root commitment that satisfies this constraint, we can explain why the physical necessity of the actual physical necessities is preserved under any non-nomic counterfactual antecedent *p* where Con(Λ,*p*). We can also explain why the lack of physical necessity of claims in U that actually lack physical necessity is likewise preserved (as long as it is taken for granted that in the closest *p*-world, there is no more than a single grade of physical necessity). It might be supposed that the best way to find a root commitment that satisfies this constraint is to construe the root commitment undertaken in believing ■*p* (where *p* is non-nomic) simply as the belief that *p* belongs to some set that nontrivially possesses non-nomic stability.

However, there are two reasons to think that this would be a mistake. First, as I sketched in chapter 1 (though do not fully explain until chapter 7), scientific practice distinguishes between laws of nature and physical necessities that are not natural laws. The above proposal regarding the root commitment cannot account for this distinction; a non-nomic truth's lawhood, over and above its physical necessity, does not affect the range of non-nomic counterfactual antecedents under which it is preserved.[21] Second, as I explained in chapter 1 and reviewed in this chapter, the above proposal regarding the root commitment cannot account for the great scientific *importance* of identifying the laws (or even the physical necessities). Why should science especially care about determining whether some non-nomic truth belongs to a set that nontrivially possesses non-nomic stability?

To answer this question and to find the resources for distinguishing the laws from the physically necessary nonlaws, we must turn to the relation between our beliefs about the laws and the inductive confirmations that we carry out. There we shall ultimately find the root commitment.

• FOUR

Inductive Confirmability and Physical Necessity

1. "Inductive" confirmation

1.1. Goodman's suggestion

Suppose we are predicting whether a given actual *F*, heretofore unexamined for its *G*ness, is *G*. We may be in a position to make a justified prediction on the basis of having already discovered that it is a law that all *F*s are *G*. But it is often maintained that we cannot make a justified prediction on the basis of having already discovered that it is an accident that all *F*s are *G*, because we cannot be justified in believing it to be an accident that all *F*s are *G* in advance of justly believing that we have already examined every *F* for its *G*ness.

In section 4 of chapter 2, I showed J. L. Mackie making this very assumption. Nelson Goodman (1983) influentially made an assertion along the same lines:

> What then does distinguish a law like "All butter melts at 150° F" from a true and general non-law like "All the coins in my pocket are silver"? Primarily, I would like to suggest, the fact that the first is accepted as true while many cases of it remain to be determined, the further, unexamined cases being predicted to conform with it. The second sentence, on the contrary, is accepted as a description of contingent fact *after* the determination of all cases, no prediction of any of its instances being based upon it. . . . A general statement is lawlike [i.e., is a law if

> it is true] if and only if it is acceptable prior to the determination of all its instances. (pp. 20–22)

Goodman then clarifies his view of how we ascertain a "true and general non-law." Goodman's view is not that with each discovery of an instance, we can justly increase our confidence that the generalization holds in a designated unexamined case, but that the rate of increase is so slow, or the cases so few, that even if we have examined every case besides the designated one and have found each to accord with the hypothesis, our confidence that the generalization holds in the designated unexamined case will still not have risen high enough to warrant our using the generalization for making a prediction about that case. Rather, Goodman's view is that we regard the discovery of an instance as unable to increase our confidence *at all* that an unexamined case accords with the generalization:

> That a given piece of copper conducts electricity increases the credibility of statements asserting that other pieces of copper conduct electricity. But the fact that a given man now in this room is a third son does not increase the credibility of statements asserting that other men now in this room are third sons, and so does not confirm the hypothesis that all men now in this room are third sons. . . . The difference is that in the former case the hypothesis is a *lawlike* statement; while in the latter case, the hypothesis is a merely contingent or accidental generality. Only a statement that is *lawlike*—regardless of its truth or falsity or its scientific importance—is capable of receiving confirmation from an instance of it; accidental statements are not. (p. 73)

I find it easy to share Goodman's intuitions in this passage. Yet his remarks could stand some refinement. Oddly, Goodman seems to regard *h*'s confirmability as dependent on whether *h is* lawlike. Whatever Goodman means precisely by "confirmation" (a subject to which I turn momentarily), it is surely an epistemic matter; whether a given discovery confirms *h* must depend on what our background beliefs are, but not on whether those background beliefs are *true*. So I take Goodman's suggestion to be roughly that whether we can "confirm" *h* by discovering one of its instances depends on whether *we believe* that *h* is lawlike.

I ultimately defend a slightly weaker thesis: to "confirm" *h*, we must believe that *h* may be physically necessary—we cannot believe that $\neg\Box h$. This represents a twofold weakening of Goodman's thesis. First, it permits us to "confirm" *h* even when we believe $\neg\blacksquare h$; I leave room for physically necessary nonlaws.[1] Second, the thesis that I defend permits us to "confirm" *h* even when we have not yet concluded that $(h \supset \Box h)$, as long as we have some intermediate degree of confidence that $\Box h$, somewhere between belief and denial. This allows us to be collecting evidence that *h* is physically necessary if it is true while we are assembling evidence that "confirms" *h*.

Also odd in Goodman's passage above is his contention that by discovering a given man now in this room to be a third son, we do not "confirm" that all of the men now in this room are third sons. We may agree with Goodman that by discovering a given man now in this room to be a third son, we do not increase our confidence that, say, the man in this room nearest to the given man is a third son. But we may insist that by discovering a given man in this room to be a third son, we increase our confidence that all of the men now in this room are third sons, since this evidence at least precludes one way in which this hypothesis might have been false: by the given man not being a third son.

Similarly, suppose *h* is that the next three tosses of this die land six. Suppose we already believe that the die is fair. Then by finding that the first toss among these three lands six, we raise our confidence in *h* (from 1/216 to 1/36) but not in either of the two other tosses landing six.

So why does Goodman (1983) deny that we "confirm" that all men now in the room are third sons when we discover a given man now in the room to be a third son?

> Consider the heterogeneous conjunction
>
> 8497 is a prime number and the other side of the moon is flat and Elizabeth the First was crowned on a Tuesday.
>
> To show that any one of the three component statements is true is to support the conjunction by reducing the net undetermined claim. But support of this kind is not confirmation; for establishment of one component endows the whole statement with no credibility that is transmitted to other component statements. Confirmation of a hypothesis occurs only when an instance imparts to the hypothesis some credibility that is conveyed to other instances. Appraisal of hypotheses, indeed, is incidental to prediction, to the judgment of new cases on the basis of old ones. (pp. 68–69)

So we "support" but do not "confirm" that all men now in the room are third sons when we discover a given man now in the room to be a third son. I prefer a different terminology. I say that evidence *e* "confirms" *h* exactly when *e*'s discovery leads us to increase our confidence in *h*—that is, exactly when $\mathrm{pr}(h \mid e) > \mathrm{pr}(h)$.[2] Hypothesis *h* undergoes *inductive* confirmation (what Goodman calls "confirmation" or "projection") exactly when, roughly speaking, we justly increase our confidence in *h and* increase our confidence in what *h* says about cases as yet unexamined.[3]

In the examples involving the third sons, the die, and the heterogeneous conjunction, a hypothesis *h* ("All *F*s are *G*") is confirmed, but not inductively. Indeed, *none* of *h*'s predictions regarding unexamined cases is confirmed.[4] Rather, an instance *e* (that *a*, already known to be *F*, is *G*) confirms *h* solely in virtue of precluding one possible exception to *h* (namely, $Fa \;\&\; \neg Ga$). In a vivid image, Ken Gemes deems *h* to have undergone "mere content cutting"; he pictures *e* as confirming *h* solely by cutting one conjunct from that "part" of *h* concerning unexamined cases ($[Fa \supset Ga] \;\&\; [Fb \supset Gb]$ & etc.).

1.2. What could "inductive" confirmation be?

Many philosophers besides Goodman maintain roughly that *h* cannot be confirmed *inductively* if *h* is believed to be an accident if true, including Mackie

(1962, pp. 71–73), John Stuart Mill (1893, p. 230; bk. 3, ch. 4, sect. 1),[5] C. S. Peirce (1934, p. 66), G. E. Moore (1962, p. 12), R. B. Braithwaite (1927, pp. 467, 473), Hans Reichenbach (1947, pp. 359–361, 368),[6] William Kneale (1952, esp. pp. 45, 52, 65), P. F. Strawson (1952, pp. 199–200), Israel Scheffler (1981, pp. 225–27), Max Black (1967, p. 171), Fred Dretske (1977, pp. 256–260), and Jaegwon Kim (1992, pp. 11–13).[7] Unfortunately, these philosophers fail to specify precisely when *h*'s confirmation qualifies as *inductive*. At best (see, e.g., Kneale 1950, p. 125, or Goodman's remarks quoted above), they say that we confirm *h* in the special manner reserved for hypotheses that we believe may be physically necessary only when we discover an instance that not only raises our confidence in *h*, but also raises our confidence in *h*'s predictions regarding unexamined cases. But although "inductive" confirmation is then distinguished from mere content cutting (where *none* of *h*'s predictions is confirmed), it remains unclear *which* of *h*'s predictions regarding unexamined cases must be confirmed in order for *h*'s confirmation to qualify as "inductive." Evidence that confirms *h*'s predictive accuracy might confirm *h*'s accuracy regarding a particular unexamined case or a limited range of unexamined cases while nevertheless remaining irrelevant to or even disconfirming *h*'s accuracy regarding certain other unexamined cases.

Notice that *e* confirms "*h* holds in all unexamined cases" if *e* confirms, of a given unexamined case, that it accords with *h*, and for each other unexamined case, *e* is irrelevant to whether it accords with *h*. For instance, suppose we believe that the crowd of people in the room includes Jones and his brother, but none of his other siblings. Suppose *h* is that each person in the room has a father named Robert. Let *e* be that Jones's father is named Robert. Then *e* confirms (indeed, convinces us), regarding one of *h*'s unexamined cases, that it accords with *h*. But for each of *h*'s other unexamined cases, suppose that *e* fails to bear confirmationwise on whether it accords with *h*. For *e* to confirm "*h* holds in all unexamined cases," it is not necessary that, for *each* unexamined case, *e* confirms that *h* holds there. What sort of confirmation is supposedly reserved for hypotheses that we believe may express physical necessities?

The problem is not merely that the notion of "inductive" confirmation is ambiguous. Suppose we set our threshold very high: *h* is confirmed "inductively" exactly when for *each* unexamined case, the discovered instance confirms *h*'s accuracy regarding that case. Then despite this high threshold, we can easily find examples where we confirm a hypothesis "inductively" even though we believe it to be accidental if true. I mentioned in chapter 1 that, by examining an appropriate sample of persons all of whose ancestors through many generations were Native Americans, anthropologists have justified believing that all such Native Americans have blood of type O or type A. Anthropologists have obviously not checked every such Native American who ever lives. They regard the examined cases as having confirmed this hypothesis "inductively" by the above definition. Yet anthropologists believe this regularity to be accidental; all such Native Americans are descended entirely from a single, small, genetically isolated company that migrated across the Siberia-Alaska land bridge, and as it happened, no one in the company possessed allele B. I also mentioned that we do not need to taste every pear on a given tree in order to justify believing that all of the pears now on the tree are ripe. Rather, we might well regard each instance as confirming, of each unexamined pear on the tree, that it is ripe, and so as confirming the generalization "inductively" (by the above definition) even though we believe that it expresses an accident if it turns out to be true.

Carl Hempel (1965) seems to have in mind examples like these, where our discovery that all of the *F*s in a random sample are *G* justifies our belief that unexamined *F*s are *G*, when he argues that even a hypothesis believed to lack physical necessity can be confirmed inductively:

> It is sometimes possible to make predictions of an inductive character exclusively on the basis of information about a finite set of particular cases, without the mediation of any laws whatever. For example, information to the effect that a large sample of instances of P has been examined, that all of its elements have characteristic Q, and that a certain case x, not included in the sample, is an instance of P, will lend high inductive support to the prediction that x, too, has the characteristic Q. (p. 176 n6; see also p. 356)

Rudolf Carnap (1950, pp. 574–75), Arthur Pap (1958, pp. 206–207), Peter Achinstein (1971, p. 40), Elliott Sober (1988, pp. 95–98), and Bas van Fraassen (1989, pp. 134–38, 163–69) share Hempel's view. Skeptics regarding natural law regard examples like these as showing that physical necessity bears no link to some special kind of confirmation.

Whether these skeptics are correct depends on whether the threshold we set above in defining "inductive" confirmation captures the intuitive sense of confirmation toward which Goodman and others gesture. I believe it does not. I now work to cash out this intuitive sense of "inductive" confirmation. It will then be evident that the above examples do not involve the inductive confirmation of hypotheses believed to lack physical necessity. In fact, I demonstrate that we cannot confirm h "inductively" (as properly defined) if we believe $\neg\Box h$. In chapter 5, I further develop these ideas and use "inductive" confirmation to account for the laws' scientific importance (as I discuss briefly in chapter 3). In chapter 7, I use "inductive" confirmation to cash out the root commitment associated with believing $\blacksquare h$, and thereby explain why the laws bear their special relation to counterfactuals.

2. Two intuitions about induction

2.1. *Induction as the projection of observed patterns: a mistaken paradigm?*

I shortly turn to two intuitions that have motivated various philosophers to regard inductive confirmation as unavailable to any hypothesis believed to lack physical necessity. This first of these intuitions is manifested in the familiar conception of induction as essentially involving the projection of observed patterns onto unexamined cases. Black (1967, p. 169), for example, refers to "the classical conception of induction as generalization from particular instances." On this conception, the evidence that confirms "All *F*s are *G*" inductively must be an instance of the pattern projected: an *F* found to be *G*. Why have a host of philosophers focused on this "classical conception," sometimes called "induction by enumeration" or "instantial induction"?

Today, this question seems typically to receive the following answer: because they were wrong, that's why! Over the last decades, the study of confirmation has thoroughly left behind the "classical conception," and for very good reasons. Whether and to what extent some evidence confirms a given hypothesis depends crucially on various background opinions that "instance confirmation" neglected to take into account: the scientist's confidence in various "auxiliary hypotheses," her estimation of the evidence's likelihood if the hypothesis is false, her prior opinions regarding various alternatives to the hypothesis, and so on. These factors may keep an instance from confirming a hypothesis, or even make it disconfirm the hypothesis. None of these influences has a place in a "principle of the uniformity of nature," which portrays the confirmed hypothesis as a straightforward "generalization" of or "projection" from some observed instances. In contrast, these factors are automatically taken into account by models of confirmation that exploit the machinery of the probability calculus. These models do not accord instances any privilege—do not include some principle giving instances special confirmatory power. Rather, these models give a unified treatment to instances and other kinds of evidence, even examples where the hypothesis does not admit of "instances" at all because it does not assume the form "All *F*s are *G*." In bringing up the "classical conception," am I suggesting that we set aside all of the progress that confirmation theory has made by abandoning "instance confirmation"?

Of course not. I do not dispute the standard history, which depicts instance confirmation as "a mistaken paradigm, a false ideal, that misguided philosophy of science for decades" (van Fraassen 1983, p. 323).[8] I agree that the now-familiar philosophical miscellany of ships lost at sea, philosophers checking their hats, and grasshoppers outside of Pitcairn Island show the inadequacy of accounts based fundamentally on "projection," "generalization," or a "principle of the uniformity of nature."[9] An instance's confirming power depends on our background beliefs in just the way that the confirming power of any other kind of evidence does. How this works is nicely revealed by quantitative confirmation theories in which instances occupy no privileged place.

However, I also believe that something valuable has been left behind with the "classical conception"—certain intuitions that, once properly refined, should find their place within the probabilistic machinery. These intuitions are closely tied to those that motivated various philosophers to contend that a special kind of confirmation is available only to hypotheses that we believe may be physically necessary.[10] After elaborating these intuitions, I will show (in appendix 1 of this chapter) that those who have purported to explicate the kind of confirmation Goodman and others had in mind have not successfully done so.[11]

What instance confirmation includes but recent confirmation theory leaves out involves our *reason* for taking certain evidence as confirming a given hypothesis. Our reason for regarding an instance as bearing on a *given* unexamined case makes a difference to which *other* unexamined cases we must *by parity of reasoning* also regard that instance as bearing upon. In other words, whatever it is about a given unexamined case that makes the instance bear upon it, that feature is reflected in the *range* of unexamined cases that the instance bears upon. For example, that feature is reflected in whether the instance bears upon unexamined *F*s that have a certain property *P*, or (if we do not believe that there are any actual *FP*s) whether the instance confirms that *had* there been unexamined *FP*s, then they *would have* been *G*. If this counterfactual is confirmed, then our background belief that no unexamined *F*s are *P* was not needed to sustain *e*'s power to bear upon each and every actual unexamined *F*. If (roughly speaking) *all* such counterfactuals are confirmed, then *no* background beliefs regarding which unexamined *F*s there happen to be helps to sustain *e*'s confirming power; no such belief about what a given unexamined case is like forms part of our reason for taking *e* to bear on that case. I believe that this is intuitively why it was thought that an account of "instance confirmation" could safely neglect certain sorts of background beliefs.

Of course, accounts of "instance confirmation" were incorrect insofar as they suggested that such background beliefs can *typically* be neglected or that all confirmations are ultimately grounded in the special sort of confirmation that these accounts aimed to treat. I deny that, whenever an instance confirms a hypothesis, it does so *inductively*. For instance, suppose we find a given gold cube that formed naturally to be smaller than a cubic mile. We might regard this discovery as confirming "All gold cubes are smaller than a cubic mile" (at least by "content cutting" it). But presumably, we would not regard this discovery as bearing upon whether a gold cube constructed by an eccentric billionaire, who aimed to fashion a gold cube exceeding a cubic mile, would be smaller than a cubic mile. So this confirmation is not "inductive." (In contrast, we might well be prepared to accord broader relevance to such an eccentric billionaire's failure to create a large gold cube despite repeated attempts to do so.) But while instance confirmation should not have been taken as paradigmatic, the concept of "inductive" confirmation turns out to be important for understanding scientific reasoning. In this light, the neglect of background beliefs by accounts of "instance confirmation" was not simply mistaken or oversimplified. In it there resides a kernel of truth: in inductive confirmation, the evidence's capacity to bear upon certain predictions is independent of certain background beliefs, as I now show.

2.2. *Inductive confirmation distinguished by its range and its rationale*

Here is one intuitive motivation for the view that *h*'s inductive confirmation requires the belief that *h* may be physically necessary: *inductive* confirmation appeals to a "principle of the uniformity of nature"—that unexamined cases are relevantly like examined cases. Evidence that confirms *h* inductively must then involve an instance of *h*, so that *h* captures what it would take for unexamined cases to be relevantly like this evidence. In other words, when we confirm *h* ("All *F*s are *G*") inductively, we regard an examined *F* as bearing upon an unexamined *F* because we believe that in virtue of their both being *F*s, they may belong to the same natural kind of a certain sort (e.g., the same chemical species), and *h* portrays the two cases as similar in a respect that is relevant to this sort of natural kind. That is, there is supposed to be a property like *G*ness common to all members of such a kind; we believe that the possession of *G*ness (e.g., being disposed under standard conditions to boil at 1300°C) by all *F*s (samples of hydrogen peroxide) would make them uniform in a respect (boiling point under standard conditions) in which they are supposed to be uniform if they all belong to a certain sort of natural kind (chemical species). There are *laws* mandating such uniformities among the *F*s exactly when the *F*s all belong to the same natural kind of the given sort; the natural kinds of the given sort are exactly the kinds figuring in certain sorts of laws. (E.g., for each chemical species, there is a law specifying the boiling point under standard conditions of all pure samples of that species; I discuss this further in chapter 7.) Only if we believe that *h* may follow from such a law can we confirm *h* inductively. So we can confirm *h* inductively only if we believe that *h* may be physically necessary. On this view, *h*'s "inductive" confirmation is distinguished by our *reason* for regarding *e* as confirming *h*'s predictive accuracy.

In contrast, suppose we believe that either none or all of the coins in my pocket are dimes. Then by discovering a dime in my pocket, we confirm (indeed, ascertain) that all of the coins in my pocket are dimes (*h*). But while we have confirmed *h*'s predictive accuracy, we have not confirmed *h inductively*. Consider our *reason* for regarding the instance as bearing upon each actual unexamined case. It does not involve a belief that the coins in my pocket may form some sort of natural kind. Rather, the evidence gets a grip on unexamined cases because we already believe that the actual cases all happen to possess a certain feature: as a matter of physically unnecessary fact, the coins in my pocket are exclusively dimes if one is a dime. Had we instead already believed that every coin in my pocket is a dime if I am wearing an orange shirt, and had we then observed me to be wearing an orange shirt, we would have confirmed (indeed, ascertained) that each unexamined coin in my pocket is a dime. This just goes to show that under the appropriate background beliefs, any evidence can confirm that each unexamined coin in my pocket is a dime. No uniformity-of-nature principle need be involved. Therefore, when the evidence consists of the discovery of a dime in my pocket, it is really incidental to the evidence's bearing upon unexamined cases that the evidence involves an instance of h. In *inductive* confirmation, on the other hand, it is essential to the evidence's relevance to unexamined cases that the evidence, too, is a case of h. Background beliefs about which physically unnecessary possibilities happen to be realized are not needed to sustain the evidence's relevance to unexamined cases—to fix the range of unexamined cases to which the evidence is relevant.[12]

Consider another example. Suppose we are about to withdraw marbles at random from an urn of marbles, each time replacing the marble we have just withdrawn. For simplicity, suppose we know that there are only two marbles in the urn (labeled A and B) and that each marble is either black or white. When we withdraw a marble, we note its color but not its label. Since we have no other information, suppose that we assign equal likelihood to each of these possibilities: both marbles are black (h_1), A is black and B is white (h_2), A is white and B is black (h_3), and both are white (h_4). Let b_1 be that the first withdrawal yields a black marble, and let b_2 be that the second withdrawal yields a black marble. Let h be that every withdrawal from the urn yields a black marble. We have: $pr(h_1) = pr(h_2) = pr(h_3) = pr(h_4) = .25$, $pr(b_1) = pr(b_2) = .5$. Now we make our first withdrawal and discover that b_1 obtains. It is easy to show this evidence has the following effect: $pr(h_1 \mid b_1) = .5$, $pr(h_2 \mid b_1) = pr(h_3 \mid b_1) = .25$, $pr(h_4 \mid b_1) = 0$.[13] Furthermore, h has been confirmed.[14] Has h's predictive accuracy been confirmed? Yes: $pr(b_2 \mid b_1) = .75$.[15] Yet I insist that h has not been confirmed *inductively*. As in the dime example, the instance's power to confirm that unexamined cases accord with h is being sustained by our background beliefs regarding which circumstances just happen, without physical necessity, to obtain. In particular, the power of h's instance to bear on the next unexamined case depends on our *not* having certain beliefs about which physically unnecessary possibilities happen to be realized. For example, the instance's bearing on the outcome of the second withdrawal depends on our *not* believing that, as a matter of happenstance, the second marble we select is distinct from the marble that we have just withdrawn. If we believed this, then the outcome of the first withdrawal would not confirm anything about the outcome of the second withdrawal.[16] Indeed, even if we have already made many withdrawals from the urn and all of them have yielded black marbles, these instances of h do not bear upon the outcome of the next withdrawal given that the marble that we will next withdraw has never been withdrawn before.[17] An instance's confirming power is being sustained by—and accordingly, as I explore below, the limits of this power are being fixed by—our beliefs about which physically unnecessary possibilities may happen to be realized. The confirmation I am calling "inductive" is set apart by our having a different kind of reason for regarding past instances as confirming that unexamined cases accord with the hypothesis.

An instance confirms, of each future withdrawal (the next one, the hundredth one, and so on), that it accords with h, since although we believe it likely that *some* future withdrawal will yield a marble distinct from the one withdrawn in the given instance, we have no idea *which* future withdrawal this will be (the next one? the hundredth one?). Regarding each future case (e.g., the hundredth withdrawal), we believe that it *may* involve the same marble as the current instance. But there is no reason why we have to identify unexamined cases by their "number." We might identify an unexamined case by its other properties, such as who makes the withdrawal, what time of day it is made, what city is the largest at the time it is made, and so on, or that the marble withdrawn is distinct from the marble withdrawn in the given instance. The unexamined withdrawal thereby posited may be one that we do not believe is ever realized; we may believe it *likely* that some future withdrawal will involve a different marble from the given instance, without yet being willing to believe that some such withdrawal actually will take place. So for each unexamined withdrawal *that we believe actual* (such as the one I am just about to make), the given instance confirms that it accords with *h*. Nevertheless, the instance does not confirm *h inductively* because, roughly speaking, there is some property *P* that we believe it physically possible for an unexamined case of *h* to possess, where we do not regard the instance as confirming that an unexamined case *would* accord with *h were* that case to possess *P*.[18] Notice that it is a subjunctive conditional whose confirmation is at issue here. Property *P*—which, in this example, could be the property of having never before been withdrawn—insulates an unexamined case from the instance's confirming power. Since there is such a property *P*, the instance bears upon each unexamined case that we believe actual only because of our beliefs about which physically unnecessary possibilities happen to be realized—in particular, only because we do not believe, of some actual unexamined case, that it is *P*.

For the same reason, we fail to confirm "All of the pears now on this tree are ripe" *inductively* in the example I discussed in chapter 1. Our discovery that a given pear from the tree is ripe confirms, of each unexamined pear on the tree, that it is ripe, but only because we believe that the examined and unexamined pears may be subject to the same environmental conditions (rainfall, temperature, etc.). Therefore, we do not regard an instance of the hypothesis as confirming that *had there been* a pear on this tree whose environmental conditions were experimentally manipulated so as to differ radically from those of the other pears on the tree, then it would now have been ripe. Here we have a counterfactual supposition positing a case that exhibits an insulating property *P*. Likewise, in the Native-American blood-type example, an insulating *P* is the property of being descended from a different band of land-bridge migrants from the Native American in the instance under examination.[19]

In contrast, when a hypothesis is confirmed inductively, an instance's power to confirm, regarding each unexamined case believed actual, that it accords with the hypothesis is independent of background beliefs about which hypothetical unexamined cases, of those that we do not believe physically impossible, happen to be actual.[20] For example, Newton regarded an instance of his proposed gravitational-force law (inferred, say, from the fact that the moon's orbital motion accords with his hypothesis) as confirming, of each actual unexamined pair of material particles at each moment, that these particles gravitationally attract each other in accordance with his hypothesis. This confirmation does not depend on Newton's background beliefs about which physically unnecessary possibilities happen to be realized; it is not the case that Newton judged the instance to be relevant only to any unexamined pair of material particles that has, say, a certain spatial arrangement, but he believed that, as it happens, all pairs of material particles throughout the universe's history are so arranged. In other words, it is not the case that there was some insulating property P but Newton believed that every actual unexamined pair of material particles happens not to be P. Rather, Newton took an instance to be relevant to each unexamined pair of material particles *there is* because he took it to be relevant to each unexamined pair of material particles *there might have happened to be.* No beliefs about the further properties that happen to be possessed by the instance and the unexamined case sustained the instance's power to confirm that the unexamined case accords with the hypothesis. That is, Newton (1971, pp. 398–399) appealed to a uniformity-of-nature principle, his Third Rule of Reasoning: the "analogy of Nature" permits the extrapolation from qualities of bodies that "are found to belong to all bodies within the reach of our experiments" to "all bodies whatsoever." I am certainly not endorsing this rule as Newton expressed it. I mention the rule because it indicates the *range* of hypothetical unexamined cases to which Newton regarded his instances as relevant, or (what amounts to the same result) because it indicates Newton's *reason* for taking the discovered instances as bearing upon certain unexamined cases.

2.3. *Are all accidental generalizations coincidences?*

I now turn to a second intuition behind the view that we cannot confirm h inductively if we believe that $\neg\Box h$. Suppose we believe that for both of us to be at the same store simultaneously would be a coincidence. Then, intuitively, we regard my being at a given store later today as failing to confirm your being there then. An accidental generalization is often thought to describe a similar coincidence, albeit perhaps on a cosmic scale. If we believe that h is accidental if true, then we believe that it would be mere happenstance—utterly coincidental—if h, having held in every case already examined, turns out also to hold in a given, as yet unexamined case. So for any two cases of h, we consider one case's according with h as unable to confirm the other's doing so. To "project" a uniformity in examined cases onto unexamined cases, we must believe that it may be "no accident" that the uniformity has held of our observations so far.

Is this intuition correct? Some accidental generalizations do indeed describe coincidences. Typically, the reason why one gold cube is smaller than a cubic mile has nothing to do with the reason why another gold cube is smaller than a cubic mile. That *both* turn out so to be is entirely coincidental. However, as the Native-American example demonstrates, we may believe that $\neg\Box h$ and yet also believe, of *any* two actual cases, that it would be no coincidence were h to hold of both, considering that both Native Americans are descended from the same small party of land-bridge migrants. The same point is brought out by the pear example. Thus, an instance of h can bear on each actual unexamined case.

Inspired by Herbert Feigl's (1961, p. 215) terming certain accidental generalizations "structure dependent," we might say that anthropologists believed that perhaps all Native Americans belong to the same "structure," that is, have in common certain properties (namely, the property of having certain ancestors) that it is not logically necessary for a Native American to possess, but where a Native American's possession of these properties would help to explain his or her lacking allele B. The anthropologists' belief that perhaps all actual Native Americans belong to a common structure is responsible for an instance's relevance to each actual unexamined case. This suggests that an instance fails to bear upon a hypothetical case posited as falling outside of this structure. That is, anthropologists believe it physically possible (though false) that *two* companies of migrants (drawn from populations isolated from each other) crossed the Siberia-Alaska land bridge. Anthropologists believe that it would be a *coincidence* for two Native Americans, descended from different parties of migrants, both to lack allele B. Hence, an instance of h fails to confirm that, had there been a second party of land-bridge migrants, then their descendants would have accorded with h.

The same considerations apply to the marble example. That a given withdrawal involves a black marble confirms that an unexamined withdrawal also involves a black marble only if we believe that the two withdrawals may have a common structure—may involve the same marble. We believe that, otherwise, it would be a coincidence for them both to involve a black marble.

To review: the second intuition standing behind the view that we can confirm a hypothesis inductively only if we believe that it may be physically necessary is that all accidental generalizations are coincidences in the manner of our both being present simultaneously in the same store. But for any two actual Native Americans, it would be no coincidence for them both to be blood type O or A, considering their common descent. The analogy between the Native-American and store examples is restored once we look in a different place for the coincidence in the Native-American example. What would be coincidental is for a certain pair of *hypothetical* Native Americans to be similar (were they realized) in their blood types—namely, two Native Americans descended from different companies of land-bridge migrants. Likewise, it is a coincidence that all actual Native Americans are descended from the same company of land-bridge migrants.

If we go beyond the actual unexamined cases to consider certain hypothetical unexamined cases, then we restore the analogy between any accidental generalization and the store coincidence. That "All *F*s are *G*" lacks physical necessity does not entail that, for *any* pair of actual *F*s, it would be a coincidence if both *F*s in the pair were *G*. It does not even entail that there actually exists *at least one* pair of *F*s for which it would be a coincidence if both *F*s in the pair were *G*. But it entails that if the examined case is *G*, then there is a kind of unexamined case where it would be a coincidence, *were there cases of that kind*, for those cases also to be *G*. Therefore, the examined instance does not confirm that were there *F*s of that kind, then they would be *G*. The only difference between the Native-American and store examples is whether we believe that the hypothetical unexamined case possessing the property *P* that insulates a case from the instance's confirming power happens to be actualized.[21]

2.4. The two intuitions upheld

I have now elaborated two intuitions that I believe prompt the thought that we can confirm *h* "inductively" only if we believe that □*h* may be true. The apparent counterexamples to this thought can be reconciled with the intuitions behind it when "inductive" confirmation is characterized properly. By working out these intuitions, we have found them to converge on the following characterization: by discovering *Ga* (where we already believed *Fa*), we confirm *h* ("All *F*s are *G*") *inductively* exactly when *Ga* confirms the subjunctive conditional "Were there an *F* possessing property *P*, then all *F*s possessing *P* would be *G*" (call this conditional "C_P") for each property *P* that is (I shall say) "suitable." The precise range of the suitable properties is a key point to which I devote the following section. Clearly, we do not want to allow *G* itself to qualify as "suitable," since otherwise no hypothesis could be confirmed "inductively"; we cannot confirm "Were there an *F* that is *G*, then all *F*s that are *G* would be *G*" because we already believe it. Likewise, even if we believe that ¬□*h* and so believe that an *F* can physically possibly be non-*G*, we do not want to allow non-*G* to be suitable. Intuitively, that would make it *too* difficult for us to confirm *h* inductively when we believe ¬□*h*. Recall that, intuitively, we do not blame the failure of "Every withdrawal from the urn yields a black marble" to be confirmed inductively on an instance's incapacity to confirm "Were some future withdrawal to yield a white marble, then all such future withdrawals (that is to say, future withdrawals of white marbles) would yield black marbles." Rather, we put the blame on an instance's incapacity to confirm "Were some future withdrawal to yield a marble distinct from any that has already been withdrawn, then every such withdrawal would yield a black marble." That property—being the withdrawal of a marble that has never before been withdrawn—should qualify as "suitable" for this hypothesis, but the property of being nonblack should not.

We might think of "suitable" unexamined cases, whether actual or counterfactual, as cases that someone entertaining *h* could with consistency believe to be realized and as yet unexamined. That "someone" might even be ourselves, either now or sometime after we have become better informed of the properties that happen to be possessed by various actual unexamined cases. An instance confirming *h* inductively not only confirms, for each of the kinds of unexamined cases that we believe there actually are, that all such cases accord with *h*, but also confirms, for each of the kinds of unexamined cases that we might later—while still entertaining *h*—come to believe there actually are, that all such cases would (were they realized) accord with *h*. So when we confirm *h* inductively by *e* and later learn more about the properties possessed by various unexamined cases, these further discoveries (as long as they allow us to continue to entertain *h*) give us no reason to reconsider whether *e* still bears favorably on those cases. When Newton confirmed his gravitational force law inductively, no further discoveries about, say, the arrangement of lunar craters insulated cases from the instance's confirming power. We thus see again why *G* and non-*G* should not qualify as "suitable." A case of *h* that we believe to be *G* is not an "unexamined" case; it has been "examined" in the relevant respect. And as long as we are entertaining *h*, we cannot believe that there is a case of *h* that is non-*G*.

After refining the range of "suitable" properties, I demonstrate that we cannot confirm *h* inductively if we believe that $\neg\Box h$. This result might initially appear trivial, considering the way that I am defining "inductive" confirmation. After all, if we believe that $\neg\Box h$, then presumably, considering the relation between laws and counterfactuals, we believe that there is some suitable *P* such that $\neg C_P$. Therefore, we cannot confirm, for *each* suitable *P*, that C_P obtains. But the final step of this argument ("therefore . . . ") does not follow. We may believe that there exists a suitable *P* for which $\neg C_P$ without having any idea of which suitable *P* this is, that is, without there being some suitable *P* such that we believe $\neg C_P$. So we have not yet understood why evidence cannot confirm, of each suitable *P*, that C_P obtains.[22]

3. Suitability

The discovery of *Ga* (where we already believe *Fa*) confirms *h* ("All *F*s are *G*") *inductively* exactly when for any "suitable" property *P*,

$$\mathrm{pr}(C_P \mid Ga) > \mathrm{pr}(C_P), \tag{4.1}$$

where C_P is "Were there an *F* possessing *P*, then all *FP*s would be *G*." We must now specify which properties qualify as "suitable." The range of suitable *P*s is supposed to be such that, in confirming *h* inductively, an instance bears favorably upon every unexamined case that we believe might have happened to be actual—and so bears favorably upon every unexamined case that we believe actual without depending on our beliefs about *which* unexamined cases that might have happened to exist actually do.

In fixing the "suitable" *P*s, we should also bear in mind that the laws' inductive confirmability is supposed to be connected to the laws' special relation to counterfactuals (I introduced this thought in section 4 of chapter 2). To capture this thought, the conditionals C_P that are confirmed when *h* is confirmed inductively must correspond to the conditionals C_P that exhibit *h*'s preservation under the right range of counterfactual suppositions—the C_Ps that intuitively are "supported" by *h* if *h* supports counterfactuals in the manner of a law. What manner is this? Recall Λ-preservation:

Λ is preserved under $\{p \in U^+$ such that $Con(\Lambda^+, p)\}$

So if we believe that *h* may be physically necessary, then in confirming *h* inductively, the C_Ps that are confirmed must be the C_Ps that have antecedents falling within Λ-preservation's scope and that depend on *h* for their "support." Then if our root commitment in believing $\blacksquare h$ requires us to treat *h* as if it has been confirmed inductively, it will require us to regard *h* as preserved in the manner that we have found to be characteristic of laws. On the other hand, if $\neg\Box h$, then, as I showed in section 1.5 of chapter 2, $[\Lambda \cup \{h\}]$-preservation (first revision),

> $[\Lambda \cup \{h\}]$ is preserved under $\{p \in U^+$ such that $Con(\Lambda^+, p)$ and $Con([\Lambda \cup \{h\}]$'s closure in $U, p)\}$,

expresses the relation that *h* must bear to counterfactuals if *h* is to support counterfactuals in the same manner as a law (which *h* fails to do). So if we believe that $\neg\Box h$, then the suitable *P*s must correspond to the C_Ps that have antecedents falling within the scope of $[\Lambda \cup \{h\}]$-preservation (first revision) and that depend on *h* for their support. Then if we believe that $\neg\Box h$, our incapacity to confirm *h* inductively will be related to our belief that *h* fails to stand in the same sort of relation to counterfactuals as laws do.

To pursue this line of thought, we must begin by demanding of any suitable *P* that $pr(C_P) \neq 1$. For if $pr(C_P) = 1$, then C_P has already been maximally confirmed, so *Ga*'s failure to confirm it further [violating (4.1)] does not intuitively reflect some limitation on the range of unexamined cases to which we take *Ga* to be relevant. Rather, it shows that other evidence, already taken into account, sufficed to justify *h*'s projection onto the *FP*s, or that no evidence was needed to warrant it.

The latter occurs when Fx & Px is logically consistent and logically necessitates Gx.[23] In that case, C_P is a logical truth, and so, automatically, $pr(C_P) = pr(C_P \mid Ga) = 1$.[24] For instance, the property of being *G* is not a suitable *P*, for C_P is then "Had there been an *x* such that *Fx* and *Gx*, then it would have been the case that all (*F* & *G*)s are *G*." This constraint on suitability codifies the intuition that, to project inductively "All emeralds are green," we are not required to do the impossible: to confirm the greenness of emeralds that we already believe to be green.[25]

Consider another example in which this constraint on suitable *P*s comes into play. Suppose that, before discovering *Ga*, we already believe that $\blacksquare j$. Let *Px* mean $(j \supset Gx)$. Then C_P is the counterfactual "Had there been an *x* such that *Fx* and $(j \supset Gx)$, then it would have been the case that all *x*s such that *Fx* and $(j \supset Gx)$ are *G*." We believe that C_P's antecedent is consistent with Λ^+, since we already believe *Fa* and our $pr(Ga) \neq 0$. So by Λ^+-preservation, we believe that *j* is preserved under C_P's antecedent. So our degree of confidence in C_P is our degree of confidence in "Had there been an *x* such that *Fx* and *Gx*, then it would have been the case that all (*F* & *G*)s are *G*," which we already believe. Therefore, this *P* is unsuitable. Again, this is loyal to the intuitions motivating my characterization of "inductive" confirmation. We believe C_P in virtue of treating *j* rather than in virtue of treating *h* as having been highly inductively confirmed. Intuitively, *Ga*'s failure to support *h*'s projection onto these cases does not reflect a limitation on *Ga*'s confirming power; the evidence for *j* amassed prior to our ascertaining *Ga* already sufficed to justify projecting *h* onto these cases. Because of Λ^+-preservation, $pr(C_P) \neq 1$ requires that we not already believe *Gx* to follow logically from *Fx* & *Px* together with Λ^+.

To continue to elaborate what it takes for *P* to be "suitable," let us express C_P as $p > q$. Then we must require of any suitable *P* that $pr(Con(\Lambda^+,p)) = 1$. This ensures that *p* falls within the scope of Λ-preservation and goes part of the way toward ensuring that *p* falls within the scope of $[\Lambda \cup \{h\}]$-preservation (first revision). (The remaining constraint on suitability goes the rest of the way.) This requirement, then, helps to ensure that the counterfactuals confirmed when *h* is confirmed inductively correspond to the counterfactuals that Λ-preservation demands if $\Box h$ holds and the counterfactuals that $[\Lambda \cup \{h\}]$-preservation (first revision) demands if *h* is an accidental generalization.

Consider an example in which this requirement comes into play. Let *Px* mean $\blacksquare j$, where we do not already believe that *j* (in U) is a law; suppose that we have confirmed *j*'s lawhood, but not sufficiently to warrant believing that $\blacksquare j$. Then we believe that Λ^+ may include $\blacksquare j$, but we also believe that Λ^+ may include $\neg\blacksquare j$. Thus, we believe that $Con(\Lambda^+,p)$ may be false. Thus, our $pr(Con(\Lambda^+,p)) \neq 1$, and so *P* is unsuitable. .

Consider another example. Suppose we believe that $\blacksquare h$ may be true. Let *Px* mean "$Gx \supset (h \;\&\; \neg\Box h)$," that is, "$Gx \supset$ (*h* is an accidental generalization)." Since we believe that *h* may be a law, we believe that Λ^+ may include $\blacksquare h$, which entails "All *F*s are *G*," which together with *Fx* entails *Gx*, which together with "$Gx \supset$ (*h* is an accidental generalization)" entails "*h* is an accidental generalization." Hence, we believe that Λ^+ together with *p* may entail both "*h* is an accidental generalization" and $\blacksquare h$. That is, we believe that perhaps $\neg Con(\Lambda^+,p)$. Hence, *P* is not suitable.

I turn now to the final constraint on suitable *P*s.[26] By requiring that $pr(Con(\Lambda^+,p)) = 1$, the range of suitable *P*s is made to correspond to Λ-preservation's scope. However, recall that the scope of $[\Lambda \cup \{h\}]$-preservation (first revision) is further restricted for the following reason. It is logically possible for all of the members of $[\Lambda \cup \{h\}]$ to be preserved under a given counterfactual antecedent *p* only if *p* is consistent with $[\Lambda \cup \{h\}]$'s closure in U. Therefore, the scope of $[\Lambda \cup \{h\}]$-preservation (first revision) is restricted to counterfactual antecedents satisfying this constraint. Hence, for the range of suitable *P*s to correspond to the scope of $[\Lambda \cup \{h\}]$-preservation (first revision), there must be this final restriction on *P*s suitability: if we believe that $\neg\Box h$ (i.e., that *h* is an accidental generalization if *h* is true), then $pr(Con([\Lambda \cup \{h\}]\text{'s closure in U},p) \mid h) = 1$. The reason for making this probability conditional on *h* is that, whereas *h* is taken by $[\Lambda \cup \{h\}]$-preservation (first revision) to be an accidental truth, we are treating *h* merely as a hypothesis when we are confirming it inductively. In particular, there is no guarantee that we already believe that $Con(\Lambda,h)$; we need only *not* believe that

$\neg$Con(Λ,*h*) in order to entertain *h*. To make this final requirement on *P*s suitability correspond to the scope of [$\Lambda \cup \{h\}$]-preservation (first revision), we must make this requirement attend to any inconsistency between *p*, Λ, and *h presumed true*. (If *h* is true, then it must be consistent with Λ.)

Here is an example where this constraint comes into play. Let *P* be $\neg G$. If we believe that *h* may be physically necessary, then $\neg G$ is unsuitable because it violates an earlier requirement: we do not believe the existence of an *F* that is $\neg G$ to be consistent with Λ^+, since it is inconsistent with *h* and we believe that *h* may be physically necessary (in which case $h \in \Lambda^+$). This codifies the intuition that, to confirm "All emeralds are green" inductively when we believe that it may be physically necessary, we are not required to confirm to be green an emerald that we already believe not to be green. But if we believe that $\neg\Box h$, then this *P* may satisfy the earlier requirement. This in no way diminishes the intuition, captured by our new requirement on suitable *P*s, that *h*'s inductive confirmation should not demand *h*'s projection onto an *F* that is specified as $\neg G$. Of course, I argue shortly that we cannot confirm *h* inductively if we believe that $\neg\Box h$. But *h*'s failure then to be inductively confirmed should not result from *h*'s inductive confirmation requiring that *Ga* confirm the *G*ness of the unexamined *F*s posited to be $\neg G$. *Ga*'s incapacity to confirm such cases does not result from our belief that $\neg\Box h$. Rather, it results simply from the logical inconsistency between what is posited regarding these cases and what would have to be confirmed of them. [This point was also made in connection with the marble (section 2.2) and moa (chapter 1) examples.]

The analogous intuition applies to preservation under counterfactual antecedents. If $\neg G$ remains suitable when we believe that $\neg\Box h$, then *h* is not preserved under all counterfactual antecedents positing suitable *P*s. But *h*'s failure to be preserved under a counterfactual antecedent positing *h*'s violation does not exhibit *h*'s lack of physical necessity. I now show how the *P* that prevents *h* from being confirmed inductively (when we believe $\neg\Box h$) corresponds to the *p* that prevents *h* (when $\neg\Box h$) from being preserved under counterfactual antecedents in the manner of laws, that is, the *p* that makes [$\Lambda \cup \{h\}$]-preservation (first revision) fail.

4. Inductive confirmability, physical necessity, and preservation

4.1. Induction's connection to physical necessity borne out

I have suggested that *h*'s "inductive" confirmation be defined as follows: Suppose we discover *Ga*, where we already believed *Fa*. We thereby confirm *h* ("All *F*s are *G*") *inductively* if and only if for any "suitable" property *P*, pr(C_P | *Ga*) > pr(C_P) (proposition 4.1), where C_P is the subjunctive conditional $p > q$:

> If there were an *x* such that *Fx* and *Px*, then it would be the case that all (*F* & *P*)s are *G*,

and P is suitable if and only if

$\mathrm{pr}(C_P) \neq 1$,
$\mathrm{pr}(\mathrm{Con}(\Lambda^+,p)) = 1$, and
if we believe that $\neg\Box h$, then $\mathrm{pr}(\mathrm{Con}([\Lambda \cup \{h\}]$'s closure in $\mathrm{U},p) \mid h) = 1$.

In chapter 5, this definition is refined and simplified considerably, but the above will do for now.[27]

It follows from this definition that if h is confirmed "inductively," then h is confirmed. (Here let P be the property of being self-identical.) Furthermore, if we confirm h inductively, then we confirm, of each unexamined case that we believe actual, that it accords with h. [Here let P be a property that we believe to be possessed only by a particular actual unexamined case. Such a property is "suitable": $\mathrm{pr}(C_P) \neq 1$, since the case is "unexamined"; $\mathrm{pr}(\mathrm{Con}(\Lambda^+,p)) = 1$, since we believe that p holds; and $\mathrm{pr}(\mathrm{Con}([\Lambda \cup \{h\}]$'s closure in $\mathrm{U},p) \mid h) = 1$, since we believe that p holds and that h may be true.] This confirmation does not depend on our background beliefs regarding which suitable kinds of unexamined cases there happen to be. Rather, our reason for regarding the instance as bearing upon each actual unexamined F requires us to regard the instance as bearing on each suitable kind of unexamined F—roughly, each kind of unexamined F that we believe there might have happened to be. This is what places inductive confirmation at the furthest remove from mere content cutting.[28]

Having thus cashed out the intuitions behind the concept of "inductive" confirmation, now I examine whether Goodman and other philosophers are correct in holding that we can confirm h "inductively" only if we believe that h may be physically necessary.

As I noted at the close of section 2, this claim might seem implausible. To appreciate why, suppose we believe that a given generalization is false—that at least one of its cases violates it. New evidence therefore cannot confirm this generalization. Nevertheless, new evidence can confirm, of each object that we believe to be an actual unexamined case, that it accords with the generalization. For example, suppose we believe that exactly two of these three marbles are the same color, though we do not have any beliefs about which two or their color. Discovery that the first marble is red might well, for each of the two unexamined marbles, confirm that it is red, even though we believe that they cannot both be red. Now believing that $\neg\Box h$, in that some *physically possible* case violates h, might seem like believing the marble generalization false, in that some marble among the three is not red. Our believing the marble generalization false does not prevent us from taking an instance as confirming, of each actual unexamined case, that it accords with the generalization. By analogy, our believing that $\neg\Box h$ should apparently not prevent us from taking an instance of h as confirming, of each suitable kind of unexamined case, that h would hold of all members of that kind, were there any. But as I now show, the analogy between the marble example and inductive confirmation breaks down.

Suppose we believe that $\neg\Box h$. How might it be shown that we cannot confirm h inductively? It would suffice to find a suitable P such that (4.1) fails. Which suitable P insulates an unexamined F from an instance's confirming power? Let me turn the question around: When *are* we able, despite believing $\neg\Box h$, to regard an instance of h as confirming an unexamined F to be G? When we believe that the instance and the unexamined case may belong to a common "structure" (recalling our intuition pump from section 2), that is, when we believe that the instance and the unexamined case may share a property whose possession by an object is *not* physically necessitated by its being F, but whose possession by an F would be partly responsible, if the F is G, for making it G. For instance, that the examined Native American has blood of type O or A confirms that the unexamined Native American does, too, because we believe that the two Native Americans may belong to a common structure—in particular, may be descended from the same band of land-bridge migrants. So intuitively, to insulate a category of unexamined Fs from Ga's confirming influence, we should make P the property of falling outside of a's structure.

How can we ensure that if an unexamined F possessing P is G, then its properties that help to make it G, but its possession of which is not physically necessitated by its being F, are not among the properties that the examined F possesses that help to make it G but its possession of which is not physically necessitated by its being F? By stipulating that if an unexamined F possessing P is G, then the properties it possesses that help to make it G are properties that it *must* possess *because it is* F. In other words, we can ensure that a and some posited unexamined F belong to no common structure by positing that if the unexamined F is G, then its Gness does not result from its belonging to *any* structure at all—from its possessing certain properties that are physically unnecessary for an unexamined F to possess. Rather, its Gness results from its Fness, since it is physically necessary that all Fs are G. Accordingly, let Px mean "$Gx \supset \Box h$."

Intuitively, then, this P should violate (4.1)—should insulate an unexamined F from an instance's confirming power. I first introduced this P in chapter 2, at the close of section 1: $[\Lambda \cup \{h\}]$-preservation (first revision) is bound to fail for this P.[29] The failure of h to be preserved under the counterfactual antecedent positing an FP betrays that h lacks physical necessary. Recall my example from chapter 2 involving this counterfactual antecedent: Had there been a moa in such good health and promising circumstances at age 49 years, 364 days, that if even this moa dies before age 50, then it must be physically necessary that all moas die before age 50. Since we believe "All moas die before age 50" to be an accidental generalization, we do not believe the moa generalization to be preserved under this counterfactual antecedent. Rather, we hold that, had there been such an "ideal test moa," it would not have died before age 50. That is, we hold that C_P is incorrect.

Since $\text{pr}(C_P) = 0$, (4.1) fails for this P. To show that inductive confirmation is unavailable when we believe that $\neg\Box h$, we need only show that P is suitable when we believe that $\neg\Box h$.[30] Obviously, pr(C_P) $\neq 1$, since it equals zero. In addition, pr(Con(Λ^+,"There exists an x such that Fx & Px")) = 1 because the only relevant member of Λ^+ is $\neg\Box h$, which is consistent with "There exists an x such that Fx & $\neg Gx$," which in turn entails "There exists an x such that Fx & Px." The final requirement for suitability is that pr(Con([$\Lambda \cup \{h\}$]'s closure in U,p) | h) = 1, that is, that we believe (given h) that h & "There exists an x such that Fx & ($Gx \supset \Box h$)" is consistent with Λ. It suffices that we believe (given h) that $\Box h$ is consistent with Λ. In section 3 of chapter 2, I argued that it is logically possible for any of the non-nomic truths in a given possible world to be its physical necessities. Hence, we believe that given h, Λ permits h to be physically necessary.

I have thus shown that we can confirm h inductively only if we believe that h may be physically necessary. Contrary to various skeptics concerning the concept of a natural law, the intuitions motivating the concept of "inductive" confirmation do indeed point to a kind of confirmation that we cannot accord a hypothesis as long as we believe it to be an accident if it is true. In chapter 5, I refine the concept of "inductive" confirmation and explore its scientific significance.

4.2. *Induction's connection to counterfactual support borne out*

The argument given in section 4.1 not only shows that, whenever we believe $\neg\Box h$, there is a suitable P for which (4.1) fails, but also identifies one such P.[31] This P distinguishes a kind of suitable unexamined case that *always* fails to be confirmed to accord with h. The existence of such a P rules out the possibility that on different occasions, different kinds of suitable unexamined cases prevent h's inductive confirmation, but for every kind of suitable unexamined case, there is some occasion on which it is confirmed to accord with h. That we deny this C_P precludes our respecting [$\Lambda \cup \{h\}$]-preservation (first revision) *and* renders us incapable of confirming h inductively. Here is a connection between our failing to confirm h inductively and our regarding h as unable to support counterfactuals in the manner of a physical necessity.

Likewise, recall from section 3 that the range of suitable Ps was constructed so that, in confirming h inductively, we confirm exactly the C_Ps that have no other obvious justification and that Λ-preservation requires us to believe if we believe that $\blacksquare h$. In the course of justifying h by confirming it inductively, we justify exactly the counterfactuals where, intuitively, our reason for believing that h holds is our reason for believing that h would still have held under those counterfactual antecedents. This suggests why we believe h to be preserved under the counterfactual antecedents within Λ-preservation's scope if we believe $\blacksquare h$. The explanation simply posits that the root commitment we adopt in believing $\blacksquare h$ (for h in U) requires us to regard our belief in h as capable of being justified by inductive confirmation.

In construing h's justification by inductive confirmation as h's projection onto actual unexamined cases and suitable counterfactual cases alike, I capture the intuitive sense of "projection" employed by Goodman and others in the "classical conception" of induction. I vindicate the strategy suggested by Goodman and Mackie (discussed in section 4 of chapter 2). (Of course, I do so while avoiding Mackie's mistaken assumption that if we believe that $\neg\Box h$, then we can confirm *h* only by mere content cutting.) On this view, counterfactuals are confirmed empirically in the same manner as (and in the course of confirming) empirical predictions regarding actual unexamined cases. That is, in elaborating Mackie's strategy, I do not follow him in using the flawed Ramsey test to express the relation between our justification for believing *h* and our justification for believing that *h* is preserved under various counterfactual antecedents. The Ramsey test, is, roughly, that we should believe *h* to be preserved under some counterfactual antecedent *p* exactly when we are still justified in believing *h* if we learn that *p* and modify our other beliefs accordingly. I have incorporated into my notion of "inductive" confirmation a different relation between our justification for believing *h* and our justification for believing that *h* is preserved under various counterfactual antecedents. The relation I am positing is direct: when we confirm *h* inductively, we confirm various counterfactuals required by *h*'s preservation. To do the work that the Ramsey test does for Mackie, I use the trivially correct premise that we should regard *h* as preserved under *p* when our evidence justifies our believing that *h* is preserved under *p*.

On this view, we do not need some *special, new* kind of reason for *going beyond* beliefs about actual unexamined cases to hold opinions regarding counterfactual cases. Rather, our evidence could not bear upon actual unexamined cases for the reason it does without bearing upon counterfactual cases as well. Which counterfactual cases these are depends on what that reason is.

In other words, this point concerning the empirical confirmation of counterfactuals applies whether or not *h* is being confirmed *inductively*. If we believe that *h* is an accidental generalization, then we regard any instances that justify our belief in *h* as failing to confirm it inductively, and yet the instances may highly confirm C_P for *certain* suitable *P*s. In that case, we take the accidental generalization as "supporting" those C_Ps. (In chapter 1, I showed that we cannot accurately explicate scientific practice by the view that accidental generalizations are incapable of "supporting" counterfactuals.) If we believe that *h* is an accidental generalization, then whether we are justified in accepting C_P for a given suitable *P* depends on which *P* this is, our other discoveries besides *h*, and (in view of counterfactuals' context sensitivity) the other properties that must be attributed to a posited *FP* in virtue of the context in which it is being contemplated.

For example, suppose we believe "All Native Americans have blood of type O or A" to be an accidental generalization. We regard various instances as having confirmed it by more than mere content cutting (though not inductively); each instance confirms C_Ps for some suitable *P*s. Suppose we believe that, in the course of confirming the generalization, we have examined a broad sample of actual Native Americans—in particular, for each member of the original company of land-bridge migrants, we have examined many Native Americans descended from that individual. Suppose we then consider the counterfactual antecedent "Had there been a Native American sitting in that chair." Context directs us to posit the Native American in the chair to have descended entirely from the actual company of land-bridge migrants. Then in justifying our belief that all Native Americans have blood of type O or A, the examined instances typically also justify the belief that had there been a Native American sitting in that chair, then someone with blood of type O or A would have been sitting in that chair.

Appendix 1: Alternative notions of inductive confirmation

Dretske is among the philosophers who maintain roughly that we can confirm a generalization that we believe may be physically necessary in a different manner from any generalization that we believe to lack physical necessity. John Earman tries to explicate formally the notion of "projection" to which Goodman appeals. In this appendix, I elaborate why I cash out "inductive" confirmation precisely as I do by contrasting my account to the approaches taken by Dretske and Earman, as well as to the notion of *h*'s inductive confirmation as the confirmation of *h*'s physical necessity.

1. Dretske: distinguishing inductive confirmation by its rate

Dretske (1977) tries to capture the intuitive distinction between inductive and noninductive confirmation:

> [L]aws are the *sort* of thing that can become well established prior to an exhaustive enumeration of the instances to which they apply. This, of course, is what gives laws their predictive utility. Our confidence in them increases at a much more rapid rate than does the ratio of favourable examined cases to total number of cases (p. 256)

Dretske's thought seems to be roughly that only a hypothesis that we believe may express a law can be confirmed by its instances in a manner that enables its holding in a given case to receive a high degree of support before that case has been examined. Dretske understands this as involving the "rate" at which the hypothesis is confirmed. He also implies that if we believe "All *F*s are *G*" to be physically unnecessary, then if all of the cases we have examined are favorable, our confidence in the generalization can rise no faster than the ratio of examined *F*s to the total number of actual *F*s. The idea behind this ratio (the fraction of "content" that has been "cut") seems to be that the generalization is susceptible only to mere content cutting, not to confirmation of its predictive accuracy. This is not true, as I have shown in many examples. So even if we believe $\neg\Box h$, we can confirm *h* in the manner that Dretske believes to be reserved for law candidates.

It is perhaps also worthwhile noting that our confidence in a hypothesis that we believe may express a law need not increase at a much more rapid rate than does Dretske's ratio of favorable examined cases to total number of cases, *even when the evidence consists of an instance that confirms* h *inductively*. As an example, let *h* be "All material particles accelerated from rest are accelerated to less than 3×10^8 meters per second (the speed of light in a vacuum)." After the proposal of special relativity but before its widespread adoption, a physicist might have held that, presumably, there exist forces large enough to accelerate particles from rest to beyond 3×10^8 m/s. So *h* is probably false, but it is a law if it is true—for if it is true, the explanation is probably that special relativity is correct: *h* is a law. Newtonian physics had been widely tested, but not with the large forces that, on Newton's theory, would result in particles accelerating from rest to beyond 3×10^8 m/s. Suppose artificially (so that we can state Dretske's ratio with definiteness) that there are *n* episodes of acceleration from rest in the history of the universe. Suppose that a physicist observes a case involving a small force; she makes the observation *e* that a stone I throw instantiates *h*. So Dretske's ratio increases by 1/*n*. But the physicist's confidence in *h* may well have increased at a slower rate. Let h_{low} be that *h* holds in cases involving small forces; let h_{high} be that *h* holds in cases involving large forces. So *h* holds if and only if h_{low} and h_{high} hold. Now

$$\mathrm{pr}(h \mid e) - \mathrm{pr}(h) = \mathrm{pr}(h_{high} \,\&\, h_{low} \mid e) - \mathrm{pr}(h_{high} \,\&\, h_{low}),$$

and so, by the definition of conditional probability,

$$\mathrm{pr}(h \mid e) - \mathrm{pr}(h) = \mathrm{pr}(h_{high} \mid h_{low}, e)\ \mathrm{pr}(h_{low} \mid e) - \mathrm{pr}(h_{high} \mid h_{low})\ \mathrm{pr}(h_{low}).$$

Plausibly, that a stone I have thrown remains below 3×10^8 m/s does not alter the likelihood of h_{high} given h_{low}; let that likelihood be *s*. So,

$$\mathrm{pr}(h \mid e) - \mathrm{pr}(h) = s\ [\mathrm{pr}(h_{low} \mid e) - \mathrm{pr}(h_{low})].$$

Now h_{low} has already been very highly confirmed, and in any event, it was largely to be expected that a stone I throw would travel slower than 3×10^8 m/s. So let $[\mathrm{pr}(h_{low} \mid e) - \mathrm{pr}(h_{low})]$ be $1/m$, where *m* is a very large positive number. Then

$$\mathrm{pr}(h \mid e) - \mathrm{pr}(h) = s/m.$$

But $1/n$ might well exceed $1/m$, and so (since $0 < s < 1$) exceed s/m. Admittedly, *e* confirms *h* (and h_{low}) at least by content cutting—that is, by removing one case (this thrown stone) from the set of cases that are unexamined and so might have turned out to violate *h* (and h_{low})—and perhaps even inductively. And this single case is $1/n$ of all cases. But the removal of this single case might well count confirmationwise for less than $1/n$, since our previous observations of other bodies had already left us highly confident that the thrown body travels slower than 3×10^8 m/s.

Here is another example in which our discovery of an instance of a hypothesis that we believe may be physically necessary increases Dretske's ratio by $1/n$ and confirms the hypothesis inductively though by a smaller increment. Consider the hypothesis "All protons have mass 1.673×10^{-27} kilograms." Suppose that there are n protons in the universe's history. When we check our first proton and find its mass to be 1.673×10^{-27} kilograms, Dretske's ratio rises from 0 to $1/n$. Our confidence in the hypothesis presumably rises to much more than $1/n$, since we not only have removed one case from the set of cases that might turn out to violate the hypothesis, but also have to some degree confirmed unexamined cases to accord with the hypothesis. Presumably, this is inductive confirmation. When we check a second proton and find that it accords with the hypothesis, Dretske's ratio again rises by $1/n$, while our confidence in the hypothesis again rises by an increment exceeding $1/n$. But eventually, we find so many instances (and no counterexamples) that our confidence in the hypothesis exceeds $(1 - 1/n)$. With each further proton we check, Dretske's ratio rises by $1/n$, whereas our confidence in the hypothesis cannot rise by more than $1/n$, even though inductive confirmation is presumably continuing to occur.

Inductive confirmation—the kind of confirmation that is reserved for candidate physical necessities—is not distinguished by its *rate*. I suggest above (section 2) that it is distinguished by the *range* of hypothetical unexamined cases that are confirmed to accord with *h*. In both its rate and its range, the initial inductive confirmation of the proton hypothesis is distinguished from mere content cutting. Apparently, Dretske saw only this rate distinction and concluded that rate distinguishes inductive confirmation from *any* noninductive confirmation, even noninductive confirmation that goes beyond mere content cutting. I argue (in section 2) that, actually, its range rather than its rate sets inductive confirmation apart.

2. What Earman's notions of "projection" fail to capture

Of course, Goodman's (1983, p. 74) most notorious example of a hypothesis that fails to be confirmed inductively by its instances is "All emeralds are grue," which I construe as "Every temporal slice of an emerald [i.e., every "emerald slice"] is grue in that it is green if it precedes the year 3000, blue otherwise." We have found exactly as many instances of "All emerald slices are green" as of "All emerald slices are grue," since an emerald slice before the year 3000 is green if and only if it is grue. But we regard instances of the green hypothesis as confirming it inductively, whereas we do not regard instances of the grue hypothesis as justifying its "projection." This is Goodman's "new riddle of induction": Why is the green hypothesis projectible, whereas the grue hypothesis is not? My immediate interest is in using this example to help me determine what *inductive* confirmation is supposed to be.

By saying that the grue hypothesis is not confirmed "inductively," Goodman surely does not mean to say that by discovering an emerald slice to be grue, we do not confirm *any* unexamined emerald slices to be grue. Goodman accepts that by discovering a given emerald slice to be grue, we confirm an unexamined emerald slice *before the year 3000* to be grue (that is, green), considering that we already know the examined slice also to precede 3000. (If Goodman did not accept this, then he could not say that the green hypothesis is confirmed inductively.) Apparently, Goodman's reason for saying that the grue hypothesis is not confirmed *inductively* is that an instance fails to confirm an unexamined emerald slice *after* 3000 to be grue (that is, blue), considering that we already know the examined slice to precede 3000. Again, what sets inductive confirmation apart is the *range* of unexamined cases, actual and counterfactual, on which the evidence bears.

That inductive confirmation is distinguished by its *range* is often obscured in discussions of Goodman's example. Earman (1992, pp. 104–13), for instance, has attempted to define a notion of "projection" that captures Goodman's intent. Earman shows that the green hypothesis and the grue hypothesis cannot both be "projected" in Earman's sense, and uses this fact to argue that his notion faithfully elaborates Goodman's idea. But Earman's sense of "projection" fails to highlight the distinctive range of hypothetical unexamined cases that are confirmed to accord with a hypothesis when it is projected.

Earman investigates several different ways of trying to express Goodman's sense of "projection." Suppose we discover Ga—that a is green—and our background beliefs K include that a is an emerald slice. Consider hypothesis h: "All emerald slices are green." Earman notes that, according to Bayesian updating, h is confirmed—that is, $pr(h \mid Ga) > pr(h)$—if h is a live hypothesis and Ga is new evidence, that is, if $0 < pr(h), pr(Ga) < 1$. The same considerations apply to the hypothesis "All emerald slices are grue." Hence, instances confirm "All emerald slices are green" *and* "All emerald slices are grue," since "it is much too draconian to suppose that . . . unprojectible hypotheses are to be initially and forever condemned to limbo by receiving zero priors" (p. 105). Therefore, Earman rightly concludes, Goodman's intentions cannot be captured by defining h's "projection" merely as h's confirmation by instances.

Earman then considers another approach. Suppose that we use the index i, taking integral values from 1 to positive infinity, to enumerate various objects x_i where our background beliefs K include that each of the x_i is an E. Suppose that we examine the objects x_i, one by one, in order of increasing i, to ascertain whether they are G; at any stage in our work, there is some number N such that by that stage, we have examined exactly all of the objects x_i for $i < N$. Consider the hypothesis h, "All Es are G," and the "grue analog" hypothesis "All Es are G^*," where G^*x_i means $((i < 3000 \;\&\; Gx_i)$ or $(i \geq 3000 \;\&\; \neg Gx_i))$. Now abbreviate Gx_i as G_i. Earman (1992, p. 106) defines two senses of h's "projection":

"h is weakly projectible in the future-moving sense, relative to K" iff $\text{pr}(G_{n+1} \mid G_1, \ldots, G_n)$ approaches 1 as n approaches infinity.

"h is strongly projectible in the future-moving sense, relative to K" iff $\text{pr}(G_{n+1}, \ldots, G_{n+m} \mid G_1, \ldots, G_n)$ approaches 1 as n and m approach infinity.

All of these conditional probabilities are "relative to K" in the obvious sense that our background beliefs K are "given," and so are reflected in each conditional probability. (Analogous senses can be defined for the projection of the "grue analog" hypothesis.)

To understand these senses of protection, remember that after we discover x_i to be G, we update our degree of confidence in whether the next object to be examined, x_{i+1}, is G. When we discover it to be G, we revise our opinion of whether the next object to be examined, x_{i+2}, is G, and so on. At each stage in our work, we have some degree of confidence that the next enumerated object to be checked will be G. Bear in mind that, at different stages, different objects are the "next" enumerated object. So "h is weakly [strongly] projected in the future-moving sense, relative to K" means that as we proceed in this fashion, our opinion of whether the next object(s) to be checked will be G rises without bound.

Such "projection" makes contact with Goodman's intuitions in that it requires more than mere content cutting; it involves confirmation of some of h's predictions. Earman notes, however, that $\text{pr}(h) > 0$ suffices to guarantee that h is "projected" in these two senses. Analogous remarks apply to the "grue analog" hypothesis. Hence, the green and grue hypotheses are *both* projected in each of these senses. Therefore, Earman concludes, neither of these two senses of "projection" is the sense that Goodman intended.

Earman (1992, p. 108) then offers a sense of "projection" that he deems able to capture Goodman's idea. Allow the index i to range infinitely in *both* directions. According to Earman,

"h is weakly projectible in the past-reaching sense, relative to K" iff for any n, $\text{pr}(G_{n+1} \mid G_n, G_{n-1}, \ldots, G_{n-j})$ approaches 1 as j approaches infinity.

(Earman also defines an analogous sense of "strong projectibility," as well as analogous notions for the projection of the grue hypothesis.) Intuitively, these "past-reaching" senses of projection involve our fixing the object x_i that we are confirming to be G—namely, the $(n + 1)$th. In the "future-moving" senses, by contrast, we allow the object whose Gness we are confirming (the "next" x_i) to vary in the course of taking the limit.[32] The limiting process in the "future-moving" senses corresponds to our passing through various stages in our empirical research, since at each stage we check the individual that at the preceding stage was the next individual to be examined. We are noting how our confidence that the next individual(s) to be examined will be G grows from one stage to the next in our research. In the "past-reaching" senses, on the other hand, we have fixed the stage in our research—the next x_i to be examined is the $(n + 1)$th—and we are noting how at that stage our degree of confidence in the Gness of x_{n+1} is greater as the number of instances that we have already examined by that stage is greater. Thus, "*h* is weakly [strongly] projected in the past-reaching sense, relative to *K*" means that, at this stage, our opinion of whether the next object(s) to be examined will be *G* grows arbitrarily high as the number of instances that have already been found by this stage grows.

Clearly, a contradiction results if "All *E*s are *G*" *and* "All *E*s are *G**" are *both* projected in the past-reaching sense; $\mathrm{pr}(G_{3000})$ and $\mathrm{pr}(G^*_{3000})$ cannot *both* go arbitrarily high, since by definition, $\mathrm{pr}(G^*_{3000}) = \mathrm{pr}(\neg G_{3000})$. Earman concludes that here, finally, is a sense of "projection" in terms of which Goodman's intuitions can be expressed: there is no background *K* such that the green and grue hypotheses are both "projected" under that *K*.

Projectibility concerns the capacity of instances to confirm *h*'s accuracy to unexamined case(s). This was Goodman's concern; Earman quotes the same remark of Goodman (1983) by which I am here guided: inductive confirmation occurs "only when the instance imparts to the hypothesis some credibility that is conveyed to other instances" (p. 69). But to *which* other instances? As I have shown, this is a crucial question to ask if we are to elaborate a sense of "inductive" confirmation that might bear out Goodman's view that inductive confirmation is unavailable to a hypothesis that we believe lacks physical necessity—or even his view that we fail to confirm the grue hypothesis inductively. Its confirmation fails to qualify as "projection" because by discovering to be grue a given emerald slice already believed to precede the year 3000, we fail to confirm the grueness of an unexamined emerald slice already believed to be (or posited as being) *after* 3000. For Goodman, "All *E*s are *G*" is projected only when an instance imparts to it some credibility that is conveyed to another *E no matter what else* (within some suitably generous range) is believed about that other *E*.

Earman's senses of "projection" obscure this crucial feature. Earman characterizes various senses of "projection *relative to* K"; Earman's concern is whether for a certain (group of) object(s) believed under some *fixed* *K* to be *E*, accumulated instances of "All *E*s are *G*" make the object(s) arbitrarily likely to be *G*. In contrast, Goodman's concern is roughly whether under *every* K (in some suitably generous range), an instance confirms, regarding each object believed under that *K* to be *E*, that it is *G*. In other words, Goodman's concern is an instance's bearing confirmationwise upon *each* kind of unexamined case that we believe *might* have happened to exist, whereas Earman's concern is its bearing upon *some* unexamined case(s) that we believe *actually* happens to exist.

The contrast between Goodman's and Earman's notions of "projection" is especially dramatic when our background beliefs *K* contain a great deal of relevant information about which physically possible kinds of unexamined cases happen to exist. Projection in Earman's sense may then become far easier to achieve than projection in Goodman's sense. To take an extreme example, suppose we believe that, as it happens, all men in the lecture audience have the same number of older brothers. Then one instance suffices to establish "All men in the lecture audience have two older brothers." So this hypothesis is projected in Earman's sense relative to our background beliefs *K*. Despite this possibility, Goodman (1983, p. 73) holds that this hypothesis is not "projectible." In particular, to confirm "All men in the lecture audience have two older brothers" inductively, an instance must confirm that had I been in the lecture hall, I would have had two older brothers. Earman, like Dretske, fails to see projection as set apart by its *range* or (what amounts to the same result) by our *reason* for regarding the instance as bearing upon unexamined cases. The instance's power to confirm that a given unexamined case accords with the hypothesis must not depend on the assistance of background beliefs (in *K*) about physically unnecessary matters of fact.

Earman demands of an adequate sense of "projection" that it be logically impossible for the green and grue hypotheses both to be "projected."[33] But he does not require that the grue hypothesis be somehow intrinsically incapable of projection. How, then, do we justify projecting the green hypothesis rather than the grue hypothesis? This is Goodman's "new riddle of induction." I argue in section 3 of this chapter that we cannot confirm a hypothesis inductively if we believe that it lacks physical necessity. In section 2 of chapter 7, after distinguishing the laws from the physically necessary nonlaws, I argue that certain physically necessary nonlaws are "coincidences" and that, if we believe that h is such a coincidence if $\Box h$, then it is very difficult for us to confirm h inductively. Therefore, our failure to confirm the grue hypothesis inductively could be explained by our justly believing that the grue hypothesis, if it holds, is either an accidental generalization or a physically necessary "coincidence." But would our according a zero prior probability to its being a law be just as "draconian" (in Earman's words) as our according its truth a zero prior probability? The answer depends on the root commitment that we undertake in science when we believe in a law. According to the account of this root commitment that I set out in chapter 7, we can justly conclude before checking any emeralds that the grue hypothesis expresses neither a law nor a physically necessary nonlaw that is not a coincidence. This accounts for our failure to confirm it inductively.

3. *Inductive confirmation as confirmation of* $\Box h$

Suppose that I had identified "inductive" confirmation simply as confirmation of $\Box h$. Then it is trivial that we cannot confirm h "inductively" if we believe that $\neg\Box h$. But this is not an illuminating explanation of why "inductive" confirmation is unavailable if we believe $\neg\Box h$, since inductive confirmation is then blocked by the wrong *P*s. (Recall the sort of *P*s that intuitively should block h's inductive confirmation in the marble, Native-American, and pear examples.)

Furthermore, under this definition, it is too easy to confirm h "inductively." By Bayesian updating, it is so confirmed (if we believe that $\Box h$ may hold) by any instance that we did not already know before. That is because by Bayes's theorem, $\text{pr}(\Box h \mid Ga) = \text{pr}(\Box h)\ \text{pr}(Ga \mid \Box h)/\text{pr}(Ga)$. Our $\text{pr}(Ga \mid \Box h) = 1$ (since our background beliefs include Fa) and $0 < \text{pr}(Ga) < 1$ (since we did not already believe case *a* to accord with *h*). Hence, pr($\Box h$ | *Ga*) equals pr($\Box h$) multiplied by some number greater than 1, and so pr($\Box h$ | *Ga*) > pr($\Box h$), which means that *Ga* confirms $\Box h$. So an instance confirms *h* "inductively," on this definition, even if it merely content-cuts *h*.[34] But mere content cutting intuitively lies at the furthest extreme from inductive confirmation, since it involves no "projection" of *h* onto unexamined cases at all. We can confirm $\Box h$ without confirming *h*'s predictive accuracy and so, intuitively, without confirming *h* inductively. Furthermore, if confirmation of $\Box h$ sufficed for "inductive" confirmation of *h*, then *h* could be confirmed "inductively" without *h*'s preservation under any counterfactual suppositions being confirmed. This would fail to cash out the intuition expressed by Goodman and Mackie that inductive confirmation is related to counterfactual support.

To disqualify mere content cutting from counting as "inductive" confirmation, we might instead define *h*'s "inductive" confirmation as confirmation of $\Box h$ *together with* confirmation of "All unexamined cases accord with *h*." But again, under this definition, the reason why *h* cannot be confirmed "inductively" if we believe $\neg\Box h$ is trivial rather than illuminating. Furthermore, confirmation that qualifies as "inductive" under this definition can occur without confirmation that *certain* suitable kinds of unexamined cases accord with *h*. As I explained at the start of section 2 of this appendix, an instance of the grue hypothesis fails intuitively to support its inductive "projection," despite confirming that certain kinds of unexamined cases accord with the hypothesis (those on the same side of the year 3000 as the examined case), because it fails to confirm that certain other kinds of unexamined cases accord with the hypothesis (those on the opposite side of the year 3000 from the examined case). This confirmation qualifies as "inductive" under the above definition. On my view, this confirmation lies at an intermediate point in the spectrum running from mere content cutting through genuine inductive confirmation. Moreover, the above sense of "inductive" confirmation again fails to ensure that *h*'s preservation under counterfactual suppositions is confirmed when *h* is confirmed "inductively."

Appendix 2: Inductive confirmation and the paradox of the ravens

My elaboration of "inductive" confirmation sheds some light on Hempel's (1965) notorious "paradox of the ravens" (pp. 14–20). The "paradox" is that instance confirmation (that "All *F*s are *G*" is confirmed by the discovery that a given *F* is *G*) together with the "equivalence condition" (that evidence confirming a given hypothesis confirms any logically equivalent hypothesis) entails that we confirm "All ravens are black" by confirming its contrapositive ("All nonblack things are nonravens") by discovering a given nonblack thing to be a nonraven (e.g., by discovering a certain red object to be a pen). This result seems highly counterintuitive. The problem is to decide whether this result is not as counterintuitive as it first appears, or whether one of the premises leading to this result ignores some consideration that is responsible for our judgment that "indoor ornithology" is relatively unprofitable.

One response to this problem is to deny instance confirmation. There is surely something correct in doing so; background beliefs, for example, can prevent a hypothesis from being confirmed by one of its instances (see note 8). However, it is insufficient to respond to the raven paradox by simply rejecting instance confirmation. An adequate response must explain why the discovery that a given nonblack thing is not a raven typically fails to confirm "All ravens are black" (or, equivalently, "All nonblack things are nonravens")—or, at least, constitutes less compelling evidence for this hypothesis than does the discovery that a given raven is black. By Bayes's theorem, $\text{pr}(h \mid e) = \text{pr}(h)\ \text{pr}(e \mid h)/\text{pr}(e)$, and so the discovery of an instance invariably confirms a hypothesis [i.e., $\text{pr}(h \mid e) > \text{pr}(h)$] as long as the instance was not already known and is entailed by the hypothesis and background beliefs [i.e., as long as $0 < \text{pr}(e) < 1$, $0 < \text{pr}(h) < 1$, and $\text{pr}(e \mid h) = 1$]. What prevents (or diminishes) instance confirmation of "All nonblack things are nonravens"?

One explanation that has frequently been offered turns on our prior belief that there are many fewer ravens than there are nonblack things. It then follows that we confirm "All ravens are black" to a much higher degree by finding a raven to be black than by finding a nonblack thing to be a nonraven.[35] More specifically, let h be $(x)(Rx \supset Bx)$, that is, "All ravens are black"; let one piece of evidence be Ba, let another be $\neg Rb$, and let our background beliefs when we make these observations include Ra, $\neg Bb$, and that there are many more nonravens than black objects [so $\text{pr}(\neg Rb) \gg \text{pr}(Ba)$]. Then h together with our background logically necessitates $(Ba\ \&\ \neg Rb)$. By Bayes's theorem, $\text{pr}(h \mid Ba)/\text{pr}(h \mid \neg Rb) = \text{pr}(\neg Rb)/\text{pr}(Ba)$, which is high. Thus, Ba's confirming effect far exceeds $\neg Rb$'s. In this way, our intuitions are supposed to be saved: our discovery that a given nonblack thing is a nonraven would raise our confidence that "All ravens are black," but the increase would be quite small when compared to the increase occasioned by our discovery of a black raven.

However, as Ken Gemes first suggested to me (when he proposed the term "mere content cutting"), this response to the raven paradox shows only that more content is cut from "All ravens are black" by an instance of it than from "All nonblack things are nonravens" by an instance of it. Since we believe 1/(the number of ravens) to be much greater than 1/(the number of nonblack things), we believe that each raven represents a larger fraction of all ravens than each nonblack thing represents of all nonblack things. Intuitively, then, the fraction of h's content that is cut by a black raven is much greater than the fraction that is cut by a nonblack nonraven. In other words, we are more likely to find a falsifier of "All ravens are black" by looking at a raven than by looking at a nonblack object because there are many more nonblack objects than there are ravens; a single falsifier is a greater "fraction" of the ravens than of the nonblack objects. This reasoning concerns the likelihood of finding a case that *falsifies* the hypothesis. But to resolve the raven paradox, we must consider the power of various instances to *confirm* the hypothesis. These are the same only if an instance's power to confirm derives entirely from its failure to falsify.[36] This occurs only if an instance confirms a hypothesis by mere content cutting, that is, by showing that one circumstance that would have made the hypothesis false is not realized. If we consider confirmation that goes beyond mere content cutting, then the above Bayesian argument for *Ba* having greater confirming power than $\neg Rb$ becomes inconclusive. Let *j* be that the *next* raven-or-nonblack thing to be checked will be black if it is a raven, or a nonraven if it is nonblack. So, *j* is what might get confirmed when *h* is not merely content-cut—that is, when *h*'s predictive accuracy is confirmed. Having turned our attention from *h*'s confirmation to *j*'s, we cannot entirely retrace the course of the above Bayesian argument since unlike *h*, *j* together with our background beliefs does *not* logically necessitate *Ba* & $\neg Rb$. Bayes's theorem yields only $\text{pr}(j \mid Ba)/\text{pr}(j \mid \neg Rb) = \text{pr}(Ba \mid j)\ \text{pr}(\neg Rb)/\text{pr}(Ba)\ \text{pr}(\neg Rb \mid j)$. This result is not enlightening.

It is important to account for our intuition that, by discovering a given nonblack thing to be a nonraven, we do much less to confirm *h* than we do by discovering a given raven to be black. But the above approach neglects the distinction between the confirmation of *h*'s predictive accuracy (as when *h* is confirmed inductively) and the mere content cutting of *h* by its instances. An alternative way to account for the intuition that *Ba* has greater confirming power than does $\neg Rb$ is to imagine ordinary examples in which we treat the discovery that a given raven is black as inductively confirming "All ravens are black."[37] In contrast, we ordinarily would not treat the discovery that a given nonblack thing is a nonraven as inductively confirming "All nonblack things are nonravens."[38] Suppose we already believe that the nonblack object under examination is a bird with a certain special kind of curved bill, and we discover it to be a nonraven. Then for any (actual or hypothetical) suitable nonblack thing that is posited also to be a bird with this special kind of curved bill, we might well confirm that it is also a nonraven. But typically, having discovered a given nonblack bird with the special kind of curved bill to be a nonraven, we take this discovery to be irrelevant as to whether a suitable nonblack bird, posited as possessing a *different* sort of bill, is a nonraven. Likewise, suppose we believe that the nonblack object under examination is *not* a bird with the special kind of bill. Then by discovering it to be a nonraven, we typically fail to confirm that a suitable nonblack bird, posited as possessing the special kind of bill, is a nonraven. Hence, by discovering a given nonblack thing to be a nonraven, we do not confirm "Were there a nonblack thing possessing property *P*, then all such things would be nonravens" where *Px* means "*x* is a bird with the special kind of curved bill" and we already believe that the thing actually under examination is not a bird possessing this kind of bill. (Presumably, *P* is suitable.) It follows that, in ordinary examples, an instance of "All nonblack things are nonravens" fails to confirm it inductively.

The point that I am trying to make is, roughly, that we regard ravens as relevant to all other ravens, but we usually do not regard nonblack things as relevant (even as far as nonravenhood is concerned) to all other nonblack things. This is one important reason why we believe it more profitable to examine ravens than to examine nonblack things in testing "All ravens are black." A raven instance typically packs a greater confirmatory wallop—I refer here not to the *increment* of confirmation that it brings to the hypothesis, but to the *range* of the hypothesis's predictions that it confirms.[39]

Compare "All ravens are black and all doves are white" (i.e., "Anything that is a raven or a dove is black if it is a raven or white if it is a dove").[40] If the discovered instance is already believed to be a raven, then we regard it as irrelevant to a typical suitable case that is posited to be a dove, and vice versa. This brings us close to a point (anticipated in chapter 2, note 10) that I raise in section 2 of chapter 7. There I argue that some physical necessities—some logical consequences of the laws—are not laws themselves, and that our believing them to be nonlaws may be reflected in our treating them as failing to have been confirmed inductively. There I touch again on the raven paradox (and also in chapter 6, at the end of section 4).

• FIVE

Why Are the Laws of Nature So Important to Science (II)?

In this chapter, I refine and generalize the concept of "inductive" confirmation that I developed in chapter 4. I use this notion to answer the question that I began to address in chapter 3: Why are the laws of nature so important to science? To do this, I require the concept of an "inductive strategy," an idea I elaborate with greater care in chapter 7; it figures importantly in the root commitment associated with belief in a law.

In chapter 3, I argued that by virtue of nontrivially possessing non-nomic stability, Λ is distinguished from any set of truths in U containing accidents. For simplicity, I assume (until further notice) that there is actually only a single grade of physical necessity. Then Λ is the only set logically closed in U that is nontrivially preserved whenever—non-nomically speaking—it logically possibly could be: under every non-nomic subjunctive antecedent that is logically consistent with each of its members. Of course, to understand why it is so important to discover which truths are laws and which are not, we still must understand why science should especially care about identifying the members of the set that nontrivially possesses non-nomic stability. The scientific importance of doing so derives (as I argue in section 2) from the special *indiscriminate* character of inductive projection—which gives inductive strategies a special place in scientific research.

1. Inductive confirmation refined

The concept of "inductive" confirmation developed in chapter 4 succeeds in capturing various intuitions about the sort of confirmation that we must reserve for hypotheses that we believe may express physical necessities. For example, our having confirmed h "inductively" is related to our treating h as bearing the special relation to counterfactuals that distinguishes physical necessities from accidental truths. However, this characterization of "inductive" confirmation also has some unfortunate features—chiefly, its restriction to hypotheses of the form "All *F*s are *G*" (and, correspondingly, to evidence of the form *Ga* and predictions of the form C_P) and the references to the laws in the constraints on "suitable" *P*s. These references make it hard to understand the scientific importance of inductive confirmation without presupposing the scientific importance of the laws. It would be better to refine our characterization of "inductive" confirmation so as to retain the successes while avoiding the deficiencies of my proposal in chapter 4. This is much like the situation with which I began chapter 3. I then removed the references to Λ and Λ^+ from Λ-preservation, arriving at the concept of "non-nomic stability." In like manner, I now refine "inductive" confirmation. This will enable me in section 2 to explain the special character of inductive confirmation without begging the question of the laws' importance.

Consider, then, a hypothesis h of any kind (it need not take the form "All *F*s are *G*"), as long as $h \in$ U. The notion of h's "predictions"—a certain range of which are confirmed by evidence that confirms h "inductively"—must be generalized from the claims C_P. Let us adopt the following definition:

> h's "predictions" are all and only the subjunctive conditionals $p > m$, $q > (p > m)$, $r > (q > (p > m))$, and so on, for any $m \in$ U, $p \in$ U, $q \in$ U, $r \in$ U, and so on, where $(p \ \& \ h) \Rightarrow m$.[1]

For example, h's prediction $p > m$ specifies what h says would happen were p the case. We might say that this is a prediction "regarding p."

Suppose, then, that e is one of h's predictions, and we discover that e obtains. This discovery might confirm each of h's predictions in a broader or narrower range. To reinforce this point, consider four brief examples, each involving the confirmation of a hypothesis but differing in the *range* of its predictions that are confirmed.

1. We believe that a given die is fair. The hypothesis is that each of its next three tosses lands six. Accordingly, our initial confidence in the hypothesis is $1/6 \times 1/6 \times 1/6 = 1/216$. We toss the die once, and it lands six. Since this evidence eliminates one way in which the hypothesis might have been falsified, the hypothesis is confirmed; our confidence in it rises to $1/6 \times 1/6 = 1/36$. But none of its predictions regarding actual unexamined cases (the outcomes of either of the next two tosses) is confirmed. (In chapter 4, I follow Ken Gemes in referring to this confirmation as "mere content cutting.")
2. The hypothesis is that all emeralds are green and all rubies are red. The first emerald we examine turns out to be green. Typically, this discovery is relevant confirmationwise to unexamined emeralds but not to unexamined rubies. In other words, this evidence typically confirms the hypothesis and, unlike example 1 above, confirms *some* of the *predictions* made by the hypothesis (such as that the next

emerald I check will be green). But typically, it fails to confirm other predictions (such as that the next ruby I check will be red).

3. The hypothesis is that all of the pears on this tree are now ripe. By checking a pear from the tree and finding it to be ripe, we might confirm, of each actual unexamined pear on the tree, that it is ripe. In that event, the evidence not only confirms the hypothesis but also, unlike example 2 above, confirms *each* of the predictions (that have not already been accepted) that the hypothesis makes regarding *ps* that (we believe) *actually obtain*. Furthermore, the evidence typically also confirms some of the predictions (that have not yet been accepted) that the hypothesis makes regarding *ps* that (we believe) do not actually obtain—for example, predictions regarding the ripeness of certain *counterfactual* pears. For instance, it confirms that had there been a pear on the third branch of the tree, then it would have now been ripe. But predictions that the hypothesis makes regarding certain other counterfactual *ps* typically fail to be confirmed—for example, that had there been a pear on the tree whose environmental conditions (e.g., temperature, length of day, plant hormones) were experimentally manipulated to differ from those experienced by the other pears on the tree, then it would have now also been ripe.
4. Newton regarded a successful prediction made by his putative gravitational force law as bearing upon each of the predictions (that had not already been accepted) that the hypothetical law makes regarding *ps* that actually obtain; the Moon's according with the hypothesis confirmed that distant stars do, too. Moreover, unlike example 3 above, Newton took this evidence as bearing upon *each* of the predictions (that had not yet been accepted) that the hypothesis makes regarding *ps* that do not actually obtain but, roughly speaking, might have happened to obtain—for example, as confirming that no matter what the Earth–Moon separation might have been, their mutual gravitational attraction would still have accorded with his hypothesis.

By discovering the truth of some prediction *e* made by *h*, we confirm *h* "inductively" exactly when *e* confirms each of *h*'s predictions in a certain range. What range? In chapter 4, I used "suitability" to delimit this range. I cannot do this now (since I am considering hypotheses and predictions of any form).

In moving from chapter 2 to chapter 3, I went from discussing Λ-preservation, where Λ and Λ^+ mark off the relevant range of counterfactual antecedents, to discussing the "stability" of a set Γ of truths, where Γ itself picks out the relevant range—namely, those *ps* that are consistent with each of Γ's members. In other words, the range of non-nomic counterfactual antecedents under which the members of some logically closed set of non-nomic truths must be preserved, for that set to be non-nomically stable, is fixed not by appealing expressly to the laws, but rather by appealing to the set's members themselves. I take this as my inspiration in refining "inductive" confirmation: let the range of *h*'s predictions that are confirmed when *h* is confirmed inductively be fixed by the set Θ containing exactly the hypotheses $h_i \in$ U that we choose to regard as capable of being confirmed inductively. (Below I refine this characterization of Θ and say more about where this set comes from.)

In particular, when I characterized inductive confirmation in chapter 4, I demanded that any suitable P be such that pr(Con(Λ^+,p)) = 1, where p expresses what is posited by C_P's antecedent "Were there an x such that Fx and Px." Let us transform this requirement so that Θ plays the role that was being played by Λ^+: each of h's predictions $p > m$ that must be confirmed, in order for h to be confirmed inductively, is such that p is consistent with Θ's closure in U.[2] Likewise, each of h's predictions $p > (q > m)$ that must be confirmed, in order for h to be confirmed inductively, is such that Con(Θ's closure in U,p) and Con(Θ's closure in U,q) hold, and so on for multiply nested subjunctive conditionals.

Of course, we cannot confirm a prediction to which we already assign a subjective probability of 1. So whereas my definition of "inductive" confirmation at the close of chapter 4 required that pr(C_P) ≠ 1 for any suitable P, I now stipulate that each of h's predictions f that must be confirmed, in order for h to be confirmed inductively, is such that pr(f) ≠ 1.

The final constraint on suitable Ps applied only when we believed that $\neg\Box h$. Since I am dropping all references to our beliefs about the laws, I drop this requirement. (I dropped the analogous constraint on suitable P's when I moved from thinking about [Λ ∪ {a}]-preservation (first revision) in chapter 2 to thinking about the "non-nomic stability" of [Λ ∪ {a}]'s closure in U in chapter 3.)

The following definition results:

> Hypothesis $h \in$ U is confirmed *inductively* by e, a prediction made by h, exactly when e confirms each of h's predictions f such that
>
> pr(f) ≠ 1;
> if f is $p > m$, then Con(Θ's closure in U,p);
> if f is $p > (q > m)$, then Con(Θ's closure in U,p) and Con(Θ's closure in U,q);
> and so on.

What does this mean, intuitively? In confirming h inductively, we confirm each of h's predictions $p > m$ (that we do not already believe) about what would happen were p the case, for each hypothetical circumstance p in which it is logically possible for all of Θ's members to be true. For each of those ps, we also confirm each of h's predictions (that we do not already believe) about what would happen, in those hypothetical circumstances p, were q the case—that is, each of h's predictions $p > (q > m)$—for each hypothetical circumstance q in which it is logically possible for all of Θ's members to be true. And so on, for h's multiply nested predictions—where m, p, q, and so on are all non-nomic. In confirming h inductively, in other words, we project h out as far onto ps as it can go without running up against the projections of the other h_i (or bumping into its own projection)—in the sense that we confirm each of h's predictions $p > m$ (that we do not already believe) for any p where the predictions $p > m$ that are made by the various h_i are mutually consistent. We are prepared to project each of the h_i out as far onto *ps* as they can *all* logically possibly go (and for each of these *ps*, we are willing to project each of the h_i out as far onto *qs* as they can *all* logically possibly go, etc.).

From this definition, it follows that we cannot inductively confirm a hypothesis in U that does not belong to Θ's closure in U. For suppose that $h \notin$ Θ's closure in U. Then Con(Θ's closure in U,$\neg h$) holds. So to confirm h inductively, we must confirm each of h's "predictions" $\neg h > m$. But by the definition of "prediction," h makes some logically impossible predictions regarding $\neg h$ [e.g., $\neg h > (h \& \neg h)$], and these cannot be confirmed.[3] Therefore, we can confirm h inductively only if $h \in$ Θ's closure in U.

So the hypotheses that we are prepared to confirm inductively determine the range of any hypothesis's inductive projection. Inductive projection is *self*-limiting; the hypotheses that we are prepared to project inductively (should the right evidence come along), rather than some background beliefs, collectively limit the range across which they would then be projected. This point will assume great importance in section 2, where I explain Θ more fully.

From the definition of "non-nomic stability," it follows immediately that

> it is nontrivially the case that every prediction $p > m$, $p > (q > m)$, . . . made by Θ's members—for each p, q, . . . where Con(Θ's closure in U,p), Con(Θ's closure in U,q), . . . all hold—is true

exactly when Θ's closure in U nontrivially possesses non-nomic stability.[4] Thus, the nontrivial non-nomic stability of Θ's closure in U is a necessary and sufficient condition for the nontrivial truth of all of the predictions, made by Θ's members, that are confirmed when Θ's members are all confirmed inductively.[5] But as I showed in chapter 3, Θ's closure in U nontrivially possesses non-nomic stability if and only if it is Λ. It follows, then, that

> it is nontrivially the case that every prediction that is confirmed, when all of Θ's members are confirmed inductively, is true

just in case Θ's closure in U is Λ. That is, Λ is the only set of non-nomic claims logically closed in U such that when claims spanning the set are each projected inductively, then nontrivially, each of the predictions thereby confirmed is true.

Perhaps inductive projection has some particular importance in science. (I pursue this possibility below, where I argue that inductive confirmation's importance derives from its unique *self*-limiting character.) If inductive confirmation has special scientific importance, then science must especially value identifying a set Θ of claims whose inductive confirmation does not mislead us, but rather leads nontrivially to the confirmation of truths alone. The closure in U of any such set Θ is Λ. So in identifying the laws in U, science identifies what would be best to confirm inductively. I thereby explain why it is so scientifically important to discover what the laws in U are.[6]

2. Inductive strategies as free electives

Why might inductive confirmation be especially important in science? Let me now offer a very rough proposal. To return to the pear example above (example 3), let h be the hypothesis that all of the pears on the tree are now ripe. By finding a pear on the tree to be ripe, we confirm h's predictions regarding certain counterfactual *ps*, such as p_1 = "There is a pear on the third branch of the tree." But we do not confirm h's predictions regarding certain other *ps*, such as p_2 = "There is a pear on the tree that has experienced environmental conditions that have been experimentally manipulated to be . . . ," which we know differ radically from the conditions actually experienced by the pears on the tree. What justifies our projecting h so far but no further? When h alone is considered, the limits of its projection appear arbitrary; h itself provides no basis for treating p_1 differently from p_2. But, of course, the limits of h's projection are not arbitrary. They are determined by our *reason* for taking the ripeness of one actual pear on the tree as relevant confirmationwise to the ripeness of another actual pear on the tree. Part of our reason is our background belief, arrived at by prior empirical work, that all of the pears on the same tree that have experienced roughly the same environmental conditions are ripe to about the same degree. Hence, our reason for taking the ripeness of one actual pear on the tree as confirming the ripeness of another actual pear on the tree carries over to the pear posited by p_1, since we have already confirmed that it would have experienced roughly the same environmental conditions as the actual pear under examination. But our reason does not apply to the pear posited by p_2, since we believe that it would not have experienced roughly the same environmental conditions as the actual pear under examination. So our prior empirical work justifies our taking the discovery that a given pear actually on the tree is ripe as confirming, of each actual or counterfactual pear *in a certain range*, that it is ripe. Without that empirical work, this range would be arbitrary, unmotivated.

In contrast, when a hypothesis h is confirmed *inductively*, the limits of its projection are fixed by h itself along with the other hypotheses that we are committed to confirming inductively (should we discover one of their predictions to be borne out). No appeal need be made to background opinions that we have arrived at through prior empirical work. From the viewpoint of the various hypotheses h_i (including h) that we are committed to confirming inductively, the range of h's projection is not arbitrary; we are prepared to project *each* of the h_i as far across the *ps* in U as they can *all* logically possibly go (and likewise in the nested case). Inductive projection is *indiscriminate*; it draws no distinction among the *ps* across which we *could* project all of the hypotheses in Θ. Therefore, it is not beholden to prior empirical work for supplying a good reason to draw such a distinction—that is, for supplying a good reason to project one of these hypotheses so far but no further. Again, the projection is *self*-limiting; nothing beyond the hypotheses themselves is needed to justify the limits of their range of projection.

Here is why this is important. We pursue various strategies for arriving justly at predictions. Part of carrying out some of these strategies is categorizing objects or situations in certain ways for the purpose of seeking regularities of certain sorts. For example, we choose to regard copper, Cepheid-type variable stars in galactic spiral arms, and autism as perhaps constituting natural kinds of certain sorts. Once we tentatively adopt such a categorization and decide provisionally to seek a regularity of some specified kind covering some category, we are prepared to take, say, the electrical conductivity of one copper object as relevant confirmationwise to another copper object's electrical conductivity, or the response of one autistic patient to a given drug as relevant confirmationwise to another autistic patient's response. But, as relevant to *which* others? As illustrated by examples 1–4 in section 1 above, the range of projection in different examples varies considerably. The limits of *inductive* projection are not fixed by considerations beyond the various hypotheses that we are prepared so to project; nothing more is needed to justify including certain of *h*'s predictions but not certain others in the range to be confirmed by *h* making some accurate prediction. Therefore, inductive confirmation is the only sort of projection that we can justly make when we have nothing more available to justify such discrimination among *h*'s predictions. The only sort of strategy that we need no prior empirical work to entitle us to carry out—that makes no discrimination requiring justification from prior empirical work—is an *inductive* strategy, so-called because we are prepared to confirm "inductively" the hypotheses generated by those strategies that we are carrying out. That is why inductive confirmation is so important in science. I devote the remainder of this section to elaborating this proposal.

To understand the scientific importance of inductive projection, I must back up a bit to consider a fundamental question: What entitles us to regard one discovery as confirming the truth of various other claims? The answer, of course (albeit rather unilluminating at this high level of generality), is that the other opinions we hold when we take our discovery into account determine what we should take that discovery as confirming. For instance, in the pear case, we start with considerable confidence that all of the pears actually on the tree have experienced roughly the same environmental conditions, and that these conditions (whatever they may be) are responsible for the pears' degrees of ripeness. It is this opinion that gives our discovery of a ripe pear on the tree the power to confirm, of every other pear actually on the tree, that it is ripe. Our earlier observations justify our holding this background opinion.

But how can we use one discovery to confirm the truth of various other claims when we *lack* any considerable body of relevant prior opinion—when we have little in the way of relevant past observations and already well-supported theories? In such circumstances, how can we be justified in taking one claim's truth as confirming another's—in holding the prior opinions (i.e., conditional probabilities) required for such confirmation?[7]

Let us look at an actual scientific example. Psychiatry today is much like chemistry was in the late eighteenth century, when Kant bemoaned its lack of any overarching theory. To be sure, psychiatry has justified various generalizations, allowing predictions to be made (with some confidence) from various observations. But in large part, it lacks any comprehensive theory systematizing or explaining these generalizations. These generalizations, then, must have been arrived at on the basis of observations—various case histories—little supplemented by sophisticated theoretical considerations. How, then, have psychiatrists managed to confirm the predictive accuracy of these generalizations when they have had no basis for the prior opinions apparently needed to justify their regarding their observations as confirming those predictions?

To appreciate the difficulty here, consider a list A, B, C, and so on, of psychiatric symptoms, such as "unreasonable insistence on following routines in precise detail," "delusions," and "absence of imaginative activity." Suppose that some patient exhibiting symptoms A, B, and C, but none of the other symptoms on the list, is discovered to respond in a certain way to some new drug. Should psychiatrists regard this discovery as confirming (by some increment) that another patient, exhibiting A, B, and C but none of the other symptoms on this list, would have the same response to the drug? What about a patient exhibiting A, B, C, and only one of the other symptoms on the list? What about a patient exhibiting at least two of A, B, and C, and no more than one of the other symptoms on the list? In the absence of considerable background information, psychiatrists apparently can have no good reason for pursuing one of these ampliative policies rather than another; the range of confirmed predictions appears to be fixed arbitrarily. Yet apparently, psychiatrists are not thereby precluded from justly regarding their evidence as confirming some range of predictions rather than another. In effect, psychiatrists appear to be justified in pursuing a hunch: in simply deciding to take a certain class of evidence as relevant confirmationwise to a certain class of predictions, even though their opinions prior to making this hunch fail to supply them with a good reason for imposing this particular limitation on the range of their projections.[8] What sort of hunch is a scientist permitted to pursue, when her prior opinions do not determine which predictions she would then be prepared to regard certain evidence as confirming?[9]

This is obviously part of the classical "problem of induction." What I am about to suggest is not meant to address this entire problem or to exclude all other approaches. But certain other approaches do not work. For example, it might be supposed that psychiatrists justify grouping together the patients exhibiting certain combinations of symptoms on the basis of having already found some similarities among these cases—perhaps in their reactions to some drugs or in other features suggestive of a common mechanism responsible for these symptoms. But to have already discovered that all patients with certain combinations of symptoms also have in common the disposition to respond in a certain way to a given drug, psychiatrists presumably must have already checked some patients in this category and then projected the result across the category. Such projection requires that psychiatrists *already* regard the examined cases as relevant confirmationwise to the unexamined cases. So it presupposes precisely the sort of categorization that it was supposed to justify. Psychiatrists do tentatively form certain categories for their patients, and then support this categorization by appealing to the results reached by projecting across these categories, but for these results to support anything, they must themselves be warranted, and that depends on the initial categorization being warranted. I argue that one way for science to get off the ground is with some "free moves." But not every sort of policy for regarding a certain class of evidence as bearing confirmation-wise on a certain class of predictions is the kind of policy that scientists need no evidence to entitle them to pursue. I suggest that only strategies involving *inductive* projection are indiscriminate in a manner that allows them to be free moves because only inductive confirmation draws no distinction requiring support from prior empirical work. That is, an inductive strategy does not depend on prior empirical work to justify regarding an observation as able to confirm certain predictions but not certain others.

I discuss inductive strategies much more fully in chapter 7. But here is a brief sketch. A psychiatrist might begin an inductive strategy by grouping together some collection of symptoms as a single diagnostic category (e.g., "autism"). Then she decides to seek a truth in U that would supply certain information regarding the cases in this category—for example, a truth specifying how any autistic patient would respond to a certain drug. Her next step is to observe various autistic patients after they have received this drug. It may happen that this evidence suggests a unique hypothesis. By this, I do not mean something psychological—for example, that the scientist's observations of these autistic patients lead her to come up with some hypothesis. Rather, I mean that scientists observe what it would be for unexamined autistic patients to respond in the same way to the drug as the examined autistic patients have done. If scientists contemplate some way in which unexamined cases might depart from the hypothesis suggested by the evidence, then scientists observe that this departure would involve the unexamined cases behaving differently from the cases heretofore examined. That this is an *observation* means that the scientists' belief possesses a certain special kind of justificatory status. In the manner characteristic of observation reports, this status requires that there be widespread agreement, among qualified observers who are shown the data, regarding which hypothesis expresses what it would be for unexamined cases to go on in the same way as the cases already examined.[10] The scientist who is gambling on this risky strategy then regards any successful prediction made by the salient hypothesis as confirming that hypothesis *inductively*—as I explain below. An inductive strategy ends when some hypothesis of the sought-after kind, having become salient, receives sufficient inductive confirmation to warrant the scientist's believing in it and in each of its predictions that she has confirmed in confirming it inductively.

The history of science is replete with the pursuit of inductive strategies. Johann Balmer (in 1885) and Johannes Rydberg (in 1889) carried out inductive strategies in confirming various regularities in the locations of lines in the spectrum of hydrogen. Henrietta Leavitt and Harlow Shapley (beginning in 1908) pursued an inductive strategy in arriving at the period–luminosity relation governing Cepheid-type variable stars. (Cepheid variables are distinguished from other variable stars by certain features of their spectra and the shape of their "light curve"—their variation in brightness over time. A "period–luminosity relation" is a function relating a variable star's intrinsic brightness to the period of its light curve. See figures 5.1 and 5.2.) They could thereby use observations of only a few Cepheids in our own galaxy's spiral arms to justify predictions regarding the period–luminosity relation of Cepheids even in the Andromeda "nebula," which allowed the nebula's extragalactic distance to be determined for the first time. Einstein employed an inductive strategy (from about 1905) in regarding the success of any prediction made by the light-quantum hypothesis (e.g., equations governing the blackbody spectrum, the photoelectric effect, the Volta effect, specific heats, etc.) as confirming each of the others. The 1987 edition of the American Psychiatric Association's *Diagnostic and Statistical Manual of Mental Disorders* (3rd ed., rev.) codifies the diagnostic categories across which U.S. psychiatrists collectively gamble in projecting their hypotheses. For instance, it classifies as "autistic" any patient possessing at least 8 of 16 designated symptoms, where at least two of the eight belong to the first category of five symptoms, one to the second category of six symptoms, and one to the third category of five symptoms. One autistic patient's response to a given drug is then supposed to be taken as bearing on what another autistic patient's response would be, even if the two patients share no particular symptoms at all. As Carl Hempel (1965)

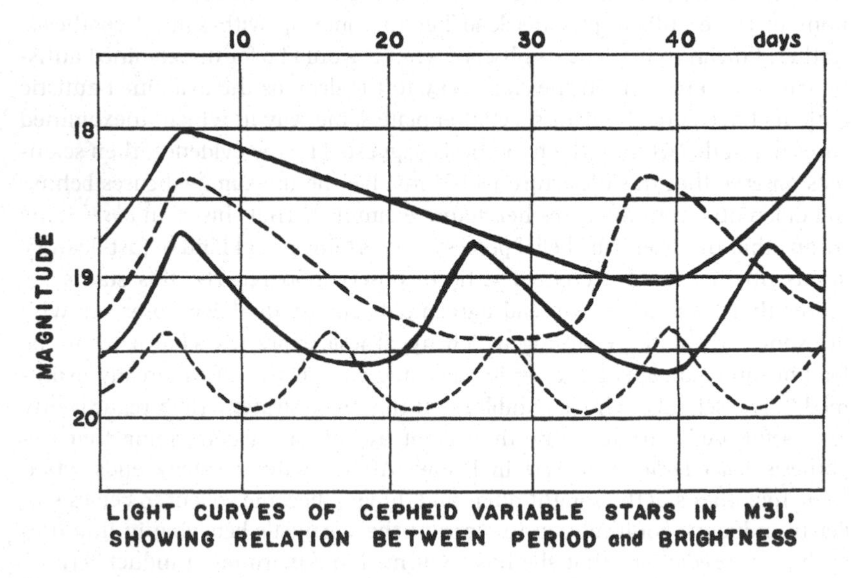

Figure 5.1. Light curves of Cepheid variables displaying characteristic shape as well as increasing period with increasing luminosity. These Cepheids are from the Andromeda Galaxy. From Burnham (1977, p. 135), with kind permission.

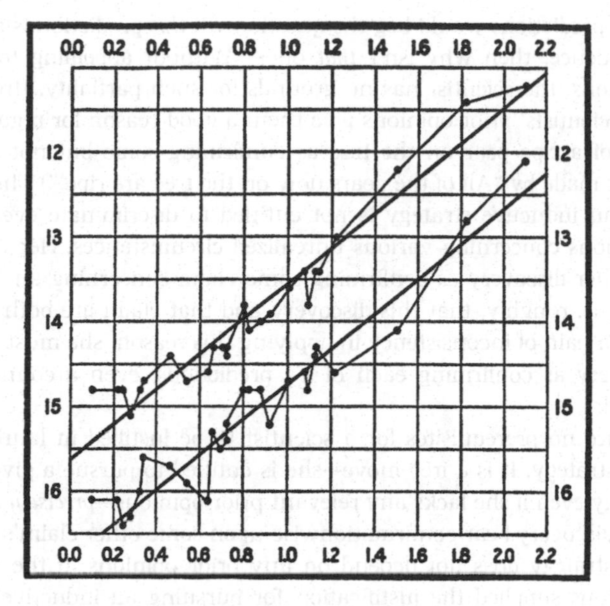

Figure 5.2. The original figure from Leavitt's key (1912) paper proposing the Cepheid period-luminosity relation. The y-axis is the variable's apparent magnitude. (Since these variables all lie in the Small Magellanic Cloud, they are all at approximately the same distance from Earth, and so their apparent magnitudes nearly directly exhibit their intrinsic luminosities.) The x-axis is the logarithm of the period in days. The two curves are for the magnitudes at maximum and minimum. The evidence suggests a certain hypothesis. As Leavitt says: "A remarkable relation between the brightness of these variables and the length of their periods will be noticed. In [a previous paper], attention was called to the fact that the brighter variables have the longer periods, but at that time it was felt that the number was too small to warrant the drawing of general conclusions. The periods of 8 additional variables which have been determined since that time, however, conform to the same law. . . . A straight line can readily be drawn among each of the two series of points corresponding to maxima and minima, thus showing that there is a simple relation between the brightness of the variables and their periods."

emphasizes in trying to resolve his "Theoretician's Dilemma," a theory supplies "inductive systematization" (p. 214); it can take what would otherwise seem an arbitrary collection of claims (e.g., the black-body equation, the photoelectric-effect equation, etc.) and make each of these claims relevant confirmation-wise to the others.[11]

When a scientist pursues an inductive strategy, the evidence suggesting *h* thereby justifies her treating one of *h*'s predictions as able to confirm another. The scientist's prior opinions play no part in this. Therefore, the scientist has no basis for taking one of *h*'s predictions rather than another as confirming *h*, or (with one restriction I explain shortly) for regarding one of *h*'s predictions rather than another as thereby confirmed. Any such discrimination among *h*'s predictions would beg the question: "If *this* prediction is confirmed by the evidence, then why isn't *that* one?" Without appealing to relevant prior opinions, the scientist has no grounds for such partiality. (In the pear example, scientists' prior opinions give them a good reason for regarding the discovery of a ripe pear on the tree as confirming some but not all of the predictions made by "All of the pears now on the tree are ripe.") The scientist pursuing an inductive strategy is not entitled to discriminate even against *h*'s predictions concerning various unrealized circumstances. Her reason for regarding her discovery as confirming some claim concerning an *actual* circumstance is, roughly, that this discovery and that claim are both predicted by *h*. So, on pain of inconsistency in applying this reason, she must recognize her discovery as confirming each of *h*'s predictions, even a counterfactual conditional.

There are no prerequisites for a scientist to be justified in launching an inductive strategy. It is a free move—she is entitled to pursue a given inductive strategy even if she lacks any relevant prior opinions—*precisely because* in making a discovery bear confirmationwise upon some other claim's truth, an inductive strategy does not depend on any prior opinions. If the scientist's prior opinions supplied the justification for pursuing an inductive strategy, then that justification would depend on the scientist's warrant for those prior opinions, which would depend on the scientist's having already done certain empirical work. The strategy would then fail to be a free elective. This is a version of the first of the two intuitions cited in chapter 4 as motivating the distinction between "inductive" and "noninductive" confirmation: background beliefs about which physically unnecessary possibilities happen to be realized are not needed in inductive confirmation to sustain the evidence's relevance to some range of unexamined cases (see chapter 4, note 12).

So the kind of confirmation figuring in an inductive strategy must draw among *h*'s predictions no distinction that could be motivated—rendered nonarbitrary—only by such prior opinions. The inductive strategy's status as a free move thereby determines the particular range of projection that characterizes inductive confirmation. Specifically, in pursuing some set of inductive strategies, we are gambling that they will succeed in taking us to accurate predictions. We are hoping, in particular, that the hypotheses h_i currently salient on these strategies (one per strategy, together forming the set Θ) will receive enough inductive confirmation to warrant their acceptance and the acceptance of all of their predictions that are confirmed in confirming them inductively, and that all of those predictions are true. So it must be at least logically possible for all of those predictions to be true, else we are hoping for what we know to be impossible. Suppose that *p* is inconsistent with Θ's closure in U. Then not all of the predictions regarding *p* that are made by the h_i can be true. So, for our hope to be realizable, it cannot be that each prediction that a salient hypothesis makes regarding *p* is confirmed when that hypothesis is projected inductively. Rather, the inductive strategies must select which of the predictions regarding *p* are to be confirmed during inductive projection. But the inductive strategies lack precisely the basis that they would need to justify any such selection; from the viewpoint of the h_i alone, any such selection would be arbitrary. So in confirming *h* inductively, none of its predictions regarding *p* is confirmed.[12] In this way, the distinctive range of *h*'s *inductive* projection [specified in section 1 above as *h*'s predictions $p > m$ where Con(Θ's closure in U,*p*) and $pr(p > m) \neq 1$, and likewise for nested subjunctives] is explained by the inductive strategy's status as a free move.

Inductive strategies are especially important in science because scientists are entitled to pursue them even before otherwise becoming entitled to hold relevant background opinions; inductive strategies can be pursued right from the outset of empirical work in some area. (This is just what happened in Balmer's and Rydberg's investigations of spectral lines, and also in the work of Leavitt and Shapley on Cepheid variables.) The range across which the salient hypotheses are to be inductively projected is fixed not by our prior opinions, but by those hypotheses themselves. The inductive strategies are collectively *self*-limiting, with each change in Θ—as when a different hypothesis becomes salient on a given strategy—affecting for each other h_i the range of its predictions that we take to have been confirmed in the course of its inductive confirmation. Recall that the range of inductive projection is what connects inductive confirmation to non-nomic stability, and hence to the laws. On this view, then, the inductive strategy's status as a free move is one source of the laws' scientific importance.

Here, then, are the morals of this story. (1) We can presume, without begging the question in this chapter's title, that science is interested in predicting non-nomic facts in the broad sense of facts in U*, as discussed in chapter 3. (2) To achieve this goal, it is especially valuable for science to find a set of inductive strategies for which it is nontrivially the case that, as these strategies are carried out, the predictions that become confirmed sufficiently to justify their acceptance are all true. (3) This occurs if and only if the logical closure in U of the various hypotheses that are ultimately generated by these strategies is Λ. In short, to discover which truths in U state laws is to discover which inductive strategies form the best set for us to carry out. Therefore, it is especially important for science to identify the laws in U.[13]

In carrying out some inductive strategies, we aim to arrive at accurate predictions—at some set whose logical closure in U possesses nontrivial non-nomic stability. When an inductive strategy that we are pursuing suggests and then leads us to accept *h*, we conclude that *h* belongs to a set that nontrivially possesses non-nomic stability, and so that $\Box h$. In the course of pursuing an inductive strategy, we must believe that the hypothesis currently salient may turn out to be the hypothesis to which the strategy ultimately leads us. Therefore, we must believe that *h* may be physically necessary.[14] Likewise, we must believe that the hypotheses currently salient on our inductive strategies (i.e., the hypotheses belonging to Θ) may turn out to be the hypotheses to which these strategies ultimately lead us—at which point we would believe each of the predictions $p > m$, $p > (q > m)$, . . . made by any of these hypotheses, where Con(Θ's closure in U,*p*), Con(Θ's closure in U,*q*), and so on. Hence, we must now believe it may be that each of these predictions is true.[15] This is the case, we know, if and only if Θ's closure in U is Λ. Therefore, we must believe that Θ's closure in U may be Λ.

I demonstrated in section 1 above that we can confirm *h* inductively only if *h* ∈ Θ's closure in U. I have just argued that we must believe that Θ's closure in U may be Λ. It follows that we can confirm *h* inductively only if we believe that *h* may belong to Λ. Thus, "inductive" confirmation as I have elaborated it in this chapter is a kind of confirmation that we must reserve for hypotheses that we believe may express physical necessities. Recall the central role that this idea played in chapter 4; by showing that some such confirmation has special scientific importance, I respond to the skeptic regarding the concept of a natural law.[16]

In chapter 7, I offer a proposal regarding the "root commitment" associated with believing ■*h*. One consequence of undertaking this root commitment is believing *h* to be the hypothesis that is reached at the conclusion of some inductive strategy that belongs to the best set of inductive strategies for us to pursue. This proposal entails that Λ possesses nontrivial non-nomic stability: roughly speaking, a set of inductive strategies would not be the best for us to pursue if it led to our making false predictions. Although epistemic considerations—concerning the indiscriminate character of inductive confirmation—are one source of the laws' scientific *importance*, I am not defending an epistemic characterization of *what it is for something to be physically necessary*. What makes a set of inductive strategies best is, in part, the correctness of certain subjunctive conditionals. Here we have an unreduced metaphysical (rather than epistemic) element.

For the sake of simplicity, I have thus far set aside the possibility that there are multiple "grades" of physical necessity (as discussed at the close of chapter 3). To leave room for this possibility, we must regard each inductive strategy as belonging to some grade. The definition of "inductive" confirmation must then be amended accordingly: in pursuing an inductive strategy *of a certain grade*, we are prepared to project the salient hypothesis across every *p* in U that is consistent with the logical closure of the hypotheses currently salient on the inductive strategies *of that grade* (and, for each of these *p*s, we are prepared to project the salient hypothesis across every *q* in U that is likewise so consistent, and so forth for multiply nested subjunctives).

3. Conclusion

Let me take stock. I began chapter 3 by suggesting that science is *interested* in discovering whether □*p* or ¬□*p* (for *p* ∈ U), insofar as this goes beyond discovering whether or not *p*, because science thereby learns something about what would have happened under various unrealized non-nomic circumstances. Since the subjunctive conditionals that science thereby ascertains to be correct have antecedents and consequents in U, we can grant science's interest in ascertaining their correctness without begging the question (i.e., without thereby merely stipulating science to be interested in ascertaining whether $\Box p$ or $\neg\Box p$).

I have now gone one step further. I have just argued that a reason why discovering whether $\Box p$ or $\neg\Box p$ (for some truth $p \in$ U) is so *important* to science is because this is discovering how best to go about discovering whether various claims in U* are correct.

This view is reminiscent of the response that Hempel offers to *his* Theoretician's Dilemma. (In chapter 3 I pointed out that the question in the title of this chapter is a nomic version of the Theoretician's Dilemma.) Hempel contends that a theory's claims about unobservables, insofar as they go beyond claims about observables, may supply "inductive systematization," enabling one observation to bear confirmationwise upon another to which it would otherwise be irrelevant. Likewise, in seeking to discover whether $\Box p$ (where $p \in$ U), not merely whether *p*, science seeks to know how best to proceed in order to predict non-nomic facts (in the broad sense of facts in U*). In aiming to learn which set of inductive strategies is best for us to carry out, science is aiming to identify the members of Λ.

If we select the right inductive strategies to carry out, then we can ascertain some of the laws' accurate predictions in U* without depending on prior empirical work to have already given us a good reason for picking out those predictions, rather than others, as the ones to be confirmed.[17] In this respect, the laws differ in their range of predictive accuracy from the non-nomic facts that lack physical necessity, such as that all of the pears on the tree are ripe. For us to confirm *exclusively* the *accurate* predictions that some accidental generalization makes, we must have already conducted enough empirical work to warrant the background opinions needed to privilege that particular range of projection. In the pear example, for instance, we must have already become confident that a pear's degree of ripeness is determined partly by the amount of sunlight received in the preceding weeks. Otherwise, in confirming the accidental generalization, we might confirm that all of the pears on the tree would still have been ripe even if they had received much less sunlight in the preceding weeks. It may, then, be far easier to justify projecting a hypothesis that turns out to be a law across a range in which it turns out to make exclusively accurate predictions than to justify projecting a hypothesis that turns out to be an accidental truth across a range in which it turns out to make exclusively accurate predictions, since the latter projection demands that we motivate drawing some otherwise arbitrary distinction among the various predictions made by the hypothesis.

Here I am taking issue with a remark of A. J. Ayer (1976) that develops a Theoretician's Dilemma for laws:

> If on the basis of the fact that all the A's hitherto observed have been B's we are seeking for an assurance that the next A we come upon will be a B, the knowledge, if we could have it, that all A's are B's would be quite sufficient; to strengthen the premise by saying that they not only are but must be B's adds nothing to the validity of the inference.

> The only way in which this move could be helpful would be if it were somehow easier to discover that all A's must be B's than that they merely were so. (pp. 149–50)

But this is not possible:

> It must be easier to discover, or at least find some good reason for believing, that such and such an association of properties always does obtain, than that it must obtain; for it requires less for the evidence to establish. (p. 150; see also van Fraassen 1989, p. 135)[18]

How can you become entitled to regard some observed *A*'s *B*ness as confirming that the next *A* to be examined will be *B*? If you are not pursuing an inductive strategy, then you need to have already done some empirical work in order to justify regarding your observation as relevant confirmationwise to the next *A*'s *B*ness. (In the pear case, you typically need to have amassed sufficient evidence to justify believing that the examined and unexamined pears developed under environmental conditions that are similar in those respects that influence the pears' ripeness.) It may be far easier—require much less prior empirical work—for the observed *A*'s *B*ness to confirm the next *A*'s *B*ness in the course of some inductive strategy on which "All *A*s are *B*" becomes salient. The decision to carry out that inductive strategy is a free move, and it may then take relatively little evidence to render "All *A*s are *B*" salient (as I discuss more extensively in chapter 7). It is hard to imagine how psychiatrists today could justly regard a given autistic patient's response to a drug as relevant confirmationwise to how any other actual autistic patient would respond to that drug except through carrying out an inductive strategy. The same could be said for any of the other historical examples of inductive strategies that I gave above (section 1). For instance, how could Leavitt and Shapley in 1908 have justified taking the periods and luminosities of nearby Cepheid-type variable stars as bearing upon the period–luminosity ratio of Cepheids in the Andromeda "nebula" except as a risky inductive gamble that they were entitled to elect in virtue of the indiscriminate character of the projection involved? Utterly lacking any theories concerning the internal structure of stars, much less of Cepheid variables, they were in no position to offer any other sort of justification for the requisite conditional probabilities. The same thought is echoed in Max Planck's 1913 recommendation that Einstein be admitted to the Prussian Academy. Concerning Einstein's inductive strategy of taking any successful prediction made by the light-quantum hypothesis (e.g., regarding the black-body spectrum and photoelectric effect) as confirming any other prediction made by the light-quantum hypothesis (e.g., regarding the Volta effect and specific heats), Planck prematurely judged that Einstein had failed. But Planck understood that a scientist is permitted to elect to take a gamble of this kind:

> That [Einstein] may sometimes have missed the target in his speculations, as, for example, in his hypothesis of light-quanta, cannot really be held too much against him, for it is not possible to introduce really

new ideas even in the most exact sciences without sometimes taking a risk. (Kirsten and Korber 1975, p. 201)

Now if "All *As* are *B*" becomes salient on one of your inductive strategies, and you confirm it inductively to a very high degree, and you ultimately adopt it, then you believe that it is physically necessary, since you believe that "All *As* are *B*" belongs to a set possessing nontrivial non-nomic stability. Contrary to Ayer, then, it can be easier to justify believing that it is physically necessary that all *As* are *B* than to justify believing that all *As* are *B* without believing that it is physically necessary that all *As* are *B*.

Of course, Ayer is correct to point out that, since the lawhood of "All *As* are *B*" entails its truth, our degree of confidence in its lawhood cannot exceed our degree of confidence in its truth. But this does not show that it is harder to justify believing that it is a law than to justify believing that it is true *without thereby justifying the belief that it is a law*.[19] Ayer's argument presupposes that some evidence would, as it were, *first* bear upon whether various unexamined *As are B*, and *then* you would need some *further reason* for regarding that evidence as bearing upon whether those unexamined *As must be B*. I suggest, on the contrary, that you may be required to believe that various unexamined *As must be B* in virtue of your reason for believing that various unexamined *As are B*.

A basic presupposition of scientific research is that we do not need to *observe* whether a claim is true in order to ascertain this fact. (Indeed, to ascertain the correctness of some nontrivial counterfactual conditional, we do not have the option—even in principle—of observing whether it is true.) Science is very much interested in knowing how it can use its observations to make accurate predictions when it begins without already having any reason to regard its observations as bearing confirmationwise upon any predictions. In seeking the best way to proceed from such ignorance to knowledge of non-nomic facts (in the broad sense of facts in U*) beyond the limited range of past observations, science seeks to identify the laws. Beliefs about the laws, over and above beliefs about non-nomic facts, must therefore be acknowledged as playing an important role in scientific investigation of the non-nomic facts.

• SIX

Laws, Regularities, and Provisos

1. Introduction

It is typically taken for granted that laws are or are invariably associated with logically contingent regularities. (E.g., on a "Humean" account, laws just *are* regularities, whereas according to an account of laws as relations of "nomic necessitation" among universals, each law *underwrites* a regularity.) The key issue is then finding a way to distinguish the regularities associated with laws from the accidental regularities. But there is a prior issue: Is every law associated straightforwardly with a regularity? In section 2, I motivate this question by way of the "provisos" to various law-statements. These are clauses, often remaining implicit, that stipulate something like "ceteris paribus" or the absence of "disturbing factors." It has proved difficult to see what provisos might mean, especially when we bear in mind that we are able to discover and to apply various laws—to recognize when there are no disturbing factors—even before we have ascertained all of the kinds of influences there are.

In section 3, I offer an account of what a proviso means. I thereby remove the worry that a proviso either (1) trivializes the claim to which it is attached, (2) deprives that claim of any determinate content, or (3) renders that claim incapable of being understood or applied as it is in scientific practice. However, I argue in section 4 that this account of a proviso's content entails that the association between a law and a regularity is often not nearly as straightforward as it has typically been made out to be. If we insist on portraying a law-statement "All *F*s are *G* (ceteris paribus)" as describing a regularity, then that regularity need not be anything like the fact that all *F*s under certain specified circumstances are *G*.

A claim with a proviso attached nevertheless possesses nontrivial content because the proviso is not a totally elastic "escape clause." To understand the putative law is to know how it is supposed to be applied, and the proviso is true exactly when it is supposed to be proper to apply the putative law. Its application can be appropriate as part of using a certain approximation, model, calculational procedure, idealization, or simplification. By applying the law "All *F*s are *G*," we arrive at conclusions *Gx* that, when plugged into the model, lead to beliefs that are close enough (often enough) for the relevant purposes. Yet to be serviceable in this way, the conclusions *Gx* need not be true or even, in any ordinary sense, approximately true.

This might be put by saying that "It is a law that all *F*s are *G*" can be true even if "All *F*s are *G*" is false. This formulation, though possessing a certain salutary shock value, is somewhat misleading. I emphasize that when "All *F*s are *G*" is used to state a law, there may be a great deal implicit in it about how it is to be applied: the other laws that should be used along with it in order to solve certain problems, the sorts of practical and theoretical applications in which this law together with those others can be used, the sorts of observations and background beliefs that can and cannot be used in certain applications of the law, and so forth. To interpret the law as simply that all *F*s are *G* is to disregard these provisos.

It is not the case that, if these implicit qualifications are included explicitly in the law-statement, then the law-statement expresses some more complicated regularity: that all *F*s under certain specified circumstances are *G*. Rather, belief in the law involves belief in the effectiveness (for certain purposes) of reasoning in accordance with a certain inference rule in the course of pursuing a certain kind of approach to solving certain sorts of problems. In other words, it involves belief in what I call the "reliability" of the inference rule "associated" with the law. Belief that this inference rule is truth-preserving, over and above belief in its reliability, is not part of the root commitment undertaken in believing in the law. Otherwise, I argue, that root commitment would not account for the relations between our beliefs about the laws and our beliefs about counterfactuals, inductions, and explanations.

Of course, belief in the inference rule's "reliability" might still be understood as belief in a *certain peculiar sort* of regularity.[1] This is not the sort of regularity that is typically identified as (underwritten by) a law. I am not arguing that the world lacks any exceptionless regularities, or that science acknowledges none, or that science is unconcerned with discovering them, or that there are no laws corresponding to such regularities. Rather, I am arguing that belief in a regularity of the sort that is typically thought to be associated with every law is not part of the root commitment undertaken when a law is believed to obtain. This is so even in those cases where the law is *also* believed to be associated with such a regularity.[2]

2. The problem of provisos

Consider the familiar statement of the law of thermal expansion: "Whenever the temperature of a metal bar of length L_0 changes by ΔT, the bar's length changes by $\Delta L = k \cdot L_0 \cdot \Delta T$, where k is a constant characteristic of that metal." The relation between L_0, ΔT, and ΔL that this statement appears to express does not obtain. It may be violated, for instance, if someone is hammering the rod inward at one end. If this statement is false, then it does not express a law, according to any account that portrays laws as invariably associated with regularities.

The obvious move to make here is to contend that the familiar statement of the law of thermal expansion is understood as implicitly containing a restriction to cases where, for instance, the rod is not being so hammered. Many other familiar law-statements would have to contain similar sorts of restrictions. For instance, Snell's law of refraction is standardly expressed as "When a beam of light passes from one medium to another, sin *i*/sin *r* is a constant," where *i* is the angle of incidence of the beam upon the second medium, *r* is the angle of refraction in that medium, and the constant is characteristic of the two types of media. This statement would have to contain implicit restrictions to cases where the media fall within a certain range of temperatures and pressures, there is no magnetic or electrical potential difference across the boundary between the media, the media have uniform optical densities and transparencies, the media lack double-refractivity, and the beam is monochromatic. Even this is not a complete list of the necessary restrictions. The law of thermal expansion must also include a great many qualifications. If a complete, explicit statement of the law must specify each of the restrictions individually, then the statement will be very long indeed. It must specify not only that no one is hammering the bar inward at one end, but also, for instance, that the bar is not encased on four of its six sides in a rigid material that fails to yield as the bar is heated. For that matter, it should not deem the law inapplicable to every case in which the bar is being hammered on. The bar may be hammered on so lightly, and be on such a frictionless surface, that the hammering produces translation rather than compression of the bar. Apparently, to state the law of thermal expansion fully would require exquisite care and detail. Ronald Giere (1988) concludes that "qualified laws are incapable of being written down explicitly simply because the number of provisos implicit in any law is indefinitely large" (p. 40). This seems an overstatement (as I discuss in more detail below); no such problem seems to arise for certain laws, such as Maxwell's equations and "All material bodies, accelerated from rest, travel slower than the speed of light." But there are a great many laws that appear very difficult to express completely.

However, as a mere practical difficulty, this does not seem terribly serious.[3] It can easily be avoided by employing some such qualification as "in the absence of disturbing factors," "ceteris paribus," or "as long as there are no other relevant influences." I term these sorts of qualifications *provisos*.[4] But once we set the practical difficulty aside in this way, we are confronted by a more fundamental problem. What does a proviso mean?

By "other relevant factors," we must mean some particular set of factors. Otherwise, "All *F*s are *G* in the absence of other relevant factors" fails to assert any determinate relation at all. This point has recently been emphasized in connection with putative laws of "folk psychology," the sort of commonsense psychology that proceeds in terms of intentional states such as beliefs, desires, and so on. Statements of putative folk-psychological laws contain ceteris paribus clauses. For instance, consider "If someone believes that *p*, and believes that if *p* then *q*, then ceteris paribus, she believes that *q*." Someone who is confused or distracted, or cognitively impaired (e.g., through injury or illness), will not satisfy this ceteris paribus clause. The same goes for the ceteris paribus clause in the claim "If someone desires that *p*, believes that if *q* then *p*, and is able to bring it about that *q*, then ceteris paribus, she brings it about that *q*." In addition, someone who believes that her other desires conflict with her desire that *p*, or believes that *r* would be a better means than *q* for bringing it about that *p*, fails to satisfy the latter ceteris paribus clause. That the proviso "in the absence of other relevant factors" must refer to some particular list of factors has been emphasized by Stephen Schiffer (1991)

> The sentence
>
> [1] If a person wants something, then, all other things being equal, she'll take steps to get it
>
> is deceptive. It looks as though it's expressing a determinate proposition, because it looks as though "all other things" is referring to some contextually determinate things and "equal" is expressing some determinate relation among them. But one would be hard pressed to say what the "other things" are and what it is for them to be "equal." Yet if "all other things being equal" doesn't make a bona fide contribution to a proposition expressed by [1], then [1] is really tantamount to
>
> If a person wants something and . . . , then she'll take steps to get it,
>
> which is good for nothing, as it expresses no complete proposition. . . . (p. 2; see also Fodor 1991)

But let's be more persistent than Schiffer in seeking an interpretation of [1] that renders it contentful. Apparently, the proviso must refer to the *complete* list of physically possible factors that could cause an *F* to fail to be *G*. If the list were incomplete, then once again, we would be faced with a "law-statement" that is false. The trouble now is that, surely, we can know the law of, say, thermal expansion without having already identified each of the other kinds of factors besides temperature change that can affect a metal bar's length. Plausibly, when scientists discovered the law, they knew about the kinds of disturbing factors that sometimes produce large deviations from the law's predictions in standard sorts of cases in which we might be interested in applying the law. But they need not have known about kinds of influences that require extreme, highly atypical, or otherwise nonstandard conditions in order to produce non-negligible deviations from the law. For instance, the law of thermal expansion was discovered long before the relativistic effects of velocity on length were known. These effects are negligible in practical applications of the law to, say, bridge construction.

Of course, scientists might later develop a complete account of the various kinds of influences there are. They might then even introduce the concept of an "ideal" case—a case where none of these influences is present. The concept of an "ideal gas" was introduced in this way and defined explicitly as a gas in which (to cut a longer story short) there are no intermolecular forces and no molecular radii. An explanation could then be given of why the gas laws are accurate (to the extent that they are) in various circumstances: their accuracy depends on how small these two influences are. But this notion of an "ideal" gas was developed *after* the gas laws had been discovered; scientists understood what statements of the gas laws meant—understood the provisos—before they had a scientific explanation of why the gas laws obtain. They knew only that the gas laws worked well (for certain purposes) under low pressures, high temperatures, and so on. Only later was it understood *why* these conditions were necessary in order for the deviations from the gas laws to be negligible.[5] To ascertain a law, it is not necessary to know why that law obtains. The concept of an "ideal case" is not implicit in the law-statement.

That we would have to know the complete list of other influences in order to know the law plagues only "derivative laws" like the law of thermal expansion—that is, laws that concern the *net effect* on some condition (such as a bar's length). A derivative law derives from various, more fundamental laws, each of which concerns only the independent contribution made by a single kind of factor. So the presence of other factors does not disturb a fundamental law. For instance, here is a typical statement of Galileo's law of freely falling bodies, a derivative law: "The distance covered in time t by a body falling to earth from rest is $(1/2)gt^2$ where g is about 9.8 m/s^2, as long as there are no disturbing factors." If this proviso must identify all of the various kinds of disturbing factors there are, then what Galileo believed to be a law was not a law; he knew that collisions with other bodies and air resistance could affect a falling body, but he was unaware of electromagnetic forces on a charged falling body, for instance. (Macroscopic falling bodies are typically so close to electrically neutral that these forces can be ignored.) But it might appear that no such problem arises for the more fundamental laws from which Galileo's law follows, such as Newton's gravitational force law.[6] The component gravitational force that one body exerts on another is not affected by the presence of other material bodies or the other kinds of forces that the one body exerts on the other.

Yet if the regularity associated with the gravitational-force law is supposed to be a regularity between the masses and separation of any two bodies and the component gravitational force that each exerts on the other, then those component forces must be real entities in order for there to be any such law. The reality of such forces is controversial; the particular contribution made by one body's gravity to the net force on another body, or to the net gravitational field, may have no independent reality.[7] It might be suggested that just as we cannot say that a given governmental program was paid for by your tax dollars rather than by mine—because our tax contributions lose their separate identities once they enter the government's coffers—so perhaps a field's net vector, its energy, and so on, at a given location are not composed of individual bits attributable to particular charged particles. Rather, there may exist simply a (net) electric field throughout space (as well as truths about what that field would have been like had a given charged particle been absent). Despite this uncertainty regarding the reality of component forces, we are confident of the nomic status of Newton's law (setting aside the fact that physics has advanced beyond classical Newtonian physics). That status, then, does not seem to depend on the reality of component gravitational forces. After all, we use "the Coriolis force law" in the manner distinctive of laws; for example, the Coriolis force is used to explain the directions of cyclones and tradewinds.[8] Yet we believe that there are no component Coriolis forces to conform to a regularity that might be (underwritten by) a law. The "Coriolis force" is merely a correction factor (having the dimensions of a force) that must be used when Newton's second law of motion is applied in accelerated reference frames. Like centrifugal force, it is commonly called a "fictitious force" or "pseudoforce." So component forces may not exist, and in any case, their existence may not be required by the fundamental force laws. Hence, even a fundamental law may be difficult to associate straightforwardly with a regularity.[9]

A proviso to a derivative law does not refer to the complete list of other influences, on pain of inconsistency with scientific practice: the law can be discovered before that complete list is known. Accordingly, perhaps the proviso says not that no other relevant factors are present, but rather that the only relevant factor present is the factor with which the given law is concerned. On this view, the proviso to the law of thermal expansion, for instance, requires that temperature change be the only factor influencing the bar's change in length. We would then not have to know what all of the other kinds of relevant factors are in order to understand what the proviso means.

Carl Hempel (1988) considers this line of thought. He says that a given theory covers a particular kind of influence, and a proviso is a kind of completeness assumption: that in the given case, only the influence covered by the theory is present (pp. 26, 29). One of Hempel's examples is Newtonian gravitational theory. Its application to a planetary system "presupposes a proviso to the effect that the constituent bodies of the system are subject to no forces other than their mutual gravitational attraction" (p. 23). By a "proviso," Hempel means a *premise* essential to the application of a theory. Neither Newton's gravitational force law nor his second law of motion contains a clause referring to the absence of other factors (as I discuss more fully in section 6). But in order for Newton's second law (which refers to the *net* force on a body) to be applicable to the gravitational force given by Newton's gravitational force law, that gravitational force must be the total force. The proviso (in Hempel's sense) states that it is.

Whereas Hempel (1988) considers provisos to be *premises*—"a proviso has to be conceived as a clause that pertains to some particular application of a given theory and asserts that in the case at hand, no effective factors are present other than those explicitly taken into account" (p. 26)—I am interested in *clauses* (implicit) in law-statements, such as the "in the absence of disturbing factors" clause in the law of thermal expansion.[10] Like Newton's second law of motion, the law of thermal expansion concerns a *net* quantity—the net change in the bar's length. But like Newton's gravitational force law, the law of thermal expansion addresses only a single influence on that quantity—the bar's change in temperature. So apparently, a complete statement of the law of thermal expansion must include a clause specifying that no factors other than this temperature change are at work, and any application of the law must include a completeness premise. What Hempel says about the content of a completeness premise should apply equally well to the content of a completeness clause in a law-statement. I refer to the latter as a "proviso," and to the former as a "proviso premise."

To state a proviso, it is not always necessary to use an apparently elastic "ceteris paribus" or "in the absence of disturbing factors" clause. As Hempel (1988, p. 30) notes, the proviso in the planetary motion example can be expressed simply by asserting that the total force acting on each body in the system equals the gravitational force exerted on it by the system's other bodies—in other words, that gravitational forces are the only relevant factors. This interpretation of a proviso (unlike the interpretation according to which the proviso refers to the complete list of possible kinds of disturbing influences) allows us to understand what the proviso means prior to discovering all of the kinds of disturbing influences there are. However, this interpretation does not succeed in eliminating the basic problem. The key point, Hempel (1988) says, is that the theory itself—in this case, Newtonian gravitational theory—does not tell us when the proviso holds:

> The proviso must . . . imply the absence, in the case at hand, of electric, magnetic, and frictional forces; of radiation pressure; and of any telekinetic, angelic, or diabolic influences. . . . Neither singly nor jointly do [celestial mechanics and other physical theories] assert that forces of the kinds they deal with are the only kinds by which the motion of a physical body can be affected. A scientific theory propounds an account of certain kinds of empirical phenomena, but it does not pronounce on what other kinds there are. The theory of gravitation neither asserts nor denies the existence of nongravitational forces, and it offers no means of characterizing or distinguishing them. (p. 30)

The problem, then, is that the theory does not tell us how to recognize whether the proviso premise is satisfied. So although we can understand this proviso-laden law-statement prior to discovering all of the kinds of disturbing influences there are, we cannot justly apply this law-statement to any actual case because we have no way to ascertain, prior to discovering all of the kinds of disturbing influences there are, whether the single factor with which the law is concerned is the only relevant factor. It then becomes impossible to discover the law—to test the theory.

Hempel (1988, pp. 25–26) emphasizes that the difficulty here is not a version of the Duhem-Quine problem: that if the theory's predictions are not borne out, we could perhaps justly avoid blaming the theory by instead deeming the proviso premise to be false. The problem posed by provisos does not concern what would happen if the theory made a false prediction, but rather is that the theory makes no predictions at all. Any prediction requires a proviso premise. The resources that we had available prior to the formulation of the theory do not allow us to infer that the proviso premise is satisfied. And the theory itself (because it concerns only a single factor) does not allow us to tell whether the proviso premise is satisfied. So we cannot generate any testable predictions from the theory.

The same problem that Hempel identifies regarding our inability to tell whether a proviso premise holds has been emphasized by Davidson in connection with putative folk-psychological laws. Ironically, Davidson (1990) contrasts the folk-psychological case with physical science in this regard:

> It is an error to compare a truism like "If a man wants to eat an acorn omelette, then he generally will if the opportunity exists and no other desire overrides" with a law that says how fast a body will fall in a vacuum. It is an error, because in the latter case, but not the former, we can tell in advance whether the condition holds. . . . (p. 233)

I return to this remark below.

Another way that Hempel (1988) puts his point is that "in general, the requisite provisos cannot be expressed in terms of V_A alone" (p. 27), where V_A is vocabulary "that is antecedently understood, that is, available and understood independently of the theory" (p. 20). The proviso in the planetary motion example, for instance, employs the term "force" from Newtonian physics. So the proviso cannot be cashed out in terms intelligible independent of the theory in question. That is not because of any mere practical limitation—say, because it would take too long for us to do so. Rather, the proviso fails to be independently intelligible because there are no pretheoretically understood terms covering all and only the possible cases in which the theory is applicable. Prior to discovering the theory, we have no means of determining when the proviso premise holds, and so no way of telling when the theory is applicable. And since the theory is devoted to the strength of a single influence, it does not tell us when the proviso premise holds—when that single influence is the only influence present, and so when the theory can be applied. Even if there were no practical limitation to listing all of the kinds of disturbing influences individually, they could not be listed in terms that we know how to apply prior to developing theories covering those influences. Yet—to repeat—scientific practice suggests that we do not need to know every influence in order to know how to apply such a law.

It might be suggested that we use the fact that our observation of a metal bar accords with $\Delta L = k \cdot L_0 \cdot \Delta T$ as evidence that the proviso holds, and then, having secured the proviso, apply the law to this case. But this "test" presupposes the truth of what it is testing. If "as long as there are no disturbing factors" just means "as long as there are no factors that cause ΔL to deviate from $k \cdot L_0 \cdot \Delta T$," then of course we can understand the proviso prior to identifying the other influences on a metal bar's change in length. But we make the statement express a logically *necessary* truth rather than a logically *contingent* law of nature. As Giere (1988) says, "The problem is to formulate the needed restrictions without rendering the law completely trivial" (p. 40).

This, then, is the problem of provisos: to find an interpretation of provisos that does not render law-statements trivial, indeterminate in meaning, inaccurate, or incapable of being understood or applied just as they are in scientific practice (i.e., prior to the discovery of all of the other kinds of relevant factors).[11] In section 3, I argue that a theory covering only a single factor can supply a proviso with determinate content that fails to trivialize the "law-statement," and that we can know how to recognize circumstances in which the law is applicable even before we have identified the complete list of other influences.

We may be tempted to reject the problem of provisos by insisting that *genuine* law-statements simply lack provisos. On this view, the fact that many familiar "law-statements" are actually not law-statements at all only goes to show that we have discovered very few genuine laws. To accept this view would, I think, be unjustified. An account of laws must accommodate the fact that scientists are not reluctant to use proviso-laden claims in the manner distinctive of law-statements in connection with, for example, explanations and counterfactuals. We would find this fact difficult to explain if we held that scientists do not regard these claims as expressing laws.[12] I am not free to disregard this fact, any more than I am at liberty to disregard the fact that scientists treat "All gold cubes are smaller than one cubic mile" as an accidental generalization and "All cubes of uranium-235 are smaller than one cubic mile" as expressing a law, even though this fact is troublesome for many simple proposals regarding the root commitment associated with believing that some law obtains. Scientific practice is the only phenomenon for an account of this root commitment to save; if an account is tested against science as idealized to conform to that account, then the "test" is pointless.

Perhaps we would be justified in setting the problem of provisos aside if we had some account of law that, except for this problem, was entirely successful. Otherwise, I think it worth investigating whether greater progress can be made by reflecting on provisos than by disregarding them. According to the response to the problem of provisos that I offer in this chapter, the relation between the laws in a given possible world and the regularities there is not as straightforward as standard reductive analyses depict. Ultimately, this insight will assist us in dealing with other problematic features of laws, such as their behavior in connection with nonbacktracking counterfactuals (originally discussed in section 2.2 of chapter 2).[13]

Additional problems are sometimes thought to be raised by provisos to folk-psychological generalizations. It is sometimes conceded (at least for the sake of argument) that the ceteris paribus clause in a folk-psychological generalization could be cashed out explicitly: for each possible physical realization (P_1, P_2, . . .) of the relevant folk-psychological states, we could in principle spell out the conditions (C_1, C_2, . . .) under which "all other things are equal." The argument against folk-psychological laws is then often made on one of two grounds. Sometimes it is argued that the folk-psychological generalization, with its ceteris paribus clause filled out ("as long as C_1 if P_1, C_2 if P_2, . . . "), would not state a law *of folk psychology* because the expression cashing out the ceteris paribus clause must import nonpsychological terms (e.g., terms from neurophysiology or from fundamental physics):

> Will *every* nomologically possible physical defeater of these physical mechanisms itself realize a *psychological state*, such as confusion, irrationality, or distraction, that could take its place in a wholly *psychological* true completion of [a putative psychological law]? . . . [N]o is the most plausible answer. . . . [B]rain-injured people of the kind observed by Oliver Sachs appear to give empirical evidence that there are breakdowns in normal cognitive processes that cannot be accounted for in psychological terms. (Schiffer 1991, p. 4; see also Fodor 1991)

Sometimes, the argument takes a different form: the folk-psychological generalization, with the ceteris paribus clause cashed out ("as long as C_1 if P_1, C_2 if P_2, . . . "), would not state a law because there is nothing common to all of the various conjuncts in the expression cashing out the ceteris paribus clause. In other words, if we listed each physically possible kind of mental hardware and what it takes for that hardware to be working properly, then this clause would be, in Paul Teller's (1984) words, too "fat" to appear in a law-statement, considering the multiple realizability of folk-psychological states; in an important sense, the various entries on the list have nothing in common (pp. 58–59). Ruth Millikan (1993) writes:

> Characteristically, the same function could, at least in principle, be performed by many differently constituted items. But . . . then the supporting conditions required for them to effect this function must differ as well. Brain cells performing the division algorithm require oxygen whereas computer chips require electric currents, and so forth. . . . The result is that there are no ceteris paribus laws governing all items having a certain function. For ceteris paribus conditions are unspecified conditions that must remain the same from case to case for the law to hold, whereas here the necessary conditions would have, precisely, to vary from case to case. A "law" applying to all such cases could say no more than that the items falling under the law could be made, by adding different circumstances tailored specifically to each case, to perform the function. But surely anything can be made to effect anything if one adds the right intervening media, if one adds enough special enough circumstances. So any such "law" would be empty. (pp. 226–27)

The following discussion of the original "problem of provisos" will better illustrate how ceteris paribus clauses function, which will allow us to assess these various arguments concerning folk psychology.

3. What a proviso means

Here is how I believe the proviso to the law of thermal expansion (which is typical of many provisos) is understood in science. When seeking this law, scientists did not know or even suspect all of the factors besides temperature change that can affect a bar's change in length. But their aim was to find a law relating ΔL to ΔT *that works for certain purposes.* The intended purposes were limited to certain sorts of practical applications (e.g., bridge construction), and in connection with those applications, the law's intended uses may have been further restricted to cases where certain sorts of models or calculational procedures are (or are not) being employed. I discuss this further below. The key point is that when scientists sought this law, they had in mind a certain range of uses for it, and these involved a range of cases with which they were already familiar—so familiar, in fact, that scientists were correct in their belief that they had already identified all of the kinds of influences (such as someone hammering hard on the bar) that are sometimes non-negligibly present in this range of cases. These are the "disturbing factors." In other words, in the range of cases with which scientists understood themselves to be concerned, all of the unknown kinds of influences on ΔL turn out to be so small that they can safely be ignored for the law's intended purposes.

That is not to say that scientists are always able to tell in advance whether a given actual case falls within the range that the law is supposed to cover and, if so, involves no disturbing factors. They might be unaware that someone is hammering on the bar or that the case involves some exotic circumstance. But they can, at least in principle, learn whether or not this is so prior to noting the relation in the given case between ΔL and $k \cdot L_0 \cdot \Delta T$. Of course, once the law is known, then that a given case displays a great difference between these two quantities may sometimes be the way that scientists learn that some as yet unidentified factors are at work there.

So in referring to "disturbing factors," the proviso refers to a determinate list of influences. Though not a complete list of the kinds of influences on ΔL besides ΔT, this list suffices to make the law work for its intended purposes, which (like the proviso) are often left unstated. To understand the law-statement, a scientist must understand not only the term "disturbing factors," but also the sorts of applications for which the law is supposed to work.

There can be general agreement on what kinds of influences qualify as "disturbing factors" even if no scientist has ever listed them exhaustively. For instance, consider this statement of Galileo's law of falling bodies: "All bodies falling to Earth accelerate at 9.8 m/s^2, in the absence of any other relevant factors." Part of understanding this law-statement is grasping what is implicitly meant by "other relevant factors." This meaning is implicit because all that scientists ever say is that these factors may consist of another falling body colliding with the given body, a parachute slowing the fall, a strong updraft or downdraft, *and so on*. Presentation of examples like these (whether to a student learning the law today, or to workers testing the proposed law at the time of its discovery) suffice to produce near unanimity concerning how to extend the sequence—concerning whether some other circumstance belongs with these examples as a "disturbing factor." As Wittgenstein (1978) puts the point:

> It only makes sense to say "and so on" [as in "no one is hammering on the bar, and so on"] when "and so on" is *understood*. I.e., when the other is as capable of going on as I am, i.e., does go on just as I do. (p. 349)
>
> For the "and so on" might on the one hand be replaced by an arrow, which indicates that the end of the series of examples is not supposed to signify an end to their application. On the other hand "and so on" also means: that's enough, you've understood; we don't need any more examples. (p. 417)

So prior to the law's discovery, all of the cases involving no "disturbing factors" must be appreciated as similar, so that some examples of these cases can suffice to suggest the rest.

That the term "disturbing factors" derives its content from a shared, tacit understanding of the occasions in which the law is applicable does not entail that the law-statement is trivialized by this proviso—that the statement means "All bodies falling to Earth accelerate at 9.8 m/s^2, except when they don't." Scientists may well once have understood the proviso in such a way that the alleged "law-statement" was not, in fact, accurate because of significant departures from it in circumstances in which the sought-after law was supposed to apply and where there are no "disturbing factors" as the phrase was then used—for instance, near the equator, where (because of Earth's oblateness) gravity is weaker. Being near the equator was perhaps not originally considered a "disturbing factor." When scientists later came to recognize these departures, they rejected their original "law" in favor of an alternative in connection with which certain additional factors qualify as "disturbing," are understood to fall within the scope of "and so on," and so forth. (Of course, scientists who have changed their minds about what the law is might nevertheless continue to use the very same expressions as before, e.g., to say simply "in the absence of other relevant factors." But which factors someone who understands the putative law-statement must recognize to be "relevant" will have changed.) That the alleged "law-statement" was rejected shows that it was not trivially true.

By the same token, there can be widespread agreement on the sorts of applications in which the sought-after law is intended to work even if this range has never been characterized in a manner that would be intelligible to someone not already familiar with this or similar research programs. It might be tacitly understood, for example, that the sorts of conditions to be encountered in terrestrial bridge construction fall within the law of thermal expansion's intended range of application, whereas the sorts of conditions that would be encountered at the bottom of the sea or above the atmosphere are not. Likewise, when Galileo's law of falling bodies was accepted, it was not known that if a falling body is electrically charged and falling through an electric field, then its fall may deviate drastically from 9.8 m/s^2. Yet the omission of this influence from the "disturbing factors" did not keep the law from working for its intended purposes—which involve medium-sized objects in natural terrestrial conditions—since the contribution of electrical influences in such cases is negligible for ordinary purposes. (Any medium-sized body carrying a large charge on its surface and falling through the air will quickly be approximately neutralized.) In exotic or artificially prepared cases (e.g., fall through a vacuum) or for small bodies, the electrical influence can be too large to ignore. But these cases are not among those that the law was originally developed to cover. That is not to say that the law invariably fails in any such case. But to justify its application to a case of some exotic kind, we must first test it in cases of precisely that kind, or have greater theoretical understanding of why the law holds (to the extent that it does) in ordinary cases than was available at the time of the law's discovery. In other words (as I discuss in chapter 7), to carry out the "inductive strategy" yielding this law, we must be prepared to take the behavior of any medium-sized falling body in natural terrestrial conditions and under no disturbing influences, and to treat that behavior as relevant confirmationwise to the behavior of any other such body in such conditions. But this inductive strategy does not involve treating the behavior of any such body in such conditions as bearing upon any case of a more exotic kind.

To contrast the provisos to law-statements in science with provisos that do indeed trivialize the claims to which they are attached, consider some fortune tellers who have made some inaccurate predictions. They claim that, in arriving at these predictions, they misapplied their theories of the occult by mistakenly failing to bear in mind some clause in the "law-statement" on which they had relied, a clause that they had neglected to make explicit earlier. If they give this excuse repeatedly, then it is doubtful that they have undertaken any determinate commitment at all by adopting this "law-statement." Or suppose someone says "I can run a four-minute mile" but with each failure reveals a proviso that she had not stated earlier: "except on this track," "except on sunny Tuesdays in March," and so on. It quickly becomes apparent that this person will not acknowledge having committed herself to any claim by asserting "I can run a four-minute mile." Science is distinguished from such bunkum *not* by explicitly including in its law-statements every qualification—that is to say, *not* by the absence of any merely implicit understanding among scientists of how to apply accepted law-statements. What is noteworthy about science is that this implicit understanding of "and so on" or "in the absence of all disturbing factors" is genuine implicit *understanding*: there is widespread agreement on which testable predictions are made by a given hypothesis in a given case, even if that hypothesis includes a ceteris paribus clause.[14]

With these considerations in mind, I briefly return to the questions I discussed at the end of section 2 regarding the provisos associated with folk-psychological generalizations. To cash out one of these provisos completely would be a daunting prospect; as Schiffer (1991) points out, we "would be hard pressed to say what the 'other things' are and what it is for them to be "equal'" without resorting to "and so on" or some other such phrase (p. 2). Still, as I have shown, that is no reason to believe that the ceteris paribus clause renders the folk-psychological generalization indeterminate in meaning or trivial. Just as we understand (if we have learned basic physics aright) what qualifies as a "disturbing factor" in connection with the law of freely falling bodies, so we understand what qualifies as a disturbing factor in connection with a folk-psychological generalization: not only confusion and distraction, but also injury, illness, and so on, along perhaps with various other factors (depending on the generalization) such as other conflicting desires.

I discussed in section 2 how Davidson (1990, p. 233) contends that the escape clause in "Anyone who wants to eat an acorn omelette and has the opportunity to do so will eat an acorn omelette, so long as no other desire overrides, and so on" is fundamentally different from the escape-clause in "Any falling body accelerates at 9.8 m/s^2 as long as there are no disturbing factors" because in the latter case, but not the former, we can tell in advance whether the proviso holds. But the two examples seem fairly similar to me. Of course, one way of confirming that the fall is influenced by no disturbing factors is to see if it conforms to the law. (This is mere confirmation, not conclusive proof, since the body would fall at 9.8 m/s^2 if various disturbing factors existed but precisely counterbalanced each other.) But this is not the only way. We could instead draw on our past observations (along perhaps with our knowledge of other laws) and thereby confirm in advance that the fall is not taking place near the equator or being slowed by a parachute. Likewise, to confirm that the individual has no overriding desires and is cognitively intact, we might observe, for example, whether she eats an omelette under certain circumstances. Alternatively, since every folk-psychological law contains a ceteris paribus clause requiring that the agent be cognitively intact, we could draw on our knowledge of these other laws (and our past observations of the individual's behavior) to confirm in advance whether she is cognitively intact. In the same way, we can use our knowledge of her past actions and relevant psychological laws to confirm in advance whether she has overriding conflicting desires.

In section 2 I noted how Schiffer (1991, p. 4) contends that if we cashed out one of these ceteris paribus clauses, we would need to use terms from neurophysiology or fundamental physics, and so the generalization cannot qualify as a *folk-psychological* law. But this remark is of dubious relevance, since there is no good reason to believe that these clauses need to be cashed out at all. Admittedly, to describe a particular illness or injury, or to explain why it occurred, we might use terms drawn from neurophysiology or fundamental physics. But "illness" and "injury" are not themselves terms from fundamental physics, and they are not restricted to neurophysiology. A generalization with the proviso "as long as the individual is cognitively intact" is not thereby rendered incapable of qualifying as a law *of folk psychology*. On the contrary, the category of "being cognitively intact" is peculiarly a folk-psychological category. The distinction between those physical or neurological states that qualify as injuries, malfunctions, illnesses, and so on, and those that do not can be drawn only by referring to how healthy individuals are expected to be, and this includes not only how they are expected to be neurophysiologically, but also how their intentional states are expected to be ordered. I agree with Millikan's (1993, pp. 226–7) remark that considering multiple realizability, the defeaters of a given folk-psychological generalization are anatomically (or microstructurally) heterogeneous; from the viewpoint of neurophysiology or fundamental physics, they may have nothing in common. But, contrary to Millikan, what they compose may constitute a real *kind*—a real *folk-psychological* kind.

Consider students coming to grasp the content of a ceteris paribus clause. Since the clause cannot be cashed out "explicitly," students have no choice but to learn its meaning by appreciating what is similar in several examples. If students did not have a grasp of the vocabulary of folk psychology (beliefs, desires, etc.), and were shown (in full neurophysiological or microphysical detail) a series of examples in which the ceteris paribus clause of a given folk-psychological generalization fails, then I daresay they would not latch onto this category; they would not all extend it to other cases in the same manner as their teachers (or even, perhaps, in the same way as one another). Because the realizations in different hardware of the same folk-psychological state are so microstructurally diverse, the particular realizations appearing in the examples will not allow students justly to anticipate the realizations that they have never encountered *if* they think about the examples from an autonomously neurophysiological (or microstructural) perspective. The category of "cognitively intact" individuals is gerrymandered from a neurophysiological or microphysical viewpoint, but is a natural folk-psychological kind; individuals that in neurophysiological (or microphysical) terms have nothing in common, save some "fat" or "wildly disjunctive" property, have something in common from a folk-psychological viewpoint. What counts as "remain[ing] the same from case to case" (Millikan 1993, p. 226) is different from a folk-psychological standpoint than from a "lower layer" standpoint.

This is a point that I develop in chapters 7 and 8. (Millikan's remark about what counts as "remain[ing] the same from case to case" anticipates the notion of a *salient* hypothesis that I elaborate, and the relation between macro and micro explanations, kinds, and laws is one of the principal topics of chapter 8.) For now, I must set this point aside in order to examine how the account of provisos that I have just sketched affects the relation between laws and regularities.

4. Provisos and regularities

4.1. For some laws, a good regularity is hard to find

I have just suggested that scientists tacitly understand a given law as intended to cover a certain range of cases, and that the kinds of influences qualifying as "disturbing factors" include (as it turns out) every kind of influence that in some of these cases is too great to be ignored for the law's intended purposes. For example, Galileo's law of falling bodies is intended to deal with macroscopic heavy bodies falling near Earth's surface—the sorts of cases that would arise in ballistics, for instance—and although the "disturbing factors" referred to by its proviso do not include all of the influences besides Earth's gravity that such a body might physically possibly feel, they include all of the influences that might cause such a body to deviate from the law's predictions by enough to be important for the law's intended purposes. Because the law's attention is understood to be restricted to a certain familiar range of cases, scientists who sought the law (or who were being shown evidence for it) already knew many of the kinds of influences at work in these cases, and so could understand "disturbing factors" as referring to exactly those recognized influences. These turn out to be all of the other influences great enough in such cases to be non-negligible for certain purposes—which include all of the purposes in the service of which the law is tacitly understood to be applicable. So in the relevant range of cases, the law is accurate enough for its intended purposes.

The expression "disturbing factors," then, need not refer to the *complete* list of other influences in order for the law to be (in the relevant range of cases) accurate enough for its intended purposes. But "disturbing factors" must be understood to include the complete list in order for "All *F*s are *G* in the absence of disturbing factors" to be true. That is, the term "disturbing factors" fails to encompass certain influences that, although so small in the relevant range of cases as to be negligible for the law's intended purposes, are nevertheless sometimes present in such cases and hence able to produce small deviations from the law's predictions. We can safely ignore these deviations when properly using the law, but they exist all the same. Macroscopic heavy bodies falling near Earth's surface can be treated for the purposes of, say, ballistics as accelerating uniformly at 9.8 m/s^2 when under no "disturbing influences." But in fact, not all such bodies undergo exactly uniform acceleration at 9.8 m/s^2. The law merely works for the relevant purposes. It is "reliable."

At this point, repelled by the imputation that a genuine law may not be strictly true, we might be tempted to say that a law holds exactly only in ideal conditions. In the case of the law of falling bodies, ideal conditions obtain when the body feels no forces other than Earth's gravitational influence (which requires its falling in a vacuum, for instance) and falls at some "ideal" latitude and height (!) where Earth's gravitational acceleration is exactly 9.8

m/s^2. (Of course, we would on this view acknowledge that the law is sometimes applied even when it is only approximately correct, as long as it is then believed to be accurate enough for the relevant purposes.) But I have already argued that this view is mistaken. An account of the "ideal conditions" is not needed to express the law fully. Rather, such an account *explains why* the law is accurate (to the extent that it is). Giving an account of the ideal conditions is not part of discovering the law, and such an account may even be unknown to those who discover the law. They need not know what all of the other influences are, or even what factor or factors would—if unaccompanied by other influences—make bodies conform exactly to the law. Again consider the ideal gas laws. When Boyle's law was ascertained, it was not known what real kinds of influences would together produce behavior that conforms exactly to this law, or what real kinds of influences would then be left out (I discuss Boyle's law in more detail in chapter 7). Instead, in discovering Boyle's law, scientists discovered that certain kinds of *actual* systems—influenced by various factors, identified and as yet unidentified—maintain their pressure–volume relation as a constant for all practical purposes (of certain sorts). The same may be said for the law of falling bodies. Admittedly, it may sound counterintuitive to say that strictly speaking, this law concerns bodies falling under ordinary conditions and *not* bodies falling under ideal conditions. But this view seems more plausible when it is borne in mind that our conception of the ideal conditions is informed by theories that were not available at the time of the law's discovery. The law was arrived at by noting that heavy macroscopic bodies observed falling under ordinary conditions approximate uniform acceleration, and then projecting this behavior onto other, similar cases. Would a case that would *now* be considered "ideal" have been considered *at the time of the law's discovery* to be similar to the cases then already examined? It is hardly evident that it would, considering how exotic a vacuum was held to be (if even physically possible at all); Torricelli's vacua were called "factitious," artificial. Without depending on a theory that purports to explain why this approximate regularity holds, scientists could not justly regard a fall through a vacuum, in conditions that we might now consider ideal, as a similar case, and so as included within the range of this projection.

In short, then, the "disturbing factors" do not need to encompass all of the other influences because application of the law is implicitly restricted to certain purposes. But then, since the proviso does not refer to all of the other influences, "All *Fs* are *G*" is not perfectly true even when the proviso holds. It is merely accurate enough for the relevant purposes. The only regularity in which we must believe in believing in some law—the only regularity figuring in the "root commitment" — is that by applying the law in the course of taking certain kinds of inferential routes to solving certain kinds of problems posed in certain kinds of contexts, we reach conclusions that often enough are accurate enough for our purposes. (The proviso refers to these restrictions on the inference rule's applicability.) But this is not the kind of regularity with which accounts of law typically take laws to be associated.

4.2. How to accommodate reliability

When we believe in the law of thermal expansion, the inference rule that we believe to be reliable specifies a quantity (namely, $k \cdot L_0 \cdot \Delta T$) that for certain purposes we can take to equal the bar's change in length (ΔL). But when we take that quantity as being ΔL, we do so only within the context of solving a particular problem in a particular way. Outside of that context, it may be inappropriate for us to take that quantity to be ΔL. There may be other laws that we could use instead to predict ΔL for these same purposes, but which lead to slightly different (but also sufficiently accurate) values for ΔL. Once we have applied one of these other laws, it becomes inappropriate for us to take ΔL also to equal the quantity given by the law of thermal expansion. The inference rule associated with the law of thermal expansion specifies that we *can* but not that we *must* take a certain inferential route in order to predict ΔL. (I made analogous points in chapter 1 in connection with Hooke's law and other spring laws with nonlinear terms.)

Some provisos to a given law restrict the inferential route in the course of which that law may be applied. These provisos might constrain the other laws that we can use along with the given law or the manner by which we can arrive at the premises to which we apply the given law. Though some law may be properly used in the course of one idealization, model, calculational procedure, approximation, or simplification, it may lead to unsatisfactory results when applied together with certain other laws.[15] Part of learning how to use the law is learning how to distinguish the inferential routes in which it can be used from those in which it cannot.

In section 1 of chapter 5, I defined *h*'s "predictions," for $h \in U$, as the conditionals $p > m$, $q > (p > m)$, and so on, where $m, p, q, \ldots \in U$, and $(p \,\&\, h) \Rightarrow m$. Having recognized that laws are associated with reliable inference rules, I must amend this definition. Rather than require that *m* be logically entailed by *p* & *h*, it must now require instead that the inference rule associated with *h* say that we can infer from *p* to *m*.

Likewise, my former definition of "Γ is preserved under Δ" (in section 1 of chapter 2) must be amended: all references to some subjunctive conditional's "correctness" must now be understood as referring instead to the conditional being close enough for the relevant purposes to being correct. For example, if Boyle's law is preserved under the counterfactual antecedent "Had the gas's pressure been halved while its temperature and number of molecules remained fixed," then " . . . its volume would have been doubled" must be close enough (for the purposes for which Boyle's law is applicable) to being correct.[16] Likewise, my definitions of "inductive confirmation" in chapters 4 and 5 must be amended. When we confirm *h* inductively, we confirm not that certain subjunctive conditionals are correct, but that they are close enough for the relevant purposes to being correct. The evidence that confirms *h* inductively is the discovery that some "prediction" made by *h* is sufficiently accurate for the relevant purposes (see chapter 5, note 6).

In addition, all of these definitions employ "Con(Γ,*p*)." This must be redefined to mean that the claims to which we would be entitled, by reasoning from *p* in accordance with the inference rules associated with the members of Γ, could all be close enough for the relevant purposes to being correct (see chapter 2, note 4). Likewise, "Γ's logical closure in U" must now be understood to be the set containing exactly those claims corresponding to inference rules (mediating inferences among claims in U) that must be reliable if the inference rules corresponding to the claims in Γ are all reliable.[17]

There are several ways that a counterfactual conditional $p > q$ may qualify as sufficiently accurate for our purposes. One way is for us to be interested in a certain feature of the closest *p*-world, and for $p > q$ to describe that feature in a manner that is close enough to the truth for our purposes. (This is presumably the case in the example involving Boyle's law that I just gave.) Alternatively, we may have no interest in discovering whether or not *q* holds in this possible world, but doing so may be one step along an inferential route by which we are to ascertain some other feature of interest regarding this world. Another possibility is that discovering whether or not $p > q$ holds is part of no inferential route by which we are to ascertain some feature of this possible world that is of interest to us. Then *trivially*, $p > q$ is accurate enough for our purposes. (This last possibility is important in section 5 below.)

In chapter 3, I required of any set possessing non-nomic stability that it contain only truths. Now I must amend this requirement: the set must contain only claims associated with reliable inference rules.[18] The physical necessities in U, then, are expressed by exactly the claims belonging to a set that nontrivially possesses non-nomic stability. There may be more than one such set, as I discussed in section 3.1 of chapter 3 in connection with "grades" of physical necessity. Having thus refined these various concepts [*h*'s "predictions," "Γ is preserved under Δ," "inductive confirmation," "Con(Γ,*p*)," "Γ's logical closure in U," "non-nomic stability"] to make room for laws that are merely "reliable," I can retain all of the conclusions involving these concepts at which I arrived earlier, such as the relation between inductive confirmation and non-nomic stability. (I.e., consider a set of claims in U where, nontrivially, each of the predictions that is confirmed, when each of these claims is confirmed inductively, is sufficiently accurate for the relevant purposes. This set's logical closure in U must have nontrivial non-nomic stability, and its members accordingly possess a distinctive grade of physical necessity.)

5. Nonbacktracking again

In chapter 2, section 2.2, I discussed our frequent failure to backtrack, as when we assert "Had Darcy asked Elizabeth for a favor, then Elizabeth would have refused to grant Darcy's request." In the relevant possible world, Darcy's request comes without the causal antecedents apparently required by the actual laws; nothing occurring prior to Darcy's request, such as his quarrel with Elizabeth, departs from the way it actually was. So the actual laws are violated in this world. Nevertheless, we find ourselves insisting that, had Darcy asked Elizabeth for a favor, the laws of nature would have been no different, just as they would have been no different had the Sun never formed or had I failed to brush my teeth this morning. I concluded that violations of a possible world's laws can take place in that world, as long as those violations remain "offstage"—that is, involve none of the events in which we are interested in that context. I emphasized that when a counterfactual supposition calls to mind a possible world, we care only about certain features of that world. For instance, in considering how Lincoln would have compared the Dole-Clinton debates to his own debates with Stephen Douglas, we would be making an utterly irrelevant remark if we noted that Lincoln could have been in a position to make this comparison only if he had not been a human being or the actual laws of nature had been violated. How Lincoln managed to be present for both of these debates is a matter that resides "offstage."

If the causal antecedents of Darcy's request are not our concern, then *trivially*, the inference rules associated with the actual laws yield predictions regarding these causal antecedents that are sufficiently accurate for our purposes—since our purposes require no predictions to be made regarding this feature of the relevant possible world. So despite being violated in this possible world, the actual world's laws can qualify in this context as "reliable" regarding this possible world, and so can be laws there. In particular, by applying the actual laws, we are led to the accurate conclusion that Elizabeth would not have granted Darcy's request, and to accurate conclusions regarding subsequent events. We often use the actual laws to ascertain what we want to know regarding a given counterfactual world.[19]

My response to the problem of provisos fits well with my account of laws in nonbacktracking contexts: each suggests that the laws in a given world can be violated there, as long as, when properly applied, they lead to conclusions all of which are sufficiently accurate for the relevant purposes. It might have been suspected that in responding to the problem of provisos by allowing the laws to be merely reliable, I mistakenly run together what are really several different notions of law: one for approximate or "derivative" laws or for laws of "special" sciences, and another for exact or fundamental laws. Or it might have been objected that the ways that laws are applied in practice (namely, to cases where they hold only approximately) have misled me into thinking that laws themselves may be merely approximately true. But the challenge posed by our failure to backtrack is not restricted to "derivative" laws or to laws of "special" sciences, and is not an artifact of scientists' practical concerns in applying a law. Even the fundamental laws are violated offstage in the *p*-world that qualifies in a nonbacktracking context as the closest. So my reply to the problems arising from nonbacktracking counterfactuals supports my response to the problem of provisos. (In chapter 8, I return to these ideas in connection with laws of particular scientific disciplines, as I anticipated in section 3 of chapter 1.) Some might acknowledge that our failure to backtrack reveals that the laws in a given possible world can be violated there, and yet deem this to be a peculiarity of *nonactual* possible worlds. But my response to the problem of provisos should dispel this suspicion. Even in the actual world, the laws are violated.

6. Laws for which there are no "disturbing factors"

James Woodward (1992) has also discussed the role of laws in models. He argues, as I have, that the relation between certain laws and regularities is not straightforward:

> Consider another typical example of a fundamental law of nature: Schrodinger's equation. This law is, in a variety of ways, extremely general and abstract. If we wish to use this law to analyse some specific system—for example, the case of a charged particle tunneling through a potential barrier—we must make a number of additional, much more specific assumptions to insure the applicability of the law. For example, we must choose a specific Hamiltonian for the system—we must make assumptions about the shape of the potential barrier, the kinetic energy of the charged particle, and so forth. We must also make a number of other specific assumptions, usually involving large elements of simplification and idealization, about initial and boundary conditions. Quite different, but equally specific assumptions must be made if we wish to use Schrodinger's equation to analyse some other paradigmatic quantum-mechanical system such as an electron in a potential well. Scientists use these assumptions to construct *models* of real physical systems, which we can think as abstract and stylized representations which make possible the application of the Schrodinger equation to the systems in question. But while scientists certainly use the Schrodinger equation together with such modeling assumptions to provide descriptions and explanations of the behavior of real physical systems, it is far from obvious that there exists some single pervasive regularity in nature involving real physical systems corresponding to the Schrodinger equation itself. The equation without the modeling assumptions needed to apply it to particular situations seems far too abstract and non-specific to play this sort of role; it is more like a general schema or constraint which can be filled in or specified in a variety of ways so as to yield descriptions of real systems than by itself a description of some regularity in the behavior of the systems to which it applies. (pp. 196–97)

Woodward apparently acknowledges the regularity that for any system, there are "modeling assumptions" (e.g., background and initial conditions, a Hamiltonian) such that the system conforms (approximately) to Schrodinger's equation together with those assumptions. But he apparently regards this regularity as "too abstract and non-specific" to qualify as the regularity associated with a natural law. I am not certain of Woodward's reason for this view, but I agree that this regularity, without some specification of how the modeling assumptions appropriate for a given system are to be determined, fails to capture the way that Schrodinger's equation is used in science.

Roughly speaking, various laws are supposed to be used to derive the potentials to be plugged into Schrodinger's equation; each of these laws takes into account one kind of influence that the system might feel. An interpretation of Schrodinger's equation as expressing the above "abstract" regularity, without provisos specifying that these other laws are supposed to be used together with Schrodinger's equation, cannot account for why the modeling assumptions that can be made for different systems must be related to one another. By regarding certain modeling assumptions as permissible for one system (e.g., a simple pendulum, whose bob feels the tension of the string and a gravitational attraction toward Earth) and certain other modeling assumptions as permissible for another system (e.g., an iron sphere near a magnet), we undertake a commitment to regarding certain modeling assumptions as permissible for a third system that combines features of the other two (e.g., a magnetically augmented pendulum: its iron bob is gravitationally attracted to Earth and magnetically attracted to the fairly strong magnet under the rest position of the bob).

Let me back up a bit in order to work my way more carefully to this result. As I noted in section 2 above, a law may lack provisos requiring the absence of disturbing factors. Consider a force law, such as Newton's for the gravitational force. It lacks an "as long as there are no other relevant factors" proviso because the presence of other influences does not affect the conclusions that we should draw regarding the component gravitational forces present. Of course, this does not entail that the gravitational force law has no provisos whatever. On the contrary, it may have provisos specifying the contexts in which it can be used.

The acceleration observed in a given case may confirm the reliability of the inference rule associated with some hypothetical force law. This confirmation may even be *inductive*; the evidence may bear upon each of the cases (actual and counterfactual) in the range across which the hypothesis is to be inductively projected. For example, we could be justified in predicting the forces on a magnetically influenced pendulum, even if we have not yet tested our hypotheses in such a case, because we have already inductively confirmed the gravitational force and magnetic force laws from examining other cases. Suppose, however, that we had inductively confirmed not these laws, but instead various narrower hypotheses—one covering only bodies in free fall, another covering only iron bodies feeling just the force exerted by a nearby magnet, another covering only bobs of simple pendulums, another covering only magnetically influenced pendulums, and so on. Then we would have to have first observed some magnetically influenced pendulums in order to have justly arrived at a reliable inference rule covering them. Because it is advantageous to be able to make accurate predictions regarding such pendulums in advance of having observed any of them, it is better for us to confirm inductively the gravitational force and magnetic force laws than to confirm inductively only various narrower, though reliable rules. Analogous points apply to the various laws used to take various influences into account in constructing the Hamiltonian for a given system to be plugged into Schrodinger's equation. (In chapter 7 I develop this point about the best "inductive strategies," as I foreshadowed in my discussion of Hooke's law in section 3 of chapter 1.)

So it is important that the gravitational force law is applicable *no matter what other influences are at work*. That certain inferential steps can likewise appropriately be taken, whatever the other influences at work, in order to arrive at the Hamiltonian to be plugged into Schrodinger's equation is a point that is missing from the "abstract and nonspecific" characterization of the regularity associated with Schrodinger's equation.

It is revealing to see how Giere's (1988) response to the problem of provisos fails to capture this point. Giere holds that certain familiar law-statements are false as claims about the world, since they lack the requisite provisos (which, Giere says, cannot be fully stated). But he thinks that, as law-statements, they must accurately describe *something*. Giere maintains that a law-statement describes a scientific model. We may hold that the relevant behavior of a given real system can be predicted by using some model. Giere terms such a claim a "theoretical hypothesis."[20] The law-statement needs no qualification by provisos to be accurate to the model.

A theoretical hypothesis refers to a single actual system, or perhaps to some narrow range of actual systems. Crucially, it does not attempt to specify the law's entire range of application, on pain of having to include the very provisos that Giere says cannot be fully stated. By interpreting law-statements as descriptions of models, Giere has made it unnecessary for law-statements to include these provisos.

Consider, then, what Giere (1988) says about two laws, each concerning a single influence, that are combined to account for a magnetically influenced pendulum:

> We have discovered a new kind of pendulum . . . in which the force of gravity is supplemented by a magnetic force directed toward a point below the point of rest. Constructing a theoretical model that does apply to such systems is a fairly easy problem in physics. (p. 44)

If a statement of the gravitational force law specifies how you are entitled to calculate the gravitational influence of one body on another no matter what the other influences present, and a statement of the magnetic force law does likewise for the magnetic influence, then this is indeed an easy problem. Having already accepted these laws, and having recognized the proper way to add forces and to use the net force on a body to infer its acceleration, we are already committed to the propriety of a particular procedure for predicting the bob's motion, even though this kind of pendulum is new to us.

But remember, Giere does not take statements of the magnetic force and gravitational force laws as specifying that certain models are properly used in certain circumstances. That is the job of theoretical hypotheses. Rather, he takes these law-statements merely as describing certain models. Thus, Giere must admit that, by accepting that the gravitational force law holds in certain models, we do not undertake any commitment to the appropriateness of using such a model to predict the motion of some pendulum's bob. Moreover, in accepting the theoretical hypothesis that such a model can be used to predict the motion of a nonmagnetic bob, we do not become committed to the theoretical hypothesis that such a model can be used to predict the motion of a magnetically influenced bob.

On my view, in accepting the gravitational force law, we become committed to a certain rule for inferring (no matter what the other factors we believe present) the component gravitational force on a bob. This view accounts for what Giere's view obscures: that we are committed to some common element in our treatments of ordinary and magnetically augmented pendulums in virtue of which we find it easy to solve the magnetically augmented pendulum. We became committed to elements of this solution when we adopted solutions to other problems.

7. A further kind of proviso

There is another kind of proviso whose role in science suggests that the root commitment we undertake when we believe in a law does not involve belief in some regularity straightforwardly associated with that law. In chapter 8, I discuss this proviso's role in connection with certain laws of biology. But I introduce this kind of proviso here in connection with a chemical law.

I start with the law of multiple proportions (Dalton's law). Here are three venerable statements of this law:

> The law . . . may be thus illustrated algebraically: If x and y be the equivalents of any two substances, their compounds must be x + y, x + 2y, x + 3y, x + 4y, etc.; sometimes we shall have 2x + 3y, and rarely 2x + 5y . . . a *simple ratio* of one to two, one to three, etc. . . . (Gray 1848, pp. 124–25)

> If two elements A and B form several compounds with each other, and we consider any fixed mass of A, then the different masses of B which combine with the fixed mass of A bear a simple ratio to one another (Remsen 1890, p. 16).

> The quantities of the various elements in their respective compounds are either in the exact ratios of their combining weights, or else in simple multiples of these. This law is the foundation of chemical investigation. . . . (Nernst 1904, p. 31)[21]

Here is one of the illustrations Ira Remsen gives, which was one of the instances that Dalton initially cited as confirming this law; the chemical analysis was originally performed by Antoine Lavoisier, Nicolas Clément, and Charles-Bernard Désormes, though, Dalton noted, they had not "taken notice of this remarkable result" (Nash 1956, p. 39). Carbon monoxide (to use anachronistic terminology) is 42.86% carbon and 57.14% oxygen by weight, which is a ratio of 6 to 8. Carbon dioxide is 27.27% carbon and 72.73%

oxygen by weight, which is a ratio of 6 to 16. So the mass of oxygen combined with a given mass of carbon in carbon dioxide is exactly twice the mass of oxygen combined with the same mass of carbon in carbon monoxide. This ratio of 2 to 1 is an instance of the law—a "simple ratio" between the different masses of *B* that combine with a fixed mass of *A* in several compounds. As another instance, Remsen notes that the three compounds that iron forms with sulfur involve 4, 6, and 8 parts of sulfur to 7 parts of iron, and 4:6:8 is the simple ratio 2:3:4. Likewise, five compounds formed by nitrogen with oxygen contain 7 parts nitrogen to 8, 16, 24, 32, and 40 parts oxygen, so the parts by weight of oxygen that combine with 7 parts of nitrogen are in the ratio 1:2:3:4:5.

It is essential that the law stipulate the ratios to be "simple," since otherwise, any ratios accord with the law, and so the law adds nothing to the law of definite proportions, also proposed by Dalton. For instance, T. M. Lowry (1926) gives this example of how the law of multiple proportions is to be applied:

> [I]f hydrogen and oxygen unite together in the ratio 1:8 to form water, they may also unite in the ratio 2:8, 3:8, 4:8, or 1:16, 1:24, 1:32 (or, generally, in the ratio $n \times 1 : m \times 8$, where n and m are whole numbers) in forming other compounds; but they cannot combine together in intermediate ratios such as 1.107:8 or 1:7.823. (p. 304)

But of course, a ratio of 1.107:8 *can* be represented as $n \times 1 : m \times 8$ where n and m are whole numbers—namely, where $n = 1107$ and $m = 1000$. But if hydrogen and oxygen united in the ratios 1:8 and 1.107:8, then the parts by weight of hydrogen that combine with 8000 parts of oxygen would stand in a ratio of 1000:1107. Apparently, Gray, Remsen, and Nernst would say that this is not a "simple" ratio. In other words, Lowry's remarks come with the implicit qualification "where n and m are small," and $n = 1107$ and $m = 1000$ are tacitly understood not to qualify.

Accordingly, the law of multiple proportions must always be expressed in terms of "simple ratios," on pain of triviality. Of course, the term "simple" does not interpret itself: Which ratios qualify as "simple"? Examples of acceptable ratios are usually provided with the law-statement (as Gray does; see likewise Mendeleev 1891, vol. 1, p. 215); these examples assist a student in latching onto how "simple ratios" is commonly understood. The expression "simple ratios" thus derives its content in the same manner as terms like "disturbing factors" and "*ceteris paribus*."[22]

The atomic theory was thought to be highly confirmed by the law. Suppose that if elements A and B form a compound, then the number of atoms of A and the number of atoms of B in each molecule of the compound are small whole numbers. Then if A and B form more than one compound, the number of atoms of B per each atom of A in one compound stands in a simple ratio to the number of atoms of B per each atom of A in the other.

But even if the notion of a "simple ratio" derives its content in the manner I have suggested, there remains a mystery: apparently, chemists who regard this as a law (indeed, in Nernst's words, as the very "foundation of chemical investigation") may nevertheless believe it to be violated. For instance, as Maureen Christie (1994, p. 619) notes, there are a myriad of ugly ratios among the hydrocarbons, since the number of hydrogen atoms and the number of carbon atoms in a given hydrocarbon molecule are not invariably *small* integers. (Christie notes that $C_{25}H_{52}$ compared with $C_{33}H_{68}$, e.g., gives the unsimple ratio 429:425.) Of course, even if Dalton and his contemporaries were utterly ignorant of these complex molecules, the same cannot be said of Nernst and his contemporaries, not to mention current chemists. Did the "law of multiple proportions" cease at some stage to be regarded as a genuine law? For that matter, did Dalton and his associates really believe that every molecule had to involve a *small* number of atoms? This does not seem required by the atomic theory.

Indeed, it is not, as Dalton and his contemporaries fully recognized (Nash 1956, p. 30). Here is the explanation of the law of multiple proportions in the first part of Dalton's *A New System of Chemical Philosophy*, originally published in 1808:

> If there are two bodies, A and B, which are disposed to combine, the following is the order in which the combinations may take place, beginning with the most simple: namely,
>
> 1 atom of A + 1 atom of B = 1 atom of C, binary.
> 1 atom of A + 2 atoms of B = 1 atom of D, ternary.
> 2 atoms of A + 1 atom of B = 1 atom of E, ternary.
> 1 atom of A + 3 atoms of B = 1 atom of F, quaternary.
> 3 atoms of A + 1 atom of B = 1 atom of G, quaternary.
> &c. &c.
>
> The following general rules may be adopted as guides in all our investigations respecting chemical synthesis.
>
> 1st. When only one combination of two bodies can be obtained, it must be assumed to be a *binary* one, unless some cause appear to the contrary.
>
> 2d. When two combinations are observed, they must be presumed to be a *binary* and a *ternary*.
>
> 3d. When three combinations are obtained, we may expect one to be a *binary*, and the other two *ternary*.
>
> 4th. When four combinations are observed, we should expect one *binary*, two *ternary* and one *quaternary*, &c. . . . (Dalton 1953, vol. 1, pp. 213–14.)

These rules are often referred to collectively as Dalton's "rule of greatest simplicity."[23] It is clear how this rule leads to the law of multiple proportions: when we have only two compounds, they must be a binary and a ternary, and so we have a ratio of 1:2; when we have only three compounds, they must be a binary and two ternaries, yielding the ratios 1:2:4, and so on.

But wait: Dalton does not contend that all compounds actually conform to the "rule of greatest simplicity." Rather, he says that when there is only a single compound composed of a given pair of elements, we should believe it to be binary "unless some cause appear to the contrary," that is, in the absence of any information suggesting otherwise. This qualification does not appear explicitly in any of the subsequent rules, but the terms Dalton uses all refer to what we should believe: "they must be presumed," "we may expect," "we should expect," the rules "as guides in all our investigations." Elsewhere he refers to the rule of greatest simplicity as "the best criterion" for assigning atomic weights (Nash 1956, pp. 64–65).

Dalton occasionally recognizes that these rules do not correspond in a straightforward manner to some regularity among the compounds. For instance, he remarks:

> As only *one* compound of oxygen and hydrogen is certainly known [hydrogen peroxide was not discovered until 1815], it is agreeable to the 1st rule, page 214, that water should be concluded a *binary* compound; or, one atom of oxygen unites with one of hydrogen to form one of water, Hence the relative weights of the atoms of oxygen and hydrogen are 7 to 1. . . . (Dalton 1953, vol. 1, pp. 275–76)

But he continues, one paragraph later:

> After all, it must be allowed to be possible that water may be a ternary compound. In this case, if two atoms of hydrogen unite to one of oxygen, then an atom of oxygen must weigh 14 times as much as one of hydrogen; if two atoms of oxygen unite to one of hydrogen, then an atom of oxygen must weigh 3½ times one of hydrogen. (Dalton 1953, vol. 1, p. 276)

It seems that this possibility is "allowed" by the 1st rule (and not because there may exist two compounds of oxygen with hydrogen), but only because this rule has the doxastic proviso that I have discussed. Likewise, when Dalton considers the three compounds of oxygen with sulfur, experimental evidence leads him to conclude that they are a binary, a ternary, and a quaternary; his third rule would have led him to believe that they are a binary and two ternaries, but the experimental evidence constitutes "some cause . . . to the contrary," and Dalton therefore cannot apply his third rule (Dalton 1953, vol. 1, pp. 383–407). Similarly, when Dalton contemplates the two oxygen-phosphorus compounds, he recognizes that there is some evidence that they are binary and quaternary, and so the proviso "unless some cause appear to the contrary" does not apply. On the basis of other evidence, Dalton eventually concludes that the two compounds are, in fact, binary and ternary, but he does not reach this conclusion by applying his second rule (Dalton 1953, vol. 1, pp. 414–15).

Dalton's contemporaries recognized this proviso in his rules. For instance, when Thomas Thomson introduced Dalton's ideas (one year before Dalton himself published them), he put the rule of greatest simplicity in these terms:

> The hypothesis upon which the whole of Mr. Dalton's notions respecting chemical elements is founded, is this. When two elements unite to form a third substance, it is to be presumed that *one* atom of one joins

> to *one* atom of the other, unless when some reason can be assigned for supposing the contrary. . . . (Nash 1956, pp. 25–26)

So the various rules that Dalton enumerates are not intended to express regularities among the various compounds. Rather, they are inference rules specifying what "should be concluded" when there is no evidence to the contrary.

While these rules are not truth-preserving, they allowed provisional values of the atomic weights to be inferred, which were the basis for further progress. In this respect, Dalton's theory was a striking advance. As Leonard Nash (1956) puts it:

> The rule [of greatest simplicity] is in no way an essential part of an atomic theory as such, but it exercised the vital function of providing the molecular formulas, however mistaken, which were absolutely essential for the operations of an atomic theory that lacked any more rational method for the evaluation of such formulas. Using the formulas provided by his rule Dalton was able to bring the simple abstract concept of atoms to bear upon the difficult concrete problem of determining atomic weights. . . . To the extent that Dalton's arbitrary "rule" was the simplest assumption permitting further progress, it was the best possible assumption. (p. 30)

As inference rules accompanied by the proviso "in the absence of evidence to the contrary," Dalton's rules may be sufficiently accurate for the relevant purposes; they provide a foothold on the atomic weights.

Moreover, where they lead to incorrect values, they provide a means of identifying them as incorrect, since the atomic weight of the same element can be inferred by applying these rules to different groups of compounds. As Dalton remarks:

> It is necessary not only to consider the combinations of A with B, but also those of A with C, D, E, etc.; as well as those of B with C, D, etc., before we can have good reason to be satisfied with our determinations as to the number of atoms which enter into the various compounds. (Dalton 1953, vol. 2, p. 350)

If these inferences lead to different weights, then there is "evidence to the contrary," and so the proviso does not obtain. Later, contrary evidence could also be found by applying Gay-Lussac's law of combining volumes (discovered in 1808) and the Dulong-Petit law (discovered in 1819).[24] As the first step in a self-correcting procedure for discovering the atomic weights, the rule of greatest simplicity is reliable. Other methods for ascertaining atomic weights were themselves originally justified, at least in part, by their yielding many of the same atomic weights as the rule of greatest simplicity.[25]

Now the law of multiple proportions was taken to follow from Dalton's rules. This suggests that the law of multiple proportions was also recognized as accompanied by the proviso "in the absence of evidence to the contrary." In this form, the law of multiple proportions is consistent with "exceptions" such as those involving large hydrocarbons, and so may still be accepted. It is associated with a rule of inference specifying the defeasible expectations we ought to have, the default assumptions we ought to make, when we are engaged in certain kinds of work.

Of course, it does not matter for my purposes whether chemists actually do still believe in the law of multiple proportions, or whether it is now merely a historical relic that is occasionally mentioned for pedagogic purposes. It suffices that, at one time, chemists believed it to be a law, and it was then understood to have the proviso "in the absence of evidence to the contrary." This suggests that belief in the law was not straightforwardly a belief in some regularity, but rather belief in the reliability of an inference rule specifying how we should reason in certain doxastic circumstances. It was the rule's reliability that William Wollaston and others confirmed inductively.

In chapter 8, I argue that certain biological laws also involve provisos of this kind. For instance, I argue that the inference rule associated with "The human gestation period is 38 weeks" licenses our expecting a given pregnancy to last 38 weeks, in the absence of any reason to expect otherwise. The policy of having such a defeasible expectation is sufficiently reliable for certain medical purposes. This "law of medicine" is associated with no regularity other than the reliability of the corresponding inference rule. I also then discuss other issues raised by laws of this sort, such as that a law of medicine may correspond to a mere accident of human evolution.

Appendix: Accounts of laws as inference-licenses

1. Schlick, Ryle, and others

I have argued in this chapter that the root commitment we undertake when believing in a law consists partly in the belief that a certain inference rule is reliable. This may call to mind various reductive accounts—proposed by Moritz Schlick (1931, p. 151), Frank Ramsey (1931, pp. 241, 251), Gilbert Ryle (1949, p. 120; 1963, p. 302), Rom Harre (1960), Norwood Russell Hanson (1969, p. 307), and Stephen Toulmin (1953, p. 84)—according to which laws are rules of inference (in their words, "inference-tickets" or "inference-licenses"). I here review very briefly where these accounts differ from mine. The most important differences arise from the fact that my concern is to account for the laws' special roles in connection with counterfactuals, inductions, and explanations, whereas these other accounts are apparently not so motivated. As far as I can tell, these accounts are not prompted by any genuine differences between laws and accidents. Accordingly, the reasons they give for regarding laws as inference rules, but accidental generalizations as describing matters of fact, do not seem to me persuasive.

The inference rules that Schlick, Ryle, and the other philosophers have in mind seem to take the form "If you believe Fx, then you are *committed* to Gx," whereas the inference rules that I discuss deem you only to be *permitted* (perhaps when certain other conditions obtain) to take x to be G. Furthermore, Schlick, Ryle, and the others regard a law-statement as neither true nor false, since it expresses an inference rule. Moreover, these accounts generally fail to distinguish the inference rules that we have decided to follow from the actual laws of nature. That an inference rule is truth-preserving or even reliable is not required, according to these philosophers, in order for it to be a law; the laws are just the inferential rules that we respect—are simply what we do (recalling Wittgenstein 1958, section 217, which inspires Toulmin and Hanson). Ryle says that the laws are conventions like those that regulate games, etiquette, style, and grammar.[26]

On my account, in contrast, the root commitment that we undertake when believing in a law involves the belief that a given inference rule possesses certain objective properties, such as reliability. This fact must be discovered empirically. We may be unaware of it, and that we all believe in it does not ensure that it is so.

Schlick, Ryle, and the others say that if we adopt "All *F*s are *G*" as a law, then we can infer from *Fx* to *Gx* without using "All *F*s are *G*" as an additional premise (see, e.g., Hanson 1969, pp. 337–44). This is supposedly the fundamental difference between adopting "All *F*s are *G*" as a law-statement and adopting it as an accidental generalization. But these accounts have rightly been criticized (e.g., by H. G. Alexander 1958a, p. 310, and by Hempel 1962, p. 112, and 1965, p. 356) for identifying no feature of actual scientific practice suggesting that scientists regard law-statements and not accidental generalizations as able to be omitted from inferences without rendering them enthymematic.

Toulmin (1953, pp. 78, 86) argues that no law-statement specifies its law's scope; a law-statement therefore is neither true nor false, and so cannot serve as a premise. Toulmin says that in this "respect, laws of nature resemble other kinds of laws, rules, and regulations" (p. 79). In reply to Toulmin, Michael Scriven (1955, p. 127) and Ernest Nagel (1954, p. 406) suggest that law-statements often specify their law's scope. And even if it were otherwise, laws need not be rules; as Alexander (1958a, p. 321) points out, law-statements might merely be "incomplete statements" that are completed when their scopes are supplied. Furthermore, even if "All *F*s are *G*" can fail to appear as a premise in an inference that depends on it without rendering that inference enthymematic, this fact seems irrelevant to that claim's capacity to function as a law-statement in connection with explanations, counterfactuals, and inductions. These special functions are not involved in any of the examples of reasoning with laws that Toulmin (1953, pp. 84–85), Hanson (1969, pp. 337–38), and Harre (1960, pp. 79–80) discuss. Rather, their examples all involve "All *F*s are *G*" simply licensing an inference from *Fx* to *Gx*. Although laws mediate inferences from one singular statement to another, this does not justify calling them "inference-tickets," for accidental generalizations can also perform this role.

No consideration invoked by Ryle and the other philosophers prevents an inference from *Fx* to *Gx* mediated by the law/inference-license "All *F*s are *G*" from being transformed into an inference from *Fx* and the corresponding regularity to *Gx*. Therefore, their accounts of laws as inference-licenses face a troublesome question: If "All *F*s are *G*" states a law, does the *premise* "All *F*s are *G*" state a law? If it does, then there is at least a prima facie problem for any argument that laws must be inference rules because law-statements do not function as premises. On the other hand, if the premise "All *F*s are *G*" does not state a law, then scientists apparently need not bother ever to discover laws; this premise will do. Accordingly, Nagel (1954, pp. 405–406; 1961, p. 138–39), following Rudolf Carnap (1937, pp. 180, 185–86), says that the choice between using law-statements to express inference rules and using them as premises "involves nothing but questions of convenience."

2. Sellars

Wilfrid Sellars also views laws as inference rules.[27] He holds that these inference rules cannot be replaced by general premises; a language must be governed by such inference rules. Some inferential rules are "formal" in that, if they deem a given inference to be good, then they would deem that inference to be good were the nonlogical terms in its premises and conclusion consistently replaced by any other nonlogical terms of the same grammatical kinds. Sellars maintains that there must also be "material" rules of inference. These rules deem a given inference to be good only because of the particular nonlogical terms that are arranged around the logical terms in its premises and conclusion. Sellars argues that without material rules of inference, the proper inferential uses of a claim would depend only on its logical form, not on the particular nonlogical terms it contains. If claims involving these nonlogical terms have only inferential uses (and not, e.g., uses as observation-reports or as justifying some behavior), then these nonlogical terms would lack any significance beyond their mere distinctness from other nonlogical terms. This Sellars believes to be inadequate. Sellars holds that to understand a term is to master its role—to know how to use it. If the only function played by claims in which this term figures is to serve as premises or conclusions of inferences, then to understand this term is to master its inferential role. But then such mastery cannot derive entirely from understanding *formal* rules of inference; there must be some rules of inference *unique to that term* to be the rules our mastery of which constitutes our understanding that term.[28] Sellars nicely summarizes his viewpoint in the title of his paper "Concepts as Involving Laws and Inconceivable without Them" (1948a).

Schlick, Ryle, and the other philosophers discussed in section 1 of this appendix ran into difficulty because they did not show that we need to understand law-statements as inference licenses in order to explain why law-statements can play their special roles in scientific reasoning. In contrast, Sellars's (1958, 1968) account of why laws "support" counterfactuals, whereas accidents do not, turns on the laws' prescriptive character.[29] The law expressed by "All *F*s are *G*" is a rule specifying that if we believe *Fx* then we should conclude *Gx*. Since this is a rule of reasoning expressing the contents of the terms involved, we are under the obligation to follow it whether we are reasoning about the actual world or some counterfactual world. In other words, the rule ranges over "the appropriate . . . senses or intensions" (Sellars 1968, pp. 117–18; see also 1958, pp. 283–84), whereas an accidental generalization "All *F*s are *G*" ranges over only the actual *F*s. Therefore, the law applies even when "*x* is *F*" is used counterfactually.

We can understand roughly what Sellars has in mind here by considering how other rules "support" counterfactuals. The law-statement "All *F*s are *G*" is prescriptive, according to Sellars, in the same way as "We should keep our promises." There is no mystery about why this "moral law" supports counterfactuals: its function is to guide our practical deliberations, so it must apply not only to the promises we actually make, but also to promises that we did not make but could have made. For example, that we ought to keep our promises explains why we would have been obliged to meet Mae at the airport, had we promised to do so. It would be absurd to interpret the fact that we ought to keep our promises as the *regularity* that, whenever someone makes a promise, the promisor ought to perform the promised action, and then to wonder why this regularity still obtains under the counterfactual antecedent "Had I promised to meet Mae at the airport." Sellars apparently believes that it is no better to interpret natural laws as regularities, and no more mysterious why natural laws are preserved under counterfactual antecedents than why this "moral law" is so preserved.[30]

Though this analogy seems to me useful for comprehending Sellars's intent, I do not find that it illuminates the laws' relation to counterfactuals. For one thing, there are counterfactual antecedents under which the laws fail to be preserved, such as "Had copper been electrically insulating, then it would not have been a good material out of which to fashion electric wires" and "Had water ice been denser than liquid water, then life as we know it would not have evolved." Sellars must concede that we can rightly fail to follow a rule that we respect as a material rule of inference. But then what determines the counterfactual antecedents in connection with which we must follow such a rule? Sellars does not say. In section 4 of chapter 2, I made a similar point regarding accounts of laws as relations among universals: they do not suffice to explain the laws' *particular* range of invariance. Sellars's explanation needs something further in order to account for the distinction between the cases where we ought to set aside material inference rules and the cases where we should not; that we respect some claim as expressing a material rule of inference does not explain why we respect its preservation precisely as we do. Whatever explains why an accident (despite its nonprescriptive character) is preserved under a given counterfactual antecedent might well explain why a law is preserved under some counterfactual antecedent. Sellars has thus not shown that the laws' special relation to counterfactuals depends on their having some prescriptive character.

SEVEN

The Root Commitment

1. The proposal

1.1. Review

In this chapter, I offer a proposal regarding the "root commitment" that we undertake when believing in a law. Whether an inference rule possesses the properties that we attribute to it in undertaking this root commitment is a matter of objective fact, in that it obtains whether or not we recognize it as obtaining. However, whether an inference rule possesses these properties depends partly on certain facts about us. In this section, I give an overview of the proposed root commitment and how it accounts for the laws' special roles in science.

In chapter 4, I defined "inductive" confirmation. Suppose our discovery that a given copper object is electrically conductive confirms that all copper objects are electrically conductive. This discovery might confirm various predictions that this hypothesis makes regarding the electrical conductivity of various unexamined copper objects, actual and counterfactual—say, that all copper objects in my hand, had there been any, would have been electrically conductive. Now evidence can confirm a hypothesis without confirming its predictive accuracy at all. Evidence can confirm the predictive accuracy of a hypothesis by confirming its accuracy regarding *some* kinds of unexamined cases, without confirming its accuracy regarding certain *other* kinds of unexamined cases. And even if the evidence confirms, for *each actual* unexamined case, that it accords with the hypothesis, there may be certain "suitable" kinds of unexamined cases that the evidence fails to confirm *would have* accorded with the hypothesis, had there been any cases of that kind. For instance, by discovering a given copper object to be smaller than a cubic mile, we might confirm that all actual copper objects are smaller than a cubic mile. We might even confirm, of *each actual* unexamined copper object, that *it* is smaller than a cubic mile (since we might believe that all actual copper objects formed under roughly similar conditions). But as I showed in chapter 4, we do not confirm that, had there been a copper object constructed by an eccentric billionaire who tried to build a copper object exceeding a cubic mile and had every advantage, so that he failed only if it is physically impossible for a copper object to exceed a cubic mile, then all such copper objects would have been smaller than a cubic mile. A hypothesis undergoes "inductive" confirmation exactly when, for each "suitable" kind of unexamined case, the evidence confirms that all cases of that kind *would* accord with the hypothesis, *were* there any cases of that kind.

Or rather, in accordance with chapter 6, I should have said that the evidence confirms, of each of these conditionals, that it is accurate enough for the relevant purposes, where we tacitly understand the purposes for which a given inference rule is supposed to be applicable. In chapter 5, I suggested that we define "inductive" confirmation in terms of an "inductive strategy." Each inductive strategy seeks a certain kind of reliable rule mediating inferences among claims in U. On each inductive strategy that we carry out, no more than a single hypothesis of the sought-after kind can at any point be "suggested" by the evidence (in a sense that I elaborate in this chapter). In pursuing an inductive strategy, we regard *h* (the salient hypothesis) as capable of being confirmed inductively by the discovery of the success of any of its predictions. The "predictions" made by *h* are the claims $p > m$, $q > (p > m)$, and so on, in U*, where the inference rule corresponding to *h* mediates an inference from *p* to *m*. To confirm *h* "inductively" is to confirm, of each of *h*'s predictions (that we have not already accepted), that it is sufficiently accurate for the relevant purposes, where the prediction's *p*, *q*, and so on, are each logically consistent with the simultaneous reliability of all of the inference rules that our evidence currently suggests on our various inductive strategies. It follows, as I showed in chapter 5, that we can confirm *h* inductively only if *h* follows from the hypotheses currently salient on our inductive strategies. I argued that, because of the indiscriminate character of inductive projection, we are entitled to gamble on any inductive strategies that we wish. Regarding the inductive strategies that we are currently carrying out, we must believe that they may constitute the best set of inductive strategies for us to pursue and that the hypotheses now salient may be exactly the hypotheses at which these strategies ultimately arrive.

1.2. Some sets of inductive strategies are better than others

Suppose, for example, that we set out to discover a reliable inference rule that we can apply whenever we believe some object to be copper, and that allows us to infer whether or not that object is electrically conductive. Suppose that we are willing to regard any successful prediction made by whatever hypothesis of this kind the evidence suggests as confirming that hypothesis inductively. Then our "inductive strategy" works out very well: the evidence quickly suggests that all copper objects are electrically conductive, and since we then confirm, of each unexamined case in a certain broad range, that it would accord with this hypothesis, we justify a great many predictions, all accurate. The result of carrying out this inductive strategy is belief in the reliability of the inference rule corresponding to "All copper objects are electrically conductive." Of course, this inductive strategy thereby also yields belief in the reliability of the inference rules corresponding to "All copper objects in Pittsburgh are electrically conductive" and "All copper objects are electrically conductive or round." But by the "result" of a given strategy, I mean only the hypothesis of the sought-after kind that ultimately becomes and remains salient on that strategy.

By saying that the evidence quickly *suggests* that all copper objects are electrically conductive, I do not mean something psychological, concerning the way our minds work. Rather, I mean that the salient hypothesis expresses what we observe it would take for unexamined cases to go on in the same way as the cases that we have already examined. In contemplating the possibility of an unexamined case that departs from the salient hypothesis (e.g., an unexamined copper object that fails to be electrically conductive), we observe that the case being posited behaves differently from the cases that we have heretofore examined. (At most, then, a single hypothesis can be suggested by the data, since any other hypothesis must involve a departure from that one.) In examining the record of past observations of copper objects, we *see* that "All copper objects are electrically conductive" expresses what it would take for unexamined copper objects to go on in the same way as the copper objects that we have already examined. This belief, in other words, possesses a certain kind of justificatory status; in the manner characteristic of observation-reports, this status requires that there be widespread agreement, among qualified observers who are shown the data, on what would count as an unexamined copper object's being relevantly the same as the copper objects already examined.

Consider, for instance, how Pierre Louis Dulong and Alexis Thérèse Petit argue for their law (roughly: that an element's atomic weight, multiplied by the quantity of heat needed to raise a given mass of that element by a given temperature, gives the same constant for all solid elements). They present a table like Boyle's (reproduced in chapter 1) and then aver that upon reviewing this table, readers will agree with them in recognizing a certain hypothesis—their putative law—as expressing what it would take for unexamined cases to be the same as examined cases. In other words, Dulong and Petit ascribe to this belief the justificatory status of an observation report:

> Mere inspection of these numbers reveals a relation so remarkable in its simplicity that in it one immediately recognizes the existence of a physical law capable of being generalized and extended to all the elements. (quoted by Nash 1956, p. 100)

To clarify this idea (evident also in Leavitt's comments quoted in the caption to Figure 5.2), I return to the copper example. Suppose that we have not yet examined the electrical conductivity of a copper object of precisely 1.234 grams. Then our observations are consistent not only with "All copper objects are electrically conductive," but also with "All copper objects are electrically conductive if they are not 1.234 grams and electrically insulating if they are 1.234 grams." But we observe that the latter hypothesis portrays certain unexamined copper objects—namely, those of 1.234 grams—as behaving differently from the copper objects that we have already examined. We might express this thought by remarking that this alternative hypothesis takes an arbitrary turn or an unmotivated bend. We might ask: "What is so special about 1.234 grams that we should expect matters to be different there?" Later in this chapter, I say more about how our justificatory practices determine which hypothesis of a given kind (if any) is suggested by some evidence.

Pursuing an inductive strategy that seeks a reliable inference rule covering the electrical conductivity of *all* copper objects turns out to be more efficient than pursuing one inductive strategy that seeks a reliable rule covering the electrical conductivity of all copper objects in Pittsburgh and another for copper objects outside Pittsburgh. In pursuing this pair of inductive strategies, we must regard observations of Pittsburgh copper objects as bearing upon each unexamined *Pittsburgh* copper object, but not upon any *other* unexamined copper object. So we must actually observe some copper objects *outside Pittsburgh* before our inductive strategies can justify our making any predictions about objects of that kind. But Pittsburgh copper objects would not mislead us about the electrical conductivity of copper objects outside Pittsburgh; the hypothesis that is suggested by observations of Pittsburgh copper objects alone is accurate to all other copper objects. So this pair of inductive strategies is inferior to a single inductive strategy that allows any examined copper object to bear upon any unexamined copper object there might be, whatever its location.

Consider a different example—the research I mentioned in chapter 5 regarding the period–luminosity relation holding of Cepheid-type variable stars. Rather than pursue a single inductive strategy seeking a period–luminosity relation covering *all* Cepheids no matter what their location, we are better off carrying out a pair of inductive strategies that divides the Cepheids by location. If the Cepheids we initially examine are all from the arms of spiral galaxies, then the hypothesis they suggest will not be accurate to other Cepheids (namely, those in globular star clusters, elliptical galaxies, and the central halos of spiral galaxies). We will be misled. Henrietta Leavitt and Harlow Shapley carried out the broader inductive strategy beginning in 1908. This strategy enabled them to use evidence from only a few nearby Cepheids (all, as it happened, from galactic spiral arms) to arrive at a period–luminosity relation governing even Cepheids in the Andromeda "nebula" (again, as it happened, from its spiral arms), with sufficient accuracy to allow that object's distance to be approximately ascertained for the first time. However, this strategy misled Leavitt and Shapley regarding other Cepheids, as Walter

Baade discovered in the 1940s. Their period–luminosity relation is different. (Today, Cepheids in galactic spiral arms are referred to as "classical" or "population I" Cepheids, whereas Cepheids outside of spiral arms are called "population II Cepheids" or "W Virginis stars.")

Here, then, is my candidate for the root commitment that we undertake in believing that *h* (in U) expresses a law in a given possible world: we believe that one result of carrying out the best set of inductive strategies for us to pursue, in order to learn what we want to know regarding this possible world, is belief in the reliability of the inference rule understood to correspond to *h*.

If we confirm *h* inductively, then *h* must follow from the hypotheses currently salient on our inductive strategies. We must believe that the adoption of those hypotheses may be what ultimately results from pursuing those strategies, which then may be the best set of inductive strategies for us to pursue. Hence, if we confirm *h* inductively, then in virtue of the root commitment, we must believe that *h* may be physically necessary. The intuitions relating lawhood to inductive confirmability are thus borne out.

What does it take for some inductive strategies to form (one grade of) the best set for us to carry out for certain purposes? It must at least be that, if we inductively confirmed the hypotheses that ultimately result from these strategies, then nontrivially all of the predictions thereby confirmed (some of which are counterfactual conditionals) are accurate enough for the relevant purposes. As I showed in chapter 5, this condition is met exactly when Θ's closure in U is nontrivially non-nomically stable (where Θ contains all and only the hypotheses that ultimately result from pursuing these strategies). So Θ's closure in U for (one grade of) the best set of inductive strategies contains exactly the claims possessing (one grade of) physical necessity. The root commitment thus accounts for the laws' special relation to counterfactuals. It also vindicates the intuition (unsuccessfully elaborated by Goodman, Strawson, and Mackie) that we regard *h* as bearing this relation to counterfactuals by virtue of having projected *h* inductively.

From this root commitment, it follows that the actual physical necessities of a given grade would still have formed such a grade had *p* obtained, for any *p* in U that is consistent with those physical necessities. (So had *p* obtained and there been no more grades of physical necessity than in the actual world, then the physical necessities would have been exactly what they actually are.) As I showed in section 3 of chapter 3, this is because, when we confirm a hypothesis inductively, we confirm various nested counterfactuals $p > (q > m)$, $p > (q > (r > m))$, and so on. For instance, in carrying out an inductive strategy on which "All copper objects are electrically conductive" becomes salient, we might (depending on our other inductive strategies) confirm that, had there been nothing at all times but a lone proton, then had the given proton been accompanied by other protons, along with various neutrons and electrons, together forming a single macroscopic copper object, then that object would have been electrically conductive. The nested counterfactuals that we thereby confirm must be sufficiently accurate for the relevant purposes in order for our inductive strategies to be the best set for us to pursue in order to learn about the actual world's facts in U*. For example, the counterfactual $q > m$ that we confirm as holding in the closest *p*-world, when we pursue the best inductive strategies for us to carry out in order to learn about the actual world, must be sufficiently accurate in the closest *p*-world. Therefore, the results of these strategies must span a set that possesses nontrivial non-nomic stability in the closest *p*-world, and thus forms a grade of physical necessities there. Even though there are no copper objects in the closest lone-proton world, "All copper objects are electrically conductive" remains physically necessary there because of what copper objects there *would have been like*.[1]

On the other hand, the actual world's physical necessities do not automatically get carried over to *every* possible lone-proton world. By employing the best set of inductive strategies for us to pursue in order to learn about the actual world, we confirm nothing at all about the counterfactual conditionals holding in some lone-proton world that is *not* the closest. So if we simply stipulate certain claims in U (e.g., "There is nothing forever but a single proton") as holding in a given possible world, then that we regard certain strategies as the best for learning about the actual world does not require us to believe that certain counterfactual conditionals obtain in the given possible world. As I argued in chapter 2, we are free to stipulate the counterfactual conditionals holding in such a world; we can posit various lone-proton worlds in which different counterfactual conditionals obtain. Again, the physical necessity or accidental character of a fact in U in a given possible world fails to supervene on the facts in U there.

1.3. Suggestions from Mill

Whereas David Lewis's account involves a best system of *generalizations* in U, mine involves a best system of *inductions*.[2] If the physical necessities (in U) were understood simply as the members of a nontrivially non-nomically stable set, then although the physical necessities' unique relation to counterfactuals would be identified, this relation's special character and the physical necessities' consequent importance (as I discussed in chapter 5) would not yet be understood. The root commitment concerning the best set of inductive strategies is also needed to distinguish the laws from the physically necessary nonlaws (such as "All copper objects in Pittsburgh are electrically conductive") and to account for this distinction's significance, as I discuss further in section 2 below. The laws are distinguished from the other sets having Λ as their closure in U by virtue of considerations such as the superiority of pursuing a single inductive strategy covering the electrical conductivity of all copper objects regardless of their location.

In emphasizing a best system of *inductions* rather than of *generalizations*, my view is akin to J. S. Mill's.[3] (Lewis also presents his view as akin to Mill's, but he apparently disregards Mill's emphasis on inductions.) Mill (1893) writes:

> [T]here are weighty scientific reasons for giving to every science as much of the character of a Deductive Science as possible; for endeavoring to construct the science from the fewest and the simplest possible inductions, and to make these, by any combinations however complicated, suffice for proving even such truths, relating to complex cases, as could be proved, if we chose, by inductions from specific experience. Every branch of natural philosophy was originally experimental; each generalization rested on a special induction, and was derived from its own distinct set of observations and experiments. . . . [A]ll these sciences have become to some extent, and some of them in nearly the whole of their extent, sciences . . . whereby multitudes of truths, already known by induction from as many different sets of experiments, have come to be exhibited as deductions or corollaries from inductive propositions of a simpler and more universal character. . . . A science is experimental, in proportion as every new case, which presents any peculiar features, stands in need of a new set of observations and experiments—a fresh induction. It is deductive, in proportion as it can draw conclusions, respecting cases of a new kind, by processes which can bring those cases under old inductions. . . . (p. 164; bk. 2, ch. 4, sect. 5)

Notice that the emphasis here is not on the best system of *truths* but on the best system of *inductions*. (Mill also emphasizes, as I do, that a law can be discovered without using one of the best inductions.) Part of what makes a system best is that it does not require "a fresh induction" for cases of a new kind. For example, even if all of the copper objects whose electrical conductivity we have already examined are in Pittsburgh, the best inductive strategy would enable us to confirm an accurate prediction regarding the electrical conductivity of a copper object outside of Pittsburgh. We would not then stand "in need of a new set of observations and experiments"—observations of copper objects outside Pittsburgh—to provide the basis of a "special induction."

Mill's own example (1893, pp. 165–66) involves the as yet "detached and mutually independent generalizations . . . that acids redden vegetable blues, and that alkalies color them green." He anticipates our discovering how to avoid using separate inductive strategies to arrive at these "by showing that the antagonistic action of acids and alkalies in producing or destroying the color blue, is the result of some one, more general law." On the best inductive strategy, evidence consisting only of the effects of acids suffice to suggest a claim making accurate predictions even regarding the effects of bases, just as evidence drawn exclusively from copper objects in Pittsburgh suffices to warrant a claim that makes accurate predictions even regarding copper objects beyond Pittsburgh. Likewise, Mill remarks that the inductions leading to the chemical laws of valence "connect a vast number of the smaller known inductions together, and change the whole method of the science at once . . . by enabling us to a certain extent to foresee the proportions in which two substances will combine, before the experiment has been tried . . . "(pp. 165–66). That is one reason why it is so important for science to find the best inductive strategies—that is, for science to discover which claims in U express laws, not merely which ones express physical necessities. This view obviously also finds *something* right in the traditional intuitions that laws of nature are general, involve no local predicates, take the form "All *F*s are *G*," and so on, without employing some ad hoc means to build these attributes into the concept of a natural law.[4]

For the "weighty scientific reasons for giving to every science as much of the character of a Deductive Science as possible," Mill refers us to the chapter of his treatise entitled "Of Laws of Nature" (bk. 3, ch. 4). He emphasizes there (in section 3) that, when one generalization is shown to be a special case of another, the evidence for the broader generalization contributes to our confidence in the narrower one. But I showed in chapter 4 that evidence for the broader generalization is guaranteed to support the narrower one only if that evidence confirms the broader generalization *inductively*. Otherwise, the evidence may confirm the broader generalization without confirming its predictions concerning certain unexamined cases. (As I noted in chapter 4, Mill believes that inductive confirmability is connected to lawhood.)

Earlier in that chapter (in section 1), Mill writes:

> [T]he following are three uniformities, or call them laws of nature: the law that air has weight, the law that pressure on a fluid is propagated equally in all directions, and the law that pressure in one direction, not opposed by equal pressure in the contrary direction, produces motion, which does not cease until equilibrium is restored. From these three uniformities we should be able to predict another uniformity, namely, the rise of the mercury in the Torricellian tube. This, in the stricter use of the phrase, is not a law of nature. It is the result of laws of nature. . . . If we knew, therefore, the three simpler laws, but had never tried the Torricellian experiment, we might deduce its result from those laws. . . . We should thus come to know the more complex uniformity, independently of specific experience (pp. 229–30)

Here "independently of specific experience" means without requiring a "fresh induction."

That a consequence of the laws may fail to be a law is a point that I explore in section 2 below. It makes sense if the laws are the hypotheses that ultimately become and remain salient on the inductive strategies in the best set. Later Mill emphasizes this point:

> [T]he expression *law of nature* has generally been employed with a sort of tacit reference to the original sense of the word law, namely, the expression of the will of a superior. When, therefore, it appeared that any of the uniformities which were observed in nature, would result spontaneously from certain other uniformities, no separate act of creative will being supposed necessary for the production of the derivative uniformities, these have not usually been spoken of as laws of nature. . . . According to this language, every well-grounded inductive generalization is either a law of nature, or a result of laws of nature, capable, if these laws are known, of being predicted from them. (pp. 230–31)

There can be little doubt that Mill sees the laws as exactly the deliverances of the best inductions.

1.4. Laws of counterfactual worlds

The root commitment I have identified ensures that the actual laws' lawhood (not merely their physical necessity) is preserved in the closest *p*-world where $p \in U$ and Con(Λ,*p*). For "All copper objects are electrically conductive" to be an actual law, the predictions that are confirmed in confirming it inductively must all be sufficiently accurate for the tacitly understood purposes. As I noted in section 1.2 above, these predictions include various nested counterfactuals, such as "Had there been nothing but a lone proton, then had the given proton been accompanied by other protons, along with various neutrons and electrons, together forming a single macroscopic copper object, then that object would have been electrically conductive." So the best set of inductive strategies for learning about the actual world must lead us to beliefs about the closest lone-proton world that are all sufficiently accurate. Moreover, this set of strategies must arrive *most efficiently* at this knowledge of the lone-proton world from observations of the actual world—more efficiently than, say, separate inductive strategies for copper objects within and outside Pittsburgh. Any other set of inductive strategies either leads to predictions regarding the closest lone-proton world that are insufficiently accurate for our purposes, or arrives at sufficiently accurate predictions less efficiently than does the best set of inductive strategies for learning about the actual world. So the best set of inductive strategies for us to pursue, in order to learn what we want to know about regarding the closest lone-proton world, must be the best set of inductive strategies for us to pursue in order to learn about the actual world. For this reason, the laws in the lone-proton world are exactly the laws in the actual world (as long as the grades of physical necessity in this world are no more numerous than the actual grades).

My view is not that the laws in a counterfactual world are the results of the inductive strategies that it would have been best for us to pursue had we lived in that world. Rather, the laws in a counterfactual world are the results of the inductive strategies that are best for *us*—living here in the actual world (i.e., "us" understood rigidly)—to carry out in order to learn what we want to know regarding that counterfactual world. The evidence on which these inductive strategies operate consists of observations of the *actual* world, even though we are aiming for knowledge of a counterfactual world.

But this applies only to those possible worlds that we can learn about by carrying out such inductive strategies. If we design a lone-proton world "from scratch" rather than pick it out with a counterfactual antecedent, then the inductive strategies that we can carry out on evidence drawn from observing the actual world cannot tell us anything about this world. (The same goes for a world called up by a counterlegal supposition.) Insofar as it is unclear just how we are to learn empirically about this world, the distinction between laws and physical necessities there has not been fixed. Along with stipulating that world's facts in U*, we are free to designate as the laws there the members of any set spanning that world's physical necessities in U (which are the facts in U forming a nontrivially non-nomically stable set there). In distinguishing this world's laws from its physical necessities, we imagine ourselves occupying some standpoint from which we are investigating this world; this is implicitly part of the science-fiction story that we are telling in designing the world. For instance, when we say "Let the gravitational force law in this world require that the gravitational force be inverse-cubed," we are imagining the result of carrying out an inductive strategy like the one yielding Newton's inverse-squared gravitational force law in the actual world. But the evidence on which we are imagining the strategy to operate now consists of facts in U concerning this possible world, facts that are somehow being supplied to us.

2. Physically necessary nonlaws

2.1. A neglected phenomenon

It is commonly presumed that a logically contingent, general truth is either a law or an accident. Since a logical consequence of the laws surely is no accident, such a generalization is commonly presumed to be a law.[5] But in fact, some physically necessary generalizations are not laws. Consider these:

1. All copper objects in Pittsburgh are electrically conductive.
2. Anything having a rest mass of 9.1×10^{-28} grams and a negative electric charge has a charge of 1.6×10^{-19} coulombs.[6]
3. All noncyclic alkane hydrocarbons differ in molecular weight by multiples of the atomic weight of nitrogen.
4. Any two Old English Sheepdogs exert on each other an attractive force that (when the dogs' separation is large compared to the sizes of the dogs) is proportional to the product of the dogs' masses divided by the square of their separation.
5. All signals travel slower than *twice* the speed of light.

Each of these generalizations except the third is currently believed to be physically necessary. But intuitively, none is a law. The third was once believed to be physically necessary, and so not accidental, but was not then believed to be a law. That it is no longer believed true (much less physically necessary) is irrelevant to my purpose, which is to show that scientific practice recognizes the category of physically necessary nonlaws.[7]

Take the copper example. Intuitively, this is not a law because it involves the wrong category: all copper objects *in Pittsburgh*. What makes "all copper objects" the right category is obviously a difficult question. For the moment, let us concentrate merely on elaborating the intuition that this is not a law.

Carl Hempel (1965, p. 268) would say that, although this truth is a law, it is not a "fundamental" law since it involves a predicate that essentially refers to a particular location. As I discussed in appendix 2 to chapter 1, difficulties with this requirement include its reliance on a robust notion of a predicate's meaning; its problematic application to biology, geology, and cosmology; and its supplying an a priori argument against Aristotelian physics. It does not seem appropriate for a logical analysis of lawhood to inform us that no law is essentially local. Surely, this could be established only by doing science. That it is logically possible for some laws to be essentially local is emphasized by Michael Tooley's (1977) imaginative example:

> All the fruit in Smith's garden at any time are apples. When one attempts to take an orange into the garden, it turns into an elephant. Bananas so treated become apples as they cross the boundary, while pears are resisted by a force that cannot be overcome. Cherry trees planted in the garden bear apples, or they bear nothing at all. If all these things were true, there would be a very strong case for its being a law that all the fruit in Smith's garden are apples. And this case would be in no way undermined if it were found that no other gardens, however similar to Smith's in all other respects, exhibited behavior of the sort just described. (p. 686)

I shall dispense with any prohibition on local predicates. In any case, it would not explain why examples 2, 3, and 5 above have been believed to be physically necessary but not full-fledged laws. Presumably, whatever is responsible for these examples also accounts for the copper generalization, leaving nothing for a prohibition on local predicates to explain.[8]

The intuition behind the nonlawhood of "All copper objects in Pittsburgh are electrically conductive" is not an intuition expressly concerning locality. It is the intuition that Lawrence Sklar (1993) expresses:

> The discovery of the right concepts with which to describe nature and the discovery of the lawlike relations among these features are processes that cannot be disentangled from one another. The appropriate concepts are the ones that allow the formulation of lawlike regularities among the phenomena so described. (p. 15)

The concept "copper object in Pittsburgh" is not one of the "right concepts with which to describe nature"; it cuts nature too narrowly. The same goes for the concept "Old English Sheepdog" in my fourth example.[9] The relevant categories could likewise be delimited too broadly. For instance, "Any object consisting either of pure iron or of pure gold rusts under . . . conditions if iron, but does not rust if gold" is a physically necessary nonlaw; the category should be subdivided. (Can a similar point be made regarding "No signal exceeds twice the speed of light?")[10] Tooley (1977, p. 681) offers an example with an excessively narrow category: "All salt, when both in water and in the vicinity of a piece of gold, dissolves." Tooley's example involves only nonlocal predicates, unlike the Pittsburgh example. Fodor (1981) offers an example with an excessively broad category:

> [O]ne may not argue from: "it's a law that P brings about R" and "it's a law that Q brings about S" to "it's a law that P or Q brings about R or S". . . . I think, for example, that it is a law that the irradiation of

green plants by sunlight causes carbohydrate synthesis, and I think that it is a law that friction causes heat, but I do not think that it is a law that (either the irradiation of green plants by sunlight or friction) causes (either carbohydrate synthesis or heat). (p. 140)[11]

Fodor seems motivated by the same thought as Sklar:

> [I]f one allows the full range of truth-functional arguments inside the context "it is a law that ____," then one gives up the possibility of identifying the kind predicates of a science with the ones which constitute the antecedents or consequents of its proper laws. (Fodor 1981, pp. 140–41)

In other words, Fodor says, the kind predicates appearing in the antecedents or consequents of law-statements could then "cross-classify the . . . natural kinds" (p. 145)—could be "a heterogeneous and unsystematic disjunction of predicates," as in "All things that are emeralds in Pittsburgh or rubies outside Pittsburgh are green if emeralds, red if rubies." Of course, talk of a kind's "naturalness," "heterogeneity," or "lack of systematicity" is merely an intuition pump, not an account of why a certain category figures in the laws.

I know of only these few brief remarks in the recent philosophical literature that take seriously the possibility that a nontrivial claim in U that (includes no local predicates and) follows entirely from laws may fail to be a law.[12] This is unfortunate. I argue that the distinction between laws and physically necessary nonlaws figures in scientific practice. (So it must, as Fodor emphasizes, in order for the laws to be bound up with the natural kinds.) It is worth noting that this distinction is suggested by the traditional conception of laws as God's commands to the natural world and so, in Hooke's words, as "prescribed by the author of things" (Zilsel 1942, p. 273).[13] The physically necessary nonlaws are decreed by God *indirectly*, as by-products of his express decrees—requiring, as Mill remarks in a passage quoted above, "no separate act of creative will." Furthermore, a physically necessary nonlaw may fail to involve, as Agassiz (1962) expresses it, "the categories of His mode of thinking" (p. 8).[14] The categories "iron-or-gold" and "copper object in Pittsburgh" do not figure in the "Divine Plan of the Creator" just as the categories "lie or covet thy neighbor's wife" and "lie in Pittsburgh" do not figure in the Ten Commandments. Whether taken literally or metaphorically, this conception of natural law leaves room for physically necessary nonlaws.

2.2. Physically necessary coincidences

Consider example 3 above: "All noncyclic alkane hydrocarbons differ in molecular weight by multiples of the atomic weight of nitrogen." The noncyclic alkanes are methane (CH_4), ethane (C_2H_6), propane (C_3H_8), and so on—each differing from the others by C_nH_{2n} (where n is a positive integer). It was once believed to be a law that carbon's atomic weight is 12 units and a law that hydrogen's is 1 unit, in which case C_nH_{2n} would weigh 14n units. It was also once believed to be a law that nitrogen's atomic weight is 14 units. Of course, we now know that an element's atomic weight depends on the relative abundances of its various isotopes, a matter of physically unnecessary fact, and so that the alkane–nitrogen generalization is not physically necessary. (In fact, the integral atomic weights of carbon, hydrogen, and nitrogen are only approximations, so the generalization is also not exactly true.) But before these facts about isotopes were known, many nineteenth-century chemists, such as Max von Pettenkofer, took the example I have given as expressing a physically necessary nonlaw.[15] They termed it "coincidental" in a sense apparently compatible with its being physically necessary. Likewise, von Pettenkofer considered it a law that all oxygen atoms have an atomic weight of 8 units (yes, he thought 8) and a law that all elements figure in "triads" (triplets of elements having many of the same properties) where members of the same triad differ in atomic weight by a multiple of 5 and/or 8 units. But he considered it a coincidence, albeit physically necessary, that these differences are multiples of 5 units and/or the atomic weight of oxygen.

Similarly, a nineteenth-century physicist might have thought it merely a physically necessary coincidence that the velocity of light in a vacuum, dictated by one law, equals the velocity by which disturbances in the electromagnetic field are propagated, as dictated by another, otherwise unrelated law. Classical astronomers, such as Hipparchus and Theon of Smyrna, apparently also recognized this category of physically necessary coincidences. If a planet's natural motions are along an epicycle and a deferent, then a theorem of geometry requires that the planet's net motion be capable of being expressed as an eccentric orbit, perhaps with a moving center (see figure 7.1). Classical astronomers held that this pair of motions then agrees "coincidentally" with the planet's net motion, even though this agreement follows entirely from the planet's natural epicyclic and deferential component motions and a theorem of geometry. That the planet's orbit can be decomposed into an eccentric, perhaps with a moving center, is physically necessary, but this decomposition does not express the planet's natural motions. (This view is discussed in Lange 1994.) Another fact that was once widely thought to be a physically necessary coincidence is that an accurate equation for the photoelectric effect can be derived by treating light as quantized. Einstein, on the other hand, believed that this may be no coincidence; he believed that light may really be quantized. This belief was reflected in his choice of inductive strategies or "heuristics" (as I mentioned in chapter 5; see Lange 1996).

The presence of two old friends in the same restaurant simultaneously is "coincidental" exactly when there is no common cause (e.g., their having arranged to meet there); the explanation of their simultaneous presence consists of an explanation of the first friend's presence plus an unrelated explanation of the second's. Likewise, chemists deemed the alkane–nitrogen claim coincidental (albeit physically necessary) because they believed that its explanation, although composed entirely of laws, contains no law concerning both nitrogen atoms and hydrocarbon radicals. Again, of course, this does not tell us why chemists failed to regard the alkane–nitrogen claim itself as a law.

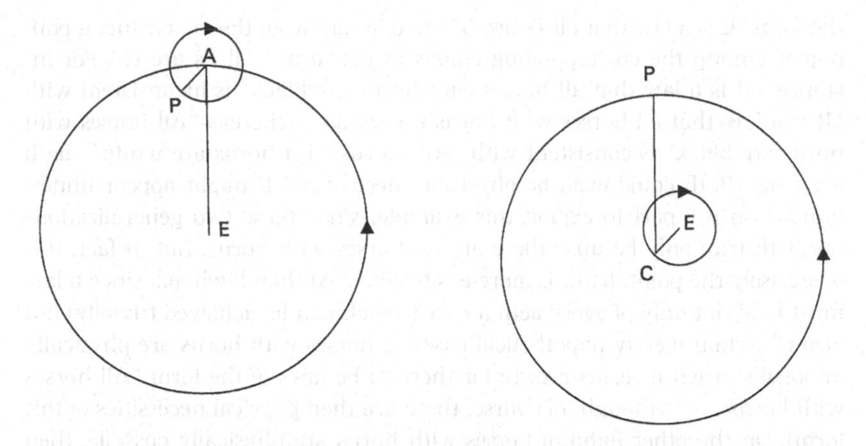

Figure 7.1. The equivalence of epicyclic and eccentric motion. The left drawing depicts a planet P orbiting clockwise along a circular path (called an "epicycle") centered at A. The center of the epicycle is in turn orbiting counterclockwise on a circular path (the "deferent") centered at Earth (E). The planet's net motion resulting from these epicyclic and deferential components could instead be decomposed as in the right drawing. There the planet orbits counterclockwise along a circle centered at C rather than centered at Earth (an "eccentric" circle). C, in turn, orbits clockwise along a circle centered at Earth. As long as the orbital distances and rates are such that AP is equal and parallel to EC, and AE is equal and parallel to PC, the epicyclic scheme yields the same net orbital motion as the moving eccentric does.

The root commitment undertaken when believing in a law should explain why this is.

The examples I have given in which lawhood fails to be preserved by truth-preserving logical operations are not simply further manifestations of the familiar fact that "It is a law that . . . " is a nonextensional context. The nonextensionality of "It is a law that . . . " is exhibited by cases such as this: it is a law that all copper objects are electrically conductive, and in fact the copper objects are exactly the objects in such-and-so spatiotemporal locations, but it is an accident that all objects in those spatiotemporal locations are electrically conductive." To say that "It is a law that . . . " is a nonextensional context is to say that the truth of a claim that includes "It is a law that . . . " is not always preserved when a term within the scope of "It is a law that" is replaced with another term that refers to exactly the same things. In contrast, the alkane–nitrogen example was believed to involve not the intersubstitution of terms that *happen* to be coreferential (such as "The copper objects" and "The objects in such-and-so spatiotemporal locations"), but of terms ("14 units" and "The atomic weight of nitrogen") that are coreferential *as a matter of physical necessity*.

The fact that a nontrivial logical consequence of laws need not be a law seems similar to the fact that the logical incompatibilities among claims of the form "It is a law that all *F*s are *G*" are different from the logical incompatibilities among the corresponding claims of the form "All *F*s are *G*." For instance, "It is a law that all horses with horns are black" is inconsistent with "It is a law that all horses with horns are white," whereas "All horses with horns are black" is consistent with "All horses with horns are white"—both are true. (Both could even be physically necessary.) It might appear underhanded on my part to exploit this example, since these two generalizations are both true only because there are no horses with horns. But in fact, this is precisely the point: truth is more easily achieved than lawhood, since a law must hold not only of every actual case (which can be achieved trivially) but also of certain merely hypothetical cases. If horses with horns are physically impossible, then it seems bizarre for there to be laws of the form "All horses with horns . . . " (though, of course, there are then physical necessities of this form). On the other hand, if horses with horns are physically possible, then it cannot be physically necessary for a horse with horns to possess two properties that logically exclude each other.

Inductive confirmation, as I characterized it in chapter 5, is bound up with inductive strategies; a hypothesis that we confirm inductively must follow from the hypotheses currently salient on our inductive strategies, which we believe may be laws. But in confirming a given salient hypothesis inductively, we may inductively confirm one of its logical consequences even though we believe that it is not a law. For example, every prediction made by "All copper objects in Pittsburgh are electrically conductive" that would have to be confirmed, for this hypothesis to be confirmed inductively, is confirmed when "All copper objects are electrically conductive" is confirmed inductively.

Alternatively, we might believe *h* to be a nonlaw, to follow from the currently salient hypotheses, but to follow from no *single* salient hypothesis. Then there is no guarantee that we are ever prepared to confirm *h* inductively. For instance, consider "Any object that is an emerald or a ruby is green if an emerald, red if a ruby," where "All emeralds are green" and "All rubies are red" have become salient on two of our inductive strategies. Some of the emerald–ruby claim's predictions that would have to be confirmed, for this claim to be confirmed inductively, are not confirmed in the course of the emerald claim's inductive confirmation. In particular, the emerald–ruby claim's predictions regarding hypothetical rubies are not among the predictions made by the emerald claim, and so are not confirmed as part of confirming the emerald claim inductively.

When *h* follows from the salient hypotheses but from no single salient hypothesis, then for each of *h*'s predictions that must be confirmed in order to confirm *h* inductively, there must be some conjunction of salient hypotheses that makes that prediction, where each hypothesis in the conjunction is indispensable to its making that prediction. Perhaps for each of *h*'s predictions, the same salient hypothesis *j* appears in a corresponding conjunction. For example, let *h* be "For each pair of noncyclic alkanes and each nitrogen atom, the molecular weights of the noncyclic alkanes differ by multiples of the atomic weight of the nitrogen atom." When scientists believed that *h* may be physically necessary, then each of *h*'s predictions that had to be confirmed, in order for *h* to be confirmed inductively, was made by the same pair of salient hypotheses: "Any two noncyclic alkanes differ in molecular weight by multiples of 14 units" and "Each nitrogen atom has an atomic weight of 14 units." When there is a single salient hypothesis *j* that is partly responsible for each of *h*'s predictions that would have to be confirmed in order to confirm *h* inductively, then it is easy to see how we might confirm *h* inductively when inductively confirming some salient hypothesis—namely, *j*.[16] But when there is no single salient hypothesis that is partly responsible for each of these predictions made by *h* (as when *h* is the emerald–ruby hypothesis), then it is much more difficult for *h* to be confirmed inductively in the course of our inductive strategies.[17] For any salient hypothesis *j* that we confirm inductively, there is a prediction made by *h* that must be confirmed, for *h* to be confirmed inductively, and that *j* is not partly responsible for making—such as when the prediction concerns hypothetical rubies and *j* is the emerald claim. So this prediction need not be confirmed when *j* is confirmed inductively.[18]

3. Inductive strategies

3.1. How an inductive strategy works

The "root commitment" associated with believing the world to be governed by a given law $p \in U$ is believing that the best set of inductive strategies for us to pursue results in *p*'s inductive confirmation. Having characterized "inductive" confirmation in chapter 5, I now explain more carefully what an "inductive strategy" is, what it is to pursue a "set" of inductive strategies, and what it is for some such set to be the "best" for us to pursue.

An "inductive strategy" is a mode of reasoning by which to justify believing in the reliability of a given inference rule. Scientists who justly believe in a given inference rule's reliability need not have arrived at this belief by carrying out an inductive strategy. But (as I argued in chapter 5) scientists are entitled to pursue inductive strategies, and they are better off pursuing certain sets of inductive strategies than others.

A given inductive strategy seeks a particular kind of inference rule: a rule to be used on certain sorts of occasions for arriving at certain sorts of information. The class of occasions in which the sought-after rule must be applicable may be designated partly in pragmatic terms—for example, the rule is to govern a certain step in a certain sort of calculational procedure to be used for certain sorts of projects. The occasions in which the sought-after rule is to be applicable influence what it takes for a rule to qualify as "reliable" (as I elaborated in chapter 6) and determine the rule's range (e.g., whether it is to cover exactly the Cepheid variable stars, or the variable stars of any sort, or only the Cepheid variables in spiral arms, etc.).

In addition, the occasions in which the sought-after rule is to be applicable determine what the premises of the inference mediated by the rule are allowed to be: they include that the case in question falls within the rule's range (e.g., is a Cepheid variable), and they may include other information that we would be expected to know (to whatever degree of accuracy we would be expected to know it) at the relevant stage in the given calculational procedure as used for such a project. For instance, the means by which we learn about a case in connection with such a project may ensure that we have certain collateral information about it, from which the premises may be drawn. (Considering the way in which we learn of the existence of some Cepheid variable star, we would be expected to know the star's location in the sky, so that we could find it again. We would also be expected to know certain features of its spectrum, since we would have used its spectral features to identify it as a Cepheid.) It may be that for the rule to be applicable, these other premises must have been arrived at by a particular means. The premises may also include other information that we would be expected to know at that stage in our line of inference or as a matter of common background knowledge. One of the premises may be that there are no other "relevant factors"; obviously, this is to be understood differently in connection with different inductive strategies, as I discussed in chapter 6.

The inductive strategy seeks a reliable rule mediating inferences from some such premises to the designated sort of conclusion. (Perhaps the conclusion must specify whether or not the object in question is electrically conductive. Or perhaps it must specify the variable star's period–luminosity ratio.) To pursue this inductive strategy, the scientist examines cases of the relevant kind, assembling information of the sort that may serve as the premises or the conclusion of an inference mediated by a rule of the sought-after kind.

Consider, for instance, Boyle's table of pressures and volumes, reproduced in chapter 1. We think of Boyle as having pursued an inductive strategy seeking a rule covering any kind of gas of a given quantity at a given temperature, in any kind of container, at any time of day, and so on. The sought-after rule was to mediate an inference to the pressure (or volume) of that gas. The premises of the inference were to state that the body in question constitutes gas in the given quantity, its volume (or pressure), and perhaps some of the other information to which we would be expected to have ready access (e.g., its temperature, the shape of the container, the chemical species of the gas, the time of day). The sought-after rule must be reliable for certain purposes. To pursue his strategy, Boyle examined his table of pressures and volumes at a given temperature, and he considered the other sorts of information that might serve as premises—until an inference rule of the sought-after kind became salient.

Once that happens, the scientist pursuing the inductive strategy regards the discovered instances of the salient rule's reliability (i.e., the rule's predictions $p > q$ that have been found empirically to be sufficiently accurate) as *inductively* confirming its reliability. Cases then continue to be accumulated, instances of the rule's reliability confirming it inductively. If evidence ultimately reveals the rule to be unreliable or the rule otherwise ceases to be salient, then data continue to be accumulated until there is again a salient rule of the relevant sort. Its reliability is then confirmed inductively, and the strategy continues until the salient rule's reliability receives sufficient inductive support to justify belief in its reliability and in the accuracy of the counterfactuals thereby confirmed. Then the strategy results in the adoption of these beliefs. This amounts to belief that the inference rule belongs to a set possessing nontrivial non-nomic stability. So we must believe that the inference rule corresponds to a physical necessity. If we continue to pursue this strategy in the belief that doing so is part of carrying out the best set of inductive strategies for us to pursue, then we must believe the inference rule to be associated with a natural law.

When carrying out Boyle's inductive strategy, those very few observations recorded in his table suffice to suggest this rule: we may (for the relevant purposes) apply $PV = k$ to any gas (of the given quantity at the given temperature). Boyle's law (in the words of a standard physical chemistry text) "holds true for the 'permanent' gases under the experimental conditions usually employed in the common laboratory courses in physics, within the precision available in such experiments" (Loeb 1934, p. 140).

Though a very few observations of atmospheric air suffice to suggest this rule, it turns out to be reliable regarding any gas to which the sought-after rule is intended to apply. (And when we confirm it inductively, the counterfactual conditionals that we confirm are all sufficiently accurate for the relevant purposes, if the other inductive strategies that we have chosen to pursue are appropriate.) Therefore, it is better to pursue a *single* inductive strategy, seeking such a rule covering *any* sample of gas, than to pursue different strategies for different gases (air, methane, etc.) or different strategies for gases in containers of different shapes.[19] There is no way of dividing samples of gas into two categories, *A* and *B*, so that observations exclusively of *A* gases are misleading concerning *B* gases (regarding a pressure–volume rule of the sought-after kind), and vice versa. Rather, if observations exclusively of *A* gases ultimately suggest a rule of the sought-after kind but covering only *A* gases (such as "$PV = k$ for methane"), a rule that belongs to a nontrivially non-nomically stable set, then that set also includes the same rule covering just *B* gases ("$PV = k$ for gases other than methane"), and vice versa. Moreover, observations of *B* gases are no more effective than observations of *A* gases at suggesting such a rule covering only *B* gases (and vice versa); it does not take more observations exclusively of *A* gases than observations exclusively of *B* gases. Therefore, it is advantageous to group all gases together for the purpose of finding such a rule. So "Boyle's law" is a law, whereas the rule applying $PV = k$ only to samples of methane is associated with a physically necessary nonlaw.[20]

As Mill noted, a very great advantage may derive from pursuing a single inductive strategy yielding a broader inference rule (as long as that rule belongs to a nontrivially non-nomically stable set). Consider the example from psychiatry that I gave in chapter 5. The 1987 edition of the American Psychiatric Association's *Diagnostic and Statistical Manual of Mental Disorders* (3rd ed., rev.) lists the combinations of symptoms covered by each psychiatric diagnostic category. For instance, the manual co-classifies 5,860 different combinations of symptoms as "autism." In adopting this classification, psychiatrists have agreed to pursue certain inductive strategies—to regard one case of autism as bearing confirmationwise on each other "suitable" kind of unexamined case of autism, no matter which one of the 5,860 combinations of autistic symptoms it displays. The significance accorded to a researcher's observations of certain autistic patients, the sorts of epidemiological studies that researchers decide to pursue, and so on, depend on the psychiatric community's commitment to these strategies. By pursuing a broad inductive strategy regarding autism, psychiatrists may accurately predict some drug's effect on a given autistic patient even if none of the autistic patients on whom the drug has already been tested exhibits this patient's combination of autistic symptoms. (Recall Mill's talk of not needing a "fresh induction" for each kind of chemical reaction.)

On the other hand, co-classifying this patient's combination of symptoms together with the others may be misleading. Then the best set of inductive strategies for psychiatrists to pursue would not include a single strategy covering all 5,860 combinations of symptoms. Autism would turn out *not* to be a species—a natural kind—of psychiatric disorder; psychiatrists would conclude that it lacks "validity" as a diagnosis. (For more on natural kinds, see the close of this chapter.) Different natural laws would then cover patients with different combinations of autistic symptoms, and the conjunction of these laws would be a physically necessary nonlaw covering all cases of autism (see section 3 of chapter 1).

Suppose there is a way to subdivide the 5,860 combinations of autistic symptoms into A combinations and B combinations such that, if we pursue the single broader inductive strategy, then observations of A cases do not contribute toward suggesting a rule that is reliable to B cases. For example, consider any rule that is reliable to B cases, and consider any collection of observations, exclusively of B cases, where these observations fail on this inductive strategy to suggest this rule. Then suppose that this rule is not rendered salient even if we add observations of some A cases, no matter which or how many. This supposition requires, in particular, that if we have observed no B cases, then no matter which or how many A cases we might observe, no rule that is reliable regarding B cases is thereby rendered salient.[21] Suppose also that observations exclusively of A cases do ultimately suggest a rule that is reliable to all A cases. Suppose in addition that observations of B cases analogously fail to help suggest a rule that is reliable regarding A cases but would ultimately suggest a rule that is reliable to all B cases. Then we are better off pursuing separate inductive strategies, one seeking a rule covering only A cases and another seeking a rule covering only B cases, than a strategy seeking a single rule covering both A and B cases.

Incidentally, this is why "All emeralds are grue"—that is, "All emerald slices before the year 3000 are green, and otherwise are blue" (see chapter 4, appendix 2)—is a nonstarter as a candidate for natural law (though it could possibly express a physical necessity). Suppose it belongs to a set that nontrivially possesses non-nomic stability. Then we are better off pursuing one inductive strategy covering emerald slices before the year 3000 and a separate inductive strategy concerning emerald slices after 3000. If we pursued a single broad strategy, then we would be misled by the emerald slices we have observed; they would suggest the green hypothesis. Even after we have observed some emerald slices postdating 3000, the earlier emerald slices do not help to suggest a rule that is reliable to later emerald slices, and vice versa. So even if "All emeralds are grue" is physically necessary, it is not the result of one of the inductive strategies in the best set for us to pursue and (recalling the close of section 2 above) it is probably not confirmed inductively in the course of our carrying out the best set of inductive strategies for us to pursue. The grue hypothesis, even if true, is a nonlaw in virtue of its most striking intuitive difference from the green hypothesis: once we have observed many instances of both and no exceptions to either, we do not see the grue hypothesis as expressing what it would take for unexamined emerald slices to be the same as examined ones insofar as their color is concerned.[22]

Because of lawhood's relation to salience, the grue hypothesis is not a candidate for lawhood. Therefore, our according a zero prior probability to its being a law would not be just as "draconian" (in Earman's words—see my discussion in chapter 4, appendix 1, section 2) as our according its truth a zero prior probability.

3.2. Salience

I now return to Boyle's inductive strategy. We observe that gases at other pressures and volumes would behave in the same way as the gases in Boyle's table if and only if they accorded with Boyle's rule. As I remarked in section 1.2 above, this point does not concern whether Boyle's rule readily crosses our minds as we peruse his table of observations. As I explain below, it may take considerable insight or creativity to hit upon the salient rule in the face of the data that suggest it. Instead, the point is normative: in pursuing this inductive strategy, if we are shown the salient rule's agreement with Boyle's data, then even if we are shown that some other rule's reliability is also consistent with these data, we recognize Boyle's rule as expressing what it would take for unexamined gases to behave like the gases already examined.

Consider, for instance, a rule that agrees with Boyle's ($PV = k$) for the kind of gas (atmospheric air) with which Boyle was experimenting, but takes a different form ($PV^2 = k$) for some second, as yet unexamined kind of gas (e.g., methane). In pursuing this inductive strategy, we would ask, "Why should we expect methane to be different in its pressure–volume behavior from atmospheric air?" Asking this question is a way of pointing to the fact that this rule is not salient. Clearly, this question presupposes some shared sense of what it would be for the two gases to be alike in their pressure–volume behavior. Analogous considerations apply to a rule that agrees with Boyle's for the range of pressures and volumes that Boyle examined when compiling his table, but makes different predictions for certain pressures and volumes outside of this range (e.g., for gases at pressures above 5 atmospheres). It is this shared sense—coupled with the working assumption that all gases (of a given quantity at a given temperature) are alike in their pressure–volume behavior—that determines which rule (if any), among those whose reliability is consistent with our observations, is salient on this strategy.

Any inductive strategy includes some working assumptions of this kind. The strategy must supply background beliefs sufficient to ensure widespread agreement on which rule of the sought-after kind becomes salient. In the case of Boyle's strategy, these background beliefs are not drawn from any particularly sophisticated scientific theory. (Nor are they supplied by any a priori principle.) Our agreement on these background beliefs is like our agreement on those beliefs that tell us how to extend various precedents to new cases, the sort of agreement that (as emphasized by Wittgenstein 1958) is presupposed by any act of ostensive definition. Different inductive strategies can seek the same kind of inference rule but employ different working assumptions; the same evidence may then render different inference rules salient on different strategies. (I discuss an example below.) While the working assumptions informing a given inductive strategy must be enough to determine which rule (if any) of the sought-after kind is suggested by certain observations, they must not be so full as to be question-begging—else scientists would not be entitled to gamble on these assumptions. As Max Planck remarked in the passage I quoted in chapter 5, rationality affords scientists the option of taking certain kinds of risks in return for the prospect of reaping the kinds of rewards that Mill discusses.[23]

What if we seek a pressure–volume rule that is reliable for certain kinds of applications for which Boyle's law is not sufficiently accurate to qualify as reliable? If we pursue an inductive strategy seeking such a rule, then Boyle's rule may initially be rendered salient. But as more evidence is accumulated, it is revealed not to be accurate enough to qualify as reliable for these more demanding applications. It takes many more observations to suggest a reliable rule on such a strategy than it takes for Boyle's rule to become salient on his strategy. For this reason, the best set of inductive strategies for chemistry contains the strategy yielding Boyle's rule, even though it also contains a strategy yielding another rule that can be applied in all cases where Boyle's rule can be applied and even in some where Boyle's rule cannot, and that typically yields more accurate predictions than Boyle's rule does. (See the example involving Hooke's law discussed in section 3 of chapter 1.)

This other rule—the standard equation of state for gases that is more accurate than Boyle's law (but, like Boyle's law, is not perfectly accurate)—is van der Waals's law: for one gram of gas at a certain temperature,

$$(P + a/V^2)(V - b) = k,$$

where a and b are constants specific to the given chemical species of gas (but holding constant at different temperatures), and k is the same constant as in Boyle's law (for one-gram gases). This equation reduces to Boyle's law for $a = b = 0$. The constants a and b have small values (e.g., van der Waals gives $a = .00874$ and $b = .0023$ for carbon dioxide); because a and b are so small, Boyle's law can be sufficiently accurate for certain purposes. Thus, we can regard van der Waals's equation as containing two small correction factors omitted from Boyle's law. Van der Waals's is "an equation of state which stands today as the most *generally* satisfactory approximation to gaseous behavior" (Loeb 1934, pp. 143–44). (Shortly, I discuss what is meant by "*generally* satisfactory.") As J. R. Partington (1949) says: "In spite of its defects it was a very great advance in the theory of gases, and still has a very important practical value. For quick and reasonably accurate orientation in situations where experimental data are lacking, and in giving a qualitative survey of the main properties of gases and liquids, the equation is indispensable, and is likely to remain so" (p. 662). He gives a sense of the equation's "practical value" by supplying the values of a and b for about 30 commonly used gases, values ascertained empirically. Regarding such values, he comments:

> [Considerable] space [in this book] . . . has been used in assembling empirical or semi-empirical formulae which are likely to be of interest to laboratory workers, or to chemists or engineers engaged in large-scale work, who often require quantitative data not available which can be calculated with sufficient approximation for their needs by means of such formulae. . . . Only those who, like the author, have had to search for such aids to research can fully appreciate their value. (p. v)

The inductive strategy yielding van der Waals's equation and belonging to the best set of inductive strategies for chemistry differs from Boyle's inductive strategy not only in seeking a rule that is reliable in certain more demanding applications, but also in being guided by more sophisticated theoretical considerations. The particular correction factors incorporated in van der Waals's equation would not be rendered salient nearly so easily (if at all) on a strategy that (like Boyle's) incorporates little in the way of theoretical motivations. For a hypothesis to be rendered salient on van der Waals's strategy, it must be capable of being derived from certain theoretical assumptions. In particular, van der Waals's strategy has us make the working assumption that a gas is composed of small, hard, roughly spherical particles ("molecules") in ceaseless random motion, occupying a small but finite fraction of the gas's total volume, whose collisions with the container walls are responsible for the gas's pressure, and exerting forces of attraction on one another that increase as the intermolecular distance is reduced; these forces are responsible, when the particles are pushed close enough together, for keeping the particles closely packed, and thereby condensing the gas into a liquid. Once science has reached a certain stage, a scientist has the right to elect an inductive strategy that is guided by this working assumption. For instance, once we are justified in believing that, for each gas, there is some low temperature at which it condenses into a liquid, we have enough reason to speculate that, if gases are composed of particles, then these particles must attract one another when they are close enough together. (Otherwise, this working assumption would have been deemed mere speculation, too question-begging to ground an inductive strategy.) This inductive strategy, then, can be pursued only after certain other inductive strategies have been pursued—for example, one yielding the law that every gas has a condensation point. The strategy yielding van der Waals's equation is part of the best *set* of inductive strategies for us to pursue, some members of which must be carried out before others.

An inference rule of the sought-after kind is a candidate for salience on this strategy only if its reliability can be explained in terms of this working assumption—only if it can be derived by using this working assumption together with the results yielded by the other inductive strategies that have already been pursued. A derivation proceeds by beginning with the reliability of Boyle's law, and then taking into account certain hypothetical influences on the pressure–volume relation that Boyle's law may neglect. Different rules result from different hypotheses about which other influences are so large that they must be taken into account in order to generate a reliable rule, as well as different hypotheses about how these influences are to be captured. In other words, to show that a given rule is a candidate for salience on this strategy, we must show how its reliability can be expressed as correcting Boyle's law for the effects of certain neglected influences—a kind of "plausibility argument."

Van der Waals gives just such a derivation of his equation; he thereby shows it to be a candidate for salience.[24] He then presents data that render it salient. To paraphrase his argument, van der Waals proposes that two further influences are so large that they need to be taken into account in order to generate a reliable rule: (1) the size of the gas particles and (2) the attractive forces that they exert on one another. (Again, other candidates for salience are derived by assuming that one or both of these influences are negligibly small and/or assuming that certain other factors must be taken into account, such as the forces of adhesion between the container wall and the gas particles, the effects of dust particles or other impurities in the gas, the shapes of the particles as they depart from perfect spheres, the inelasticity of the collisions between the gas particles, the shape of the vessel, Earth's gravity, etc.) Van der Waals then derives his equation by showing how it results from making certain approximations for these correction factors:

(i) The gas's volume is mostly empty space, but a small part is occupied by the gas particles themselves. The particles' contribution to the gas's volume can be approximated (for the purposes of generating a reliable rule) by a constant, *b*, that is independent of the pressure, temperature, and so on—that is, that depends only on the chemical species of the gas. (Again, a different candidate for salience would result from making a different approximation for this correction factor—for example, from assuming that a particle's effective radius depends on the temperature, since at higher temperatures, particle collisions are more violent, and so the particles' centers approach one another more closely. Van der Waals assumes in his derivation that this influence can safely be neglected.) It is the total volume minus the gas particles' constant contribution, $V - b$, that reflects the pressure exerted by the gas. Hence, to produce a reliable rule, V in Boyle's law should be replaced by $V - b$.

(ii) Van der Waals assumes that the intermolecular attractive force falls off with distance very quickly, compared to the thickness of the container walls. (Again, different candidates for salience are generated by making different assumptions.) From this assumption, it follows that a particle at the outer surface of the gas feels only negligible attractive forces exerted by gas particles outside of the vessel. Therefore, a particle at the gas's surface feels an unbalanced force back toward the interior of the gas; a particle about to collide with the container wall is slowed by being attracted to the gas molecules behind it, and so hits the container wall with diminished force. The result of this back attraction is thus to diminish the gas's pressure. The back attraction felt by a single particle, under this assumption, is proportional to the density of the gas behind it, and so (assuming that any deviation from uniform density is negligible) to $1/V$. The back attraction's effect on the pressure is its effect on a single particle colliding with the container wall multiplied by the frequency of particle collisions with the container wall, which is again proportional to the density of the gas at the surface layer, and so again approximately proportional to $1/V$. This argument would explain why the component of the total pressure that reflects the gas's volume is $P + a/V^2$—the gas's pressure prior to the diminution resulting from the forces of attraction exerted by the gas particles on one another.

In this way, van der Waals explains why his equation, which replaces P in Boyle's law with $P + a/V^2$ and V with $V - b$, might be reliable: because the two influences that he considers may be the only non-negligible correction factors, and they may be capable of being approximated in this manner. Alternative candidates for salience can be generated by approximating these factors differently, or by taking different influences to be those that must be taken into account in order to generate a first-order correction to Boyle's law. The candidates for salience are the equations having such derivations; observations reveal whether, for the relevant purposes, you can get away with neglecting certain influences and with making certain other approximations.

Because the working assumption behind van der Waals's strategy limits the candidates for salience, fewer observations are needed to suggest van der Waals's equation than would be required to suggest this rule on a more purely "empirical" strategy like Boyle's. Let me show just what happens if we use a strategy like Boyle's in seeking an inference rule reliable for certain applications where Boyle's law cannot be used. In 1847, Victor Regnault performed the most thorough measurements yet on the pressure–volume behavior of fixed quantities of gases at fixed temperatures. He carried his investigations of carbon dioxide up to pressures of 27 atmospheres. Van der Waals later used Regnault's observations in carrying out his own inductive strategy.[25] But these same data suggested a different equation to Regnault because he (like Boyle) proceeded without depending on any particularly sophisticated theoretical framework. Regnault's equation says that for a gram of gas at a given temperature,

$$PV = 1 - A[(1 - V)/V] + B[(1 - V)/V)]^2,$$

where A and B (like van der Waals's a and b) are constants specific to the given chemical species of gas (but holding across different temperatures). Following Regnault, I have expressed his equation in so-called "normal" units—that is, where V is measured in fractions of the volume occupied by one gram of the given species of gas at the given temperature and under pressure of 1 unit (i.e., 1 meter of mercury). (Notice that, in these units, Boyle's law simplifies to $PV = 1$, since if $P = 1$ unit, then by definition, $V = 1$. Thus, Regnault's equation simplifies to Boyle's law when $A = B = 0$.) We have here two inductive strategies, seeking the same kind of inference rule, but proceeding from different working assumptions, and consequently finding the same observations to suggest different inference rules.

Let me show precisely why Regnault's equation depicts gases as going on "in the same way" when this notion is not informed by the theoretical working assumptions of van der Waals's inductive strategy. In other words, I now demonstrate that the appropriate response to van der Waals's equation, in the absence of van der Waals's theoretical motivation, would be to point out that it takes an unmotivated bend (in the same manner as "All copper objects are electrically conductive if they are not 1.234 grams and electrically insulating if they are 1.234 grams" did in our earlier example).

To compare Regnault's equation to van der Waals's equation, we can begin by writing van der Waals's equation in normal units:

$$(P + a/V^2)(V - b) = k',$$

where k' is a new constant. Now we can massage the two equations into similar forms, to render the contrast more perspicuous. First van der Waals's equation:

$$(P + a/V^2)\ (V - b) = k'$$

$$PV - ab/V^2 + a/V - bP = k'$$

$$PV = k' + bP - a/V + ab/V^2$$

Now for Regnault's equation:

$$PV = 1 - A[(1 - V)/V] + B\ [(1 - V)/V)]^2$$

$$= 1 - A/V + A + B\ [(1+V^2 - 2V)/V^2]$$

$$= 1 + A - A/V + B/V^2 + B - 2B/V$$
$$= 1 + A + B - (A+2B)/V + B/V^2$$

These are fairly similar: if we use Boyle's law ($P = 1/V$ in these units) as an approximation good at low pressures, then van der Waals's equation becomes

$$PV = k' - (a - b)/V + ab/V^2,$$

which is made identical to Regnault's equation by setting

$$1 + A + B = k',$$
$$A + 2B = a - b,$$
$$B = ab.$$

It is therefore easy to see how the reliability of both Regnault's and van der Waals's equations could be consistent with Regnault's observations.

Since a and b are small quantities, the ab/V^2 term in van der Waals's equation is a small quantity of higher order than the others, and so for present purposes may be neglected. Since ab is approximately B, the same may be said for the B/V^2 term in Regnault's equation. So we are essentially comparing $PV = k' - (A + 2B)/V$ (Regnault) to $PV = k' + bP - a/V$ (van der Waals). For most gases, a is greater than b (e.g., recall van der Waals's values for carbon dioxide above). For such a gas at relatively low pressures (e.g., carbon dioxide below 27 atmospheres), where P is approximately $1/V$ (since Boyle's law approximately holds), the two equations approximately coincide. As P increases and so V decreases, the a/V term increases more rapidly than the bP (approximately b/V) term. Therefore $bP - a/V$ decreases, and so according to both equations, PV (at a given temperature) decreases as P increases (i.e., V decreases more rapidly than P increases).

Now if we were pursuing the more "empirical" strategy, then (as noted by Loeb 1934, p. 146) we would expect this decrease in PV to continue indefinitely as P increases. This is what it would take for unexamined cases to be *like* examined cases in their pressure–volume behavior—what it would take for the PV curve *to go on in the same way*. This is exactly what Regnault's equation predicts. In contrast, van der Waals's equation does not project this trend indefinitely. So from the viewpoint of the more "empirical" strategy, van der Waals's equation takes an unmotivated bend at higher pressures, and therefore is not rendered salient by Regnault's data (drawn from the relatively low pressure regime). Whereas Regnault's equation portrays PV as approximately proportional to $-1/V$, and so portrays PV as decreasing indefinitely as V diminishes, van der Waals's equation portrays PV as proportional to $bP - a/V$. As the pressure rises beyond the regime in which P is approximately $1/V$, bP increases more rapidly than a/V, and so van der Waals's equation eventually portrays PV as ceasing to fall and beginning to rise with increasing *P*. This bend cannot be anticipated from within the more "empirical" framework on the basis of Regnault's data.

In contrast, the molecular framework informing van der Waals's inductive strategy can explain why this bend occurs. In the relatively low-pressure regime, PV falls with increasing *P*, deviating from Boyle's constant *PV*, because some "pressure" other than *P* is active in affecting the volume—namely, "pressure" from intermolecular attraction. When increased pressure pushes the gas molecules closer together, the volume occupied by the gas decreases, but then there is increased intermolecular attraction, which diminishes the gas volume even further. But when the pressure becomes high enough, the other correction factor identified by van der Waals becomes more pronounced: the gas particles become so crowded together that the space between them is of the same order of magnitude as the size of the molecules themselves. Beyond that point, *V* cannot fall fast enough to keep pace with increasing *P*, and so *PV* begins to rise with increasing *P*. What seems like a change in the gas's behavior, from the perspective of the relatively empirical framework, seems like the gas's going on in the same way, from the viewpoint of the molecular framework. Van der Waals (1890) himself terms Regnault's equation "empirical" (pp. 399, 402), thereby noting that it has no molecular derivation, and so is not a candidate for salience on the molecular strategy.

Thus, to arrive at a reliable rule of this kind, it is better to pursue an inductive strategy that is informed by van der Waals's theoretical working assumption than a more "empirical" strategy: if the data are drawn from relatively low pressures, then the more "empirical" strategy will be led to a rule that is not reliable for relatively high pressures, whereas van der Waals's strategy will not be so misled. In the absence of certain theoretical considerations, it would not be reasonable for us to use the concepts in van der Waals's equation.[26]

The best set of inductive strategies, then, includes one relatively "empirical" strategy seeking a pressure–volume rule, covering any gas, that is reliable for fairly crude purposes. This strategy yields Boyle's law. The best set also includes a strategy proceeding from more theoretical working hypotheses and seeking a pressure–volume rule, covering any gas, that is reliable for certain more refined purposes. This strategy yields van der Waals's law. What if we seek a pressure–volume rule that is more accurate than even van der Waals's law, one that can be used in certain very exacting applications?

Any inductive strategy seeking a single such rule *covering all gases* will be repeatedly misled; observations of one species of gas will suggest a rule that is not reliable for other species of gas. That is because, in order to achieve greater accuracy than van der Waals's law, the rule must include some term reflecting how the intermolecular attractive forces depend on distance. There are many different kinds of forces acting between molecules, and they differ in their dependence on distance. For instance, there are (1) electrostatic forces, diminishing with the square of the distance, between ionized molecules; (2) electrostatic forces, diminishing with the fifth power of the distance, between an ion and a neutral molecule polarized by the field of the ion; (3) electrostatic forces, diminishing with the fourth power of the distance, between molecules possessing permanent dipole moments, such as water molecules in steam; (4) forces, diminishing with the cube of the distance, arising from the magnetic moments of the molecules; (5) electrostatic forces, diminishing with the cube of the distance, between electric dipoles and ions; (6) electrostatic forces, diminishing with the seventh power of the distance, between the dipole moment of one molecule and a second molecule possessing no natural moment but polarized by the field of the first molecule; (7) attractive forces, diminishing roughly with the seventh power of the distance, resulting from the electrostatic interactions of one molecule's electrons with another molecule's nucleus (and contributing to the so-called "van der Waals" force); and (8) forces of quantum origin, accounting both for the formation of chemical compounds (when these forces are attractive) and atoms' "impenetrability" (when these forces are repulsive). For different molecular species, different intermolecular forces predominate. That is, which of these kinds of forces can be neglected and how these forces can be approximated, at the next level of approximation beyond van der Waals's law, varies with the molecular species. Therefore, the pressure–volume behavior of a gas consisting of one molecular species, considered to a degree of precision beyond that of van der Waals's law, will mislead us as we try to predict the precise behavior of a gas consisting of another molecular species.

For this reason, the best set of inductive strategies contains no strategy seeking a single equation covering all species of gas with greater accuracy than van der Waals's law. Rather, for each species of gas, the best set of inductive strategies contains a separate strategy seeking such a rule covering just that species. That is why van der Waals's equation is referred to as "an equation of state which stands today as the most *generally* satisfactory approximation to gaseous behavior . . . [though] it breaks down quite seriously when compared with more accurate results. . . . [O]ne must depart from a single equation of state for real gases applicable in *general* form to any gas and use individual equations differing in detail and applicable only to single gases" (Loeb 1934, pp. 143–44, 140, 177; see also pp. 141, 194–95). Of course, there is an inference rule that combines all of these individual equations: reason in accordance with the equation . . . if the gas is carbon dioxide, and in accordance with the equation . . . if the gas is hydrogen, and so on. But as I said, observations of carbon dioxide do not help to suggest a reliable rule covering hydrogen, for instance, when reliability requires this extreme degree of accuracy. The inference rule that combines all of these equations is associated with a physically necessary nonlaw. (I made a similar point in chapter 1, section 3, in connection with spring laws.)

On the view I am proposing, the laws of nature are associated with the reliable rules rendered salient on certain inductive strategies. That our justificatory practices distinguish the laws from the physically necessary nonlaws may seem to make the laws refer inappropriately to ourselves. But that the concept of a natural law has this token reflexive element does not entail that, had our justificatory practices been different, then the laws of nature would have been different. Nor does it entail that had we not existed, then there would have been no natural laws. These counterfactual conditionals are incorrect—the laws would still have been laws under these counterfactual suppositions—because (as I discussed in section 1.4 of this chapter) the actual laws are associated with the best inductive strategies for us (here, in the actual world, with our actual justificatory practices) to pursue in order to learn about the closest possible worlds where these counterfactual suppositions hold.

4. Natural kinds and laws about laws

Near the start of chapter 2, I introduced the hierarchy of laws governing laws governing . . . governing laws governing non-nomic facts. So far, I have been concerned exclusively with laws governing non-nomic facts—that is, with truths $\blacksquare p$ where $p \in$ U. For instance, I stipulated that only members of U can belong to a set possessing non-nomic stability, and that $\Box p$ holds for $p \in$ U exactly when p belongs to a set that nontrivially possesses non-nomic stability, and that h is confirmed inductively only if $h \in$ U. But science sometimes purports to discover laws $\blacksquare p$ where $p \in U^+$ but $p \notin$ U—in other words, laws *governing laws* governing non-nomic facts. (Perhaps there are laws at even higher levels in the hierarchy.) Here I briefly consider how such laws should be understood, how they are related to natural kinds, and what they have to do with the best inductive strategies for us to carry out.

Some laws about laws may specify formal properties that all laws (in certain categories) possess as a matter of natural law. The following may be meta-laws: that all laws in U of elementary-particle physics are Lorentz-invariant, that they all respect charge-conjugation, parity, time-reversal (CPT) inversion, and that they all involve essentially only nonlocal predicates. (Of course, I have argued that it is difficult to cash out "involve essentially" and "nonlocal," but my concern here is not with whether this is indeed a law, or even well defined, but with the category of laws to which it may sometimes be thought to belong.) The "principle of special relativity" (that the laws of physics cannot provide a way to distinguish one inertial reference frame from another) and the "correspondence principle" (that the results of quantum-mechanical laws approach the results of classical laws at the limit of large quantum numbers) sometimes appear to be considered meta-laws.[27]

There are other sorts of patterns in the laws governing non-nomic facts, patterns that might be thought to exist as a matter of natural law. Consider a property that only a material particle can logically possibly possess, such as the property of being an electron, or the property of being an electron or a muon, or the property of being a material particle in a certain spatiotemporal region. It might be a law that for any such property *F*ness, if there are laws "All *F*s are . . . " attributing some common electric charge, rest mass, and baryon number to all *F*s, then there are also laws attributing to them some common spin, gyromagnetic ratio, lifetime, and set of decay mode probabilities. In other words, it might be a law that for any property *F*ness that only material particles can possess, if there are laws "All *F*s are . . ." of at least *n* of some *m* specified sorts (attributing to *F*s a common rest mass, electric charge, etc.), then there are laws of all *m* sorts.

The electron is a natural kind of material particle, as contrasted with the particles that are electrons or muons, or the particles that are electrons and exist in a certain spatiotemporal region, or the particles that possess electric charge greater than 5×10^{-23} coulombs; these latter are not categories that "carve nature at its joints." To say that the electron is a natural kind of material particle is to say that there exist laws "All electrons are . . ." of each of the *m* sorts stipulated by such a meta-law. That the electron is such a natural kind is then a matter of physical necessity.

There are natural kinds of various sorts: chemical elements, biological species, kinds of disease, minerals, kinds of elementary particle, kinds of force, and so on. For each sort of natural kind, there is a meta-law specifying the *m* sorts of laws of the form "All *F*s are . . ." that must exist in order for *F*s (e.g., gypsum specimens, robins, cases of autism, samples of hydrogen peroxide, component gravitational forces) to be a natural kind of that sort. For instance, it might be that for *F*s to constitute a biological species, there must be laws (with ceteris paribus clauses) specifying the *F*s characteristic anatomy, ecological niche, appearance, diet, and so on. (Laws concerning particular biological species raise certain special difficulties, however, as I discuss in chapter 8.) For *F*s to constitute a kind of disease (i.e., for *F* to have "diagnostic validity"), there must be laws (with ceteris paribus clauses) specifying *F*s characteristic symptoms, prognosis if left untreated, etiology, responses to various drugs and other stimuli, and so on. For *F*s to constitute a chemical species, there must be laws specifying *F*s characteristic elemental composition (with certain classes of exceptions that I discussed in chapter 6, note 13—in connection with the law of definite proportions), melting point and boiling point (in various conditions) or subliming point, reactivities with various other chemical species in various conditions, and so on. That is, for *F*s to constitute a natural kind of a certain sort, there must be a host of different respects (e.g., elemental composition, boiling point, reactivities) in which the laws require all *F*s to resemble one another. I mean hereby to honor the spirit of William Whewell's (1840) classic statement of this conception of natural kinds: "The Maxim by which all Systems professing to be natural must be tested is this:—that the *arrangement obtained from one set of characters coincides with the arrangement obtained from another set*" (vol. 1, p. 521).

For there to be natural kinds of, say, material particles, there must be a meta-law specifying that whenever there are laws "All *F*s are . . ." of at least *n* of some *m* specified sorts (where $n < m$), then there exist laws of all *m* of these sorts. In discovering such a law, we also discover what it takes for *F*s to be a natural kind of material particle. This law says, in other words, that there is a sharp threshold distinguishing the natural kinds of material particles from the other categories that might artificially be carved out of the elementary particles: for some *F*s, there exist none or only a few of the requisite laws, but for any *F* for which there exists a "critical mass" *n* of such laws, there must exist all *m* laws.

Consider another example: perhaps a real kind of force must have a "force law" (which specifies the strength and direction of a component of this force under various conditions), along with laws concerning the characteristic means by which the force acts (e.g., specifying the corresponding kind of virtual exchange particle), the energy possessed characteristically by such fields, and so on. As I mentioned in chapter 6, the Coriolis force has a force law but no laws of the other kinds associated with real forces. That is because it manifests the force accelerating the reference frame, which may be of any kind. Consequently, it is not one of the real forces; it is a pseudoforce.[28]

Let *h* be a meta-law. Now *h* holds in every logically possible world with exactly the same laws in U as the actual world. But in such a world, *h* could be a mere coincidence (albeit a physically necessary one), unlike in the actual world. Consider this example. Suppose it is a law that all material particles are protons or electrons or . . . or muons. There are laws specifying the rest masses of these kinds of particles; suppose *b* is the greatest of these rest masses. It might be a meta-law that, for each property *F*ness covering only particles, if there exist laws "All *F*s are . . . " of at least *n* out of some *m* specified sorts (one sort being a law attributing a common rest mass to all *F*s), then there are laws of all *m* sorts. In that event, it is a physically necessary *nonlaw* that, for each property *F*ness covering only particles, if there exist laws "All *F*s are . . . " of at least *n* out of some *m* specified sorts (one sort being a law attributing a common rest mass *not exceeding b* to all *F*s), then there are laws of all *m* sorts. Likewise, consider a claim "For each property *F*ness covering only particles or only organisms . . . " that goes on to specify the kinds of laws associated with natural kinds of particles and biological species. This claim expresses a physically necessary nonlaw about laws. (It is like "All emeralds or rubies are green if emeralds, red if rubies.") Although any truth is physically necessary if it holds in every possible world with exactly the same laws as the actual world, not every such truth is a law.

Let the set Ψ be the logical closure in U^+ of all of the laws governing laws governing non-nomic facts (such as "All laws of motion are Lorentz-invariant")—that is, all of the claims *h* such that $\blacksquare h$, $h \in U^+$, and $h \notin U$. (So Ψ is a subset of Λ^+. If $\blacksquare j$ and $j \in U$, then $\blacksquare j \in \Lambda^+$ but $\blacksquare j \notin \Psi$.) I suggest that Ψ is preserved under each counterfactual antecedent $p \in U^+$ such that Con(Ψ,*p*). In other words, I suggest that Ψ is stable.[29] For example, had the force laws been different (e.g., had gravity been stronger than it is), the force laws would still have been Lorentz-invariant. The laws about laws are preserved under "Had gravity been an inverse-cubed force," "Had there been a component force that is not ultimately the result of gravitational, electromagnetic, weak nuclear, or strong nuclear interactions," and other such counterlegals—counter, that is, to the laws $\blacksquare p$ where $p \in U$, but not to the laws about laws. Likewise, consider the counterfactual supposition "Had there been hypermesons," that is, mesons more massive than any of the kinds of mesons that actually exist. This counterfactual supposition runs counter to the law that all particles are protons or electrons or . . . or muons—the actual natural kinds of particles. This counterfactual supposition also does not suppose that the hypermeson forms a natural kind of particle; perhaps there are various different kinds of hypermesons. It supposes only that there exist some such particles. Under this supposition, the meta-laws are preserved; had there been hypermesons, then had there been *n* laws "All hypermesons . . . " out of the *m* sorts, there would have been all *m* such laws, and so the hypermeson would have been a natural kind of particle. On the other hand, recall this physically necessary nonlaw about laws: for each property *F*ness covering only particles, if there are laws "All *F*s are . . . " of at least *n* out of some *m* specified sorts (one sort being a law that attributes a common rest mass *not exceeding b* to all *F*s), then there are laws of all *m* sorts. Had there been hypermesons, their rest masses might have exceeded *b*. So this physically necessary nonlaw about laws is not preserved under every $p \in U^+$ such that Con(Ψ,*p*). This nonlaw does not belong to Ψ since the meta-laws alone do not suffice to entail it; they would have to be supplemented by the laws governing non-nomic facts (such as those governing the rest masses of the various particles).[30]

If we believe that there exists a meta-law associated with some pattern in the laws governing non-nomic facts, then by discovering certain laws governing non-nomic facts, we may confirm the existence of certain other such laws. For example, if we believe that we have discovered *n* laws "All cases of autism . . . " of the requisite sorts, then (if we already believe the relevant meta-law) we thereby come to believe that there are laws "All cases of autism . . . " of the remaining ($m - n$) kinds. These "cross-inductions" can bring great benefits. For example, we might be very confident, in advance of having observed any cases of carbon melting, that there is a law specifying carbon's melting point (under standard conditions); our confidence could derive from our having already found many other laws of the form "All carbon objects . . . " and our having already ascertained the relevant meta-law (as a result of having already discovered various laws governing iron, copper, tin, aluminum, etc., including those that specify their melting points). If we are already very confident that there is a law specifying carbon's melting point (under standard conditions), then in carrying out the inductive strategy seeking this law, we will need fewer observations of carbon in order to arrive justly at that law. The best set of inductive strategies takes advantage of the benefits that these cross-inductions can bring.[31] It includes inductive strategies leading to meta-laws along with inductive strategies leading to laws governing non-nomic facts. (I suggest that the root commitment associated with the belief that *h* is a meta-law is the belief that *h* results from carrying out some inductive strategy in the best set.) In carrying out a strategy seeking one of the remaining ($m - n$) kinds of laws, we regard the salient hypothesis "All *F*s are . . . " as confirmed by the discovery of laws "All *F*s are . . . " of at least *n* of the requisite sorts (when we have already adopted the relevant meta-law). Roughly speaking, we regard the discovery of laws confirming that, say, "dissociative disorder" is a genuinely "valid" psychiatric diagnostic category as confirming that various inductive strategies seeking other laws concerning dissociative disorder will succeed.[32]

But in order to treat the "inductive" confirmation of meta-laws, we must expand our original definition of "inductive" confirmation (from chapter 5) to encompass the confirmation of hypotheses that are in U^+ but not in U. To begin with, we can expand our definition of *h*'s "predictions" to cover any *h* in U^+: its predictions are the conditionals $p > m$, $q > (p > m)$, and so on, for any $m, p, q, \ldots \in U^+$ where the inference from *p* to *m* is mediated by the inference rule associated with *h*. Now what is the range of *h*'s inductive projection? That is, which of *h*'s predictions must be confirmed in order for *h*'s confirmation to qualify as "inductive"? When we inductively project the hypotheses ultimately rendered salient on the best set of inductive strategies, the predictions we confirm are all sufficiently accurate for the relevant purposes—that is part of what makes this set the best. I showed in section 1.4 above that if *j* in U is one of these hypotheses, then for any supposition *p* falling within the scope of inductive projection, we believe that these inductive strategies belong to the best set for learning about the closest *p*-world, and so that $p > \blacksquare j$. Hence, if *p* were inconsistent with $\blacksquare j$, we would thereby become committed to a logical impossibility. In my definition of "inductive" confirmation in chapter 5, *p* was restricted to U, and so the requirement that *p* be consistent with the closure in U of the set Θ of salient hypotheses in U sufficed to ensure *p*'s consistency with $\blacksquare j$ (since *j* is salient). But this will no longer suffice now that *p* can belong to U^+ but not to U. Likewise, suppose that *p* is "Had *j* possessed physical necessity of every grade there is," where *j* belongs to U but falls outside Θ's closure in U. Suppose that *p* falls within the scope of the inductive projection of Θ's members where Θ contains exactly the hypotheses ultimately rendered salient on one grade of the best set of inductive strategies. Then Θ contains exactly the laws of one actual grade of physical necessity, and this grade remains a grade of physical necessity in the closest *p*-world. But this grade does not include *j*. So again, *p* had better not fall within the scope of inductive projection.

I now slightly amend the range of "inductive" projection to permit the inductive confirmation of hypotheses in U^+, not merely of hypotheses in U. Consider the candidate meta-laws that are currently salient on our inductive strategies. The predictions $p > m$ made by *h* that must be confirmed, in order for *h* to be confirmed inductively, cannot include *p*s that are inconsistent with those candidate meta-laws. (Otherwise, to confirm inductively one of those candidate meta-laws *h*, we would have to confirm $\neg h > (h \mathbin{\&} \neg h)$.) So let the set Φ include every logical consequence in U^+ of the candidate meta-laws that are currently salient on our inductive strategies—that is, each salient claim in U^+ but not in U—and for any prediction $p > m$ in the range of *h*'s inductive projection, require that *p* be consistent with Φ. (Analogous remarks apply when *h*'s prediction is a nested subjective conditional $q > (p > m)$, and so on.) If the hypothesis *h* belongs to U, then the range of its predictions $p > m$ (where *p* and *m* belong to U) that must be confirmed, in order for *h*'s confirmation to be inductive, is given by the definition of "inductive" confirmation adopted in chapter 5: p must be consistent with the currently salient hypotheses h_i in U. So when $h \in$ U, the h_i must be added to Φ. What if h's prediction $p > m$ involves a subjunctive antecedent p that is in U^+ but not in U? Then for the reasons discussed in the previous paragraph, we must add to Φ the claim that the laws in U are exactly the h_i. Actually, this must be done a bit more carefully to accommodate the existence of various *grades* of physical necessity (and the corresponding grades into which the inductive strategies are organized). Among the hypotheses h_i that are currently salient on the inductive strategies of a given grade, some group form the j_i that are currently salient on those of the above inductive strategies that form a higher grade (if there be one), and some group of the j_i form the k_i that are currently salient on the inductive strategies of an even higher grade (if there be one), and so on. The j_i are candidate laws of a higher grade of physical necessity than the h_i, so when one of the j_i is confirmed inductively, some of its predictions $p > m$ are confirmed where p is inconsistent with the h_i that are not among the j_i. The range of some j_i's inductive projection is restricted by the j_i but not by the other h_i: if $p > m$ is a prediction falling within the scope of some j_i's inductive projection, then p must be consistent with the claim that one grade of laws in U consists of exactly the j_i, another consists of exactly the k_i, and so on.

Accordingly, to define the range of h's inductive projection, let the set Φ be the logical closure in U^+ of each claim in U^+ but not in U that is salient on our inductive strategies (i.e., each candidate meta-law) along with (if $h \in$ U) the claim that the laws belonging to the various grades of physical necessity at or above a given grade are exactly the h_i, the j_i, and so on, where the h_i are the claims currently salient on the inductive strategies of the given grade, and the j_i and so on are the claims among the h_i that are currently salient on the inductive strategies of the next higher grade and so on. (A "lower" grade of physical necessity is closer to the zeroth grade: no necessity at all.) Then hypothesis $h \in U^+$ is confirmed *inductively* by e, a prediction made by h, exactly when e confirms each of h's predictions f such that

pr(f) $\neq$ 1;
if f is $p > m$, then Con(Φ,p);
if f is $p > (q > m)$, then Con(Φ,p) and Con(Φ,q);
and so on.

Notice, then, that because the predictions ultimately confirmed by carrying out the best set of inductive strategies must all be sufficiently accurate for the relevant purposes, the closure in U of the hypotheses in U resulting from these strategies must form a non-nomically stable set, and so form Λ.[33] Furthermore, I have already explained that, because a set's non-nomic stability requires that its members be preserved under various *nested* counterfactual suppositions, the results in U of the best set of inductive strategies for learning about the actual world must form a non-nomically stable set *in the closest p-world* for any p falling within the scope of these inductive projections. Therefore, by the relation between physical necessity and non-nomic stability (discussed in chapter 3), the actual laws in U must be physically necessary in the closest *p*-world—and, moreover, must be laws there (since these inductive strategies are most "efficient" in leading us to correct conclusions regarding that world).

Now the range of inductive projection as just amended includes certain *p*'s in U^+ but outside U. So it follows from the root commitment that the lawhood of the actual laws in U and the physical necessity of the actual physical necessities in U are preserved in the closest *p*-world for any of these *p*s; the actual grades of physical necessity all remain grades of physical necessity in the closest *p*-world. If it is understood (or stated explicitly by *p*) that the grades of physical necessity in the closest *p*-world are no more numerous than in the actual world, then it follows that the laws and physical necessities in that possible world are exactly the same as in the actual world. It follows that had there been an ideal test moa (as discussed in chapter 2), then since "All moas die before age 50" must be physically necessary in order for this moa to die before age 50, it would have lived beyond age 50. [In order for Λ^+'s stability (i.e., Λ^+-preservation) to follow, it would need to be the case that for each of the *p*s in the range of inductive projection, the grades of physical necessity in the closest *p*-world are understood to be no more numerous than in the actual world.]

The ideal test moa (discussed in chapter 2) is posited by the antecedent of the counterfactual conditional whose incorrectness blocks $[\Lambda \cup \{h\}]$-preservation (first revision); this counterfactual conditional requires that *h* ("All moas die before age 50") rather than $\neg\Box h$ be preserved under the supposition of an ideal test moa. This counterfactual conditional's incorrectness was used in chapter 4 to show that a hypothesis can be confirmed "inductively" (in the sense defined in that chapter) only if we believe that it may possess physical necessity. The definition of "inductive confirmation" at which we have just arrived accounts for the intuition that counterfactual suppositions like the one positing an ideal test moa preclude the inductive confirmation of a hypothesis that is believed to lack physical necessity.

Notice that, in inductively confirming a candidate *meta-law*, the set Φ contains none of the candidate laws in U. That the predictions thereby confirmed are all correct means that the meta-laws generate a stable set. So the actual meta-laws remain physically necessary in the closest *p*-world for any *p* within the scope of their inductive projection, including those positing violations of the actual laws in U.

This account can be extended in parallel fashion to laws of even higher levels in the hierarchy of laws governing laws governing . . . governing non-nomic facts—should any such levels arise in scientific practice. The laws governing laws governing non-nomic facts (such as that all laws of motion are Lorentz-invariant) correspond to regularities in the distribution of ■s among the truths in U. The truth but not the lawhood of the laws governing laws governing non-nomic facts belongs to Λ^+—since these truths hold in any possible world with exactly the actual distribution of ■s among the members of U—and these truths generate a stable set. There may be regularities in the distribution of ■s among the truths in U^+ but not U. Some of these regularities may generate a set that possesses a higher order analog of stability involving preservation under the various claims *p* in U^{++} with which that set is consistent (where U^{++} encompasses claims where the scope of a "■" can include any expression in U^+). Inductive strategies (in a sense of "induction" that is revised in the above manner to encompass hypotheses and their predictions in U^{++}) yielding results that collectively span the set possessing higher order stability might permit advantageous cross-inductions and so belong to the best set of inductive strategies. Their results would be the laws at the next higher level of the hierarchy, and so on.

• EIGHT

The Autonomy of Scientific Disciplines and Levels of Scientific Explanation

1. Introduction: every science is "special"

The laws of a given scientific discipline are associated with the inference rules that result from the best inductive strategies for workers in that field to pursue, considering the kinds of evidence that the field has available, the kinds of predictions that it aims to make, and the kinds of applications in the service of which its predictions are to be used. These considerations are common to all of the laws of a given field, but differ for different fields. Accordingly, there are separate laws of physics, of population ecology, of cardiology, of "folk psychology," and so on (see appendix 1 in chapter 1). That different disciplines are "autonomous" (have their own laws) is perhaps most evident when contrasting the laws of a typical "special science," such as human medicine or ballistics or island biogeography, with the laws of fundamental physics. But these latter laws likewise depend for their lawhood on the particular range of concerns characteristic of fundamental physics. In this sense, every science is a "special science."

A discipline's concerns affect what it takes for an inference rule to qualify as "reliable" there. They limit the error that can be tolerated in a certain prediction (as I discussed in chapter 6 in connection with provisos) as well as deem certain facts to be entirely outside the field's range of interests (just as "offstage miracles" fall beyond our range of concern in nonbacktracking contexts). With regard to a fact with which a discipline is *not* concerned, *any* inference rule is *trivially* accurate enough for that discipline's purposes.

Since different disciplines have different concerns, different sets qualify in different disciplines as nontrivially non-nomically stable. (As discussed in chapter 6, "stability" requires only that various counterfactuals be good enough for the relevant purposes.) So the physical necessities differ in different scientific fields. For example, the best inductive strategies for research in cardiology to employ might involve regarding epinephrine's effect on one human heart (where ceteris paribus holds) as bearing upon its effect on every other human heart (ceteris paribus). By pursuing that strategy, a cardiologist justifies a reliable rule licensing a prediction about epinephrine's effect on a human heart to which epinephrine has never before been administered. This is useful in just the way that it is useful for an astronomer to have a means of arriving at accurate predictions regarding Cepheid variables in the Andromeda galaxy on the basis of observations only of nearby Cepheids. The cardiologist also confirms the rule's reliability regarding various counterfactual cases—for example, perhaps she confirms that, had the patient taken epinephrine along with aspirin, it would have had the same effect. Of course, there are limits to the range of counterfactual circumstances with which cardiology is concerned. For instance (as discussed in section 3 of chapter 1 and discussed further below), cardiologists are not concerned with how the human heart would have responded to epinephrine had the human heart evolved differently (just as when we contemplate how Julius Caesar would have conducted Operation Desert Storm in light of his experience in the Gallic Wars, we are not concerned with how Caesar would have managed to be in command in Kuwait in the first place). Accordingly, it is *trivial* for an inference rule to be sufficiently accurate in cardiology regarding what would have happened had evolutionary history been different. Evolutionary biology may tell us that, had certain creatures ancestral to human beings evolved in environmental conditions different from those in which they actually evolved, then epinephrine might later have come to have a different effect on the human heart. But a counterfactual conditional asserting that epinephrine would then have had the same effect is sufficiently accurate for cardiological purposes. So a law of cardiology may fail to be a law of evolutionary biology.

It has often been argued that there are no natural laws in such disciplines as cardiology, neurology, and so on, because facts about human physiology are the products of evolutionary accidents. John Beatty (1981) expresses the concern well:

> "Weber's law" . . . states that the change in stimulation necessary to excite a nerve cell is proportional to the initial value of the stimulation parameter. For instance, if at an illumination of 100 footcandles, an increase of 5 footcandles is necessary to excite an optic nerve cell, then at an illumination of 1000 footcandles, a 10 times greater increase of 50 footcandles will be necessary to excite the cell. . . . But a characteristic which owes its *presence* to natural selection or some other evolutionary agent may, in the future, also owe its *replacement* to natural selection or some other evolutionary agent. . . . Thus, Weber's "law" cannot

> really be a law, since it is evolutionarily possible that different nerve-stimulation mechanisms prevailed in the past, and will prevail in the course of the evolution of species. . . . In short, there can be no law of nature to the effect that a genetically based trait is universal within a species or among all species. All genetically based traits are subject to evolutionary forces. . . . [Similarly] the normal meiotic mechanisms described by Mendel's law, and upon which the Hardy-Weinberg law rests, is itself a genetically based trait subject to evolutionary change, [so its] *universality* is not physically necessary. (pp. 406–407; see also Beatty 1995)

However, that an evolutionary accident is responsible for the reliability of Weber's "law" does not demonstrate that Weber's fails to qualify as a law *of neurology*. Because unrealized evolutionary possibilities are outside of neurology's concerns, Weber's "law" may belong to a nontrivially non-nomically stable set in neurology.

Likewise, though Mendel's laws have exceptions, they are reliable for certain biological purposes. Moreover, they apply to such a wide range of organisms and are rendered salient so readily on certain inductive strategies that they may well result from the best set of inductive strategies for certain disciplines. As Michael Ruse writes:

> [B]ecause [Mendel's] Laws have such a wide range of applications, and because the Laws seem to complement so strongly the physical activities of the cell, it seems reasonable to suppose, as biologists do in fact suppose, that even in unexamined instances, Mendel's Laws are obeyed. (1971, p. 779)

> [T]here can be few statements of sciences which have been found to satisfy the limited condition of unrestricted universality more fully. Since Mendel first proposed his laws, they have been found to hold for a range of organisms from elephants to cod fish, from sea-weed to oak trees. (1973, p. 29)

In sections 2–4 of this chapter, I discuss the character of such laws and the sorts of provisos they possess. This will lead me to examine whether there are laws concerning particular biological species and whether biological species are natural kinds. Ultimately, I contend that Beatty's argument begs the question: it *presupposes* that a biological law must have the same range of invariance—must possess the same kind of necessity—as a law of fundamental physics. I argue that, compared to the laws of fundamental physics, the laws of a special science possess a different kind of necessity—corresponding to a range of invariance that is more inclusive in some respects, less inclusive in others. This distinctive range of invariance enables explanations in special sciences to contribute something that could not be supplied, even in principle, by micro explanations—as I discuss in sections 5–7. That discussion requires an account of the laws' relation to scientific explanations, the third of the laws' special scientific roles introduced in chapter 1 (section 2.3). (Of course, it might in fact be the case that *no* set possesses non-trivial non-nomic stability in cardiology, say. In that event, there are no laws of cardiology, no corresponding natural kinds, and no successful cardiological inductive strategies. Science aims to define fields to which there correspond genuine senses of physical necessity.)

Consider the "laws" of classical physics. Here is one way for them to belong to a set possessing nontrivial non-nomic stability. To begin with, the corresponding inference rules must have provisos limiting their application to relatively low velocities, medium-sized masses, and certain sorts of practical applications. This renders the rules reliable. Yet the non-nomic stability of their closure would require that their reliability be preserved under the counterfactual supposition "Had a material body been accelerated from rest to beyond the speed of light." But the "laws" of classical physics (even with provisos limiting their application to those bodies not so accelerated) might not then still have held.[1] So to the set we must add the laws of relativity theory, quantum mechanics, and so on. The resultant set is inconsistent with the counterfactual supposition "Had a material body been accelerated from rest to beyond the speed of light." So that set's non-nomic stability does not require its preservation under this supposition.

However, there is an alternative way for the "laws" of classical physics to belong to a set possessing nontrivial non-nomic stability. For the purposes of some scientific field that is not concerned with the sorts of applications (involving extreme velocities and masses) under which the "laws" of classical physics break down, such as embryology, it is *trivial* that "Had a material body been accelerated from rest to beyond the speed of light, then classical physics would still have held" is sufficiently accurate. When the discipline itself limits the range of potential applications under consideration, the provisos and the set's other members—which would otherwise be called on to limit this range—can be dropped. In this way, a set can omit some of the laws of fundamental physics and nevertheless be nontrivially non-nomically stable *in embryology*. The physical necessities *of embryology* are what can function *in embryology* as physical necessities in connection with counterfactuals, inductive confirmations, and scientific explanations, and these are what belong to a set that possesses nontrivial non-nomic stability *in embryology*.

Let me be a bit more precise about what it takes for a set to possess nontrivial non-nomic stability for the purposes of a certain discipline D. Each member of the set must be preserved under every subjunctive supposition in U that is consistent with the set *and* that is of interest to D. Of course, under any subjunctive supposition that is irrelevant to D, the set's members are *trivially* preserved. So another way to express the requirement is simply that each member *m* of the set must be preserved under every subjunctive supposition in U that is consistent with the set, but what it takes for *m* to be "preserved" under a given supposition depends on D's purposes. For instance, for the purposes of neurology, Weber's law is *trivially* close enough to the truth regarding the closest possible world in which human evolutionary history is different. Human medicine is not interested in what would have happened had human evolutionary history been different. That is *not* because the

ıave failed to obtain had human evolutionary his-
ıer words, the range of possible worlds in which
ıot expressly limited to possible worlds in which
rwise, it would be *trivial* that Weber's "law" is ac-
e's purposes. In that event, I take it, Weber's "law"
of medicine.
oint. Consider an inference rule that for D's pur-
egarding every possible world. So if a given possible
ge of interests, the rule is accurate enough for D's
ırd to any possible world outside of D's purposes,
e enough for D's purposes there.) But suppose that
poses the rule is accurate enough regarding every
world in which the rule is inaccurate (to a degree
thereby *by definition* outside D's range of interests.
ery possible world, the rule is accurate enough for
hat, with regard to any possible world in which D
ested in those features of the world to which the
those features are "offstage." (Perhaps, e.g., ballis-
ose matters to which the laws of embryology are
ı, it is *trivial* that in every possible world, the rule
purposes. In that event, the rule does not qualify
ıcal necessity in D. That is, for a claim to possess
ıat claim must be of interest to D. Accordingly, let
for a set to possess *nontrivial* non-nomic stability
lude any claims that are *trivially* accurate enough

at a counterfactual supposition positing a different
ry falls outside of medicine's range of interests is
upposition, Weber's law might not still have held.
edicine's practical concerns. It is not quite right to
nedicine is uninterested in what would have hap-
ionary history been different is because medicine
at possible world. Medicine treats no patients from
Still, some counterfactual suppositions are relevant
as others are not. It is of medical interest to know
ght not have been so serious had epinephrine been
ıad the patient long been engaged in a vigorous
he patient been wearing a shirt of a different color,
vaxing. But it is not of medical interest to know
ight not have been so serious had human beings
ent selection pressure. A physician might blame a
on her smoking, but not on human evolutionary
t might not have died on that occasion had human
different in a certain respect. In medicine, human
ver taken as a variable.

Suppose tha
(along with
this supposit
important is
net force and
alizations spe
possessed by
closure in U
ther that the
nomic stabili
held had the
different—ha
tric charge.)
ally possesse
held had the
electric force
now have a
hence three
I have alr
nomic stabili
world's Λ is
subset of Λ
of its physica
are two such
larger set co
be the possib
physical nec
necessity" si
there is a *u*
there is a *uni*
sity at all?
It seems t
cepts of "nat
than one set
laws' *necessit*
argue in cha
rives (at leas
we regard s
necessity—of
I have bee
ally possesse
ing section s
physical nece
than one. Pe
whose preser
the other law

So a supposition that posits a different human evolutionary history is *by definition* outside of medicine's concerns. That is not to say that any possible world where human evolutionary history is different is by definition outside of medicine's concerns. A physician might say that a patient could not have survived his injuries; the gunshot punctured his aorta. That he could not have survived his injuries entails, for instance, that he would have died even if his family had transported him to the hospital sooner. (So a physician might say "He could not have survived his injuries" in the course of reassuring the victim's family that his death was not their fault.) To say that he could not have survived the gunshot, since it punctured his aorta, is to say that in order for him to have survived the gunshot, considering its trajectory, he would have to have been different in a manner that is inconsistent with the laws of human anatomy—for example, the human aorta would have to have been constructed out of very different material, steel perhaps. (Compare this to the bus driver in chapter 2 who could not have known that there was a bomb aboard because, in order for him to have known, considering that nothing suspicious took place, he would have to have had X-ray vision or some such thing. Both counterfactual antecedents have antecedents with implicit clauses rendering them contrary to the laws of medicine.) The counterfactual conditional "Had he survived his injuries, his aorta would have to have been constructed of very different material" is of medical interest, even though its antecedent posits a possible world in which the actual laws of medicine do not hold. In the possible world called forth by this antecedent, human evolutionary history departed from its actual course. But although this possible world is relevant to medicine, it is not the case that physicians are interested in those features of this world that involve human evolutionary history. Those features are offstage; it is irrelevant just *how* the aorta managed to evolve so as to be made of different material, just as in the nonbacktracking context to which I have frequently alluded, it is irrelevant how Caesar managed to be in command of Operation Desert Storm. (Obviously, the same point could be made regarding the actual world; human beings have an actual evolutionary history, but it is of no interest to human medicine.) So facts about human evolutionary history do not possess physical necessity in medicine.[2]

Consider another example. One way to build a non-nomically stable set that includes Galileo's "law" of freely falling bodies is to include in the set *all* of the facts in U. (As I showed in chapter 3, Galileo's "law" is an accident of physics; that Earth's mass happens to be 5.98×10^{24} kg is responsible for the gravitational acceleration of 9.8 m/s^2 in Galileo's "law.") This creates a set that *trivially* possesses non-nomic stability. But we do not have to add to Galileo's "law" *all* of the other facts in U if our concerns are restricted. For example (recalling Galileo's work for fusilliers of the Venetian artillery), perhaps ballistics is not concerned with the trajectory that a body falling to Earth would have taken had Earth's mass been thus and so, or with its trajectory had the primordial nebula that eventually coalesced into the solar system possessed certain properties. That is not because under these suppositions Galileo's "law" fails to be preserved. In other words, it is not the case that a possible world in which Galileo's "law" fails to hold is for this reason outside of ballistics' range of interests. On the contrary, ballistics' interests embrace such counterfactuals as "The projectile would have landed at the desired place and moment had it been the case that bodies in free fall accelerate at 19.8 m/s^2." Galileo's "law" is reliable for the purposes of ballistics since the falls with which ballistics is concerned are from a relatively low altitude, and its practical applications demand only a certain degree of accuracy. Ballistics is not interested in the fundamental reasons for this inference rule's reliability. Ballistics might blame an errant cannon shot on the gun's angle of elevation or the air's humidity, but not on Earth's mass, even though the shot would have reached its target had Earth been, say, 10% less massive, just as it would have reached its target had the cannon's angle of elevation been a certain degree lower.

Even though "Earth's mass is 5.98×10^{24} kg" is for ballistics' purposes accurate enough in every possible world, it is not physically necessary in ballistics. It is offstage. Under any counterfactual supposition, ballistics may with sufficient accuracy say that Earth's mass would still have been 5.98×10^{24} kg.

So to construct around Galileo's "law" a set possessing non-nomic stability for ballistics, we need not add to Galileo's "law" the fact that Earth's mass is 5.98×10^{24} kg. Consider a set that omits "Earth's mass is 5.98×10^{24} kg" and any claims about the solar system's development, and is non-nomically stable for ballistics' purposes. Suppose we add "Earth's mass is 5.98×10^{24} kg" and make the result closed in U. Since the original set is non-nomically stable for ballistics, its members are all preserved for ballistics under any counterfactual supposition with which they are consistent, and so in particular under any supposition that is consistent with them and "Earth's mass is 5.98×10^{24} kg." Furthermore, this claim about Earth's mass is (trivially) preserved (considering ballistics' purposes) under all counterfactual suppositions. So the resultant set is also non-nomically stable.

Recall that I stipulated earlier in this section that the physical necessities in U for ballistics are exactly the claims that belong to a set that possesses non-nomic stability for ballistics' purposes, that contains more than the logical truths and less than all of the truths in U, *and* that contains only claims that are of interest to ballistics. Clearly, this last conjunct makes a difference; without it, claims about Earth's mass would qualify as physically necessary for ballistics.

Notice also that, without this conjunct, there would be no well-defined sense of "grades" of physical necessity in ballistics. The sets possessing non-nomic stability for ballistics' purposes are not well ordered; one need not be a proper subset of the other. For instance, one such set might contain "Earth's mass is 5.98×10^{24} kg" but omit a certain truth concerning the solar system's formation, whereas another such set might contain that truth about the solar system's developmental history but omit "Earth's mass is 5.98×10^{24} kg." In contrast, the sets possessing *nontrivial* non-nomic stability for ballistics (in the sense I have just specified) *are* well ordered; for any two, one must be a proper subset of the other. This can be shown by an argument very like the *reductio* that I originally gave in chapter 3. If r belongs to the first nontrivially non-nomically stable set but not to the second, and r' belongs to the second but not to the first, then ($\neg r$ or $\neg r'$) is consistent with both sets and is within ballistics' range of interest. Again, ballistics must be interested in whether or not r holds, since the first set is *nontrivially* non-nomically stable for ballistics, and similarly for r'. So the fact that one of these fails to hold in the closest possible world where ($\neg r$ or $\neg r'$) does not place the possible world outside of ballistics' concerns. (Recall that ballistics is interested in what would have happened had Galileo's law failed—at least, in *certain features* of that possible world, though not in whether Earth would then have had a different mass or history of formation.) Since one of these sets is not preserved under ($\neg r$ or $\neg r'$), they cannot both be nontrivially non-nomically stable for ballistics purposes.

These ideas allow us to make sense of the concept of a "derivative law." As I remarked in note 8 to chapter 7, typical examples of "derivative laws" in the philosophical literature, such as Galileo's and Kepler's laws, follow from "fundamental laws" only when the latter are supplemented by certain facts that lack physical necessity. (In the case of Kepler's laws, these facts include that the planets' masses are a small fraction of the Sun's.) But though these "derivative laws" are accidents in fundamental physics, they are physically necessary in certain other scientific disciplines (as Galileo's law is a law of ballistics).[3]

Consider another example. Island biogeography is concerned with how the abundance, distribution, and evolutionary tendencies of island species are affected by various factors. It is not interested in how island species would have been distributed had islands (small bodies of land now surrounded by water) not constituted isolated pockets of suitable habitat for those species that live there. For instance, island biogeography is not concerned with how species would have been distributed had a certain kind of creature possessed a tail that when swished automatically propels it discontinuously to the nearest island (no matter how distant it may be) with little expenditure of effort. So it is not a law of island biogeography that there are no creatures with such a tail, nor do laws of island biogeography require provisos specifying that there are no such creatures. Less remotely, suppose that, throughout Earth's history, all islands were frequently for brief periods linked by land bridges to continents. This sort of fluctuation in sea levels is consistent with the laws of physics (unlike the swishing tail); it might be brought on by mini ice ages. Island biogeography is not interested in discovering patterns of species distribution that would still have held had islands lost their isolation so frequently. Furthermore, island biogeography is not concerned with how species would have been distributed had the ranges of island species been originally set and ever since maintained by Divine will (expressed through acts of special creation, among other means).

One of the laws of island biogeography is thought to be the "area law": that the equilibrium number S of species on a given island is an increasing function of the island's area *A*, ceteris paribus. In particular, $S = cA^z$ (to the required degree of accuracy), with constants *c* and *z* specific to a given taxonomic group in a given island group (e.g., Indonesian land birds, Antillean beetles). A 10-fold increase in area is typically correlated with about double the number of species.[4] For the purposes of island biogeography, it is *trivially* accurate enough to say that the area law would still have held had dramatic changes in worldwide sea levels repeatedly taken place in the recent past. Under counterfactual suppositions with which island biogeography *is* concerned, it is *nontrivial* that the area law would still have been reliable. For example, had the island of Mauritius (actual area, 709 square miles) been 7090 square miles in area, then the area law would still have held (and so the number of species inhabiting Mauritius would have been roughly twice as great).

Of course, a counterfactual conditional's correctness is context dependent; although this Mauritius counterfactual conditional is correct in any context with which island biogeography is concerned, there are presumably other contexts in which it is not correct. For instance, the counterfactual supposition "Had either Mauritius not been 709 square miles or $S = cA^z$ failed to hold" could conceivably arise in a discussion in which the distribution of island species is not of special concern—perhaps in the context of some problem in physics, or in connection with the flow of air or water currents near Mauritius. In such a context, it is not at all evident that the correct counterfactual says that the area law would still have held. A law of island biogeography is not a law of fundamental physics; it does not have a special resilience under counterfactual suppositions in every context.

The island-biogeographical laws do not require provisos limiting their application to cases where, say, worldwide sea levels have not fallen and then risen roughly every century, since any case that this proviso would exclude already falls outside of island biogeography's scope. Nevertheless, the laws of island biogeography have *some* provisos. These refer to "disturbing factors" that sometimes arise in actual applications of island biogeography. For instance, species diversity on islands increases with area *ceteris paribus*—as long as, for instance, there is no difference in the islands' distance from a source pool of potential immigrants (namely, a continent).

The physical necessities of island biogeography are nontrivially preserved for the purposes of island biogeography under any counterfactual supposition with which they are logically consistent and that is of interest to island biogeography. Some of these suppositions posit different laws of fundamental physics (as I emphasize in section 6 below). A set that nontrivially possesses nonnomic stability *in island biogeography* does not include, for example, the law of universal gravitation. The area law, for example, would still have held under those violations of the gravity law that belong to island-biogeography's scope and that are logically consistent with the laws of island biogeography. For instance, suppose there had been birds with antigravity organs that assist them a little in becoming airborne. Unlike the creatures with the amazing tails, these birds must fly continuously from region to region, and their flights are not guaranteed to lead them safely to some land. So this supposition, though departing from the actual laws of fundamental physics, belongs to island biogeography's scope. Accordingly, had there existed such creatures, the island-biogeographical laws would still have held. After all, the factors affecting species dispersal would have been no different. For instance, smaller islands would still have presented smaller targets to off-course birds or driftwood-borne seeds and so would have picked up fewer stray creatures as migrants.[5]

It might be objected that on my view, the logical closure of *any* generalization qualifies as nontrivially non-nomically stable in *some* field—one having an appropriately circumscribed range of interests. For example, suppose that all of the human beings who ever drink from this bottle are poisoned. As it happens, the bottle contains arsenic whenever it contains any liquid. Let the "field" be concerned only with whether human beings who actually drink from this bottle are actually poisoned; in other words, its range of concerns encompasses no counterfactual conditionals at all. Of course, had the bottle been emptied, cleaned, refilled with water, and drunk from by someone, then some of the human beings who drink from this bottle would not have been poisoned. But if we are concerned only with whether human beings who actually drink from this bottle are actually poisoned, then the counterfactual "Had the bottle been emptied, cleaned, refilled with water, and drunk from by someone, then all of the human beings who drink from this bottle would still have been poisoned" is *trivially* close enough to the truth. For certain purposes that do not encompass whether this generalization is preserved under counterfactual suppositions, this generalization's logical closure in U qualifies as nontrivially non-nomically stable. So this set's members are all physical necessities of that field!

Although this result may appear counterintuitive, I am not sure that my intuitions about physical necessity are terribly strong when it comes to "fields" like this artificial example. When a field's range of interests encompasses no counterfactuals, the distinction between nontrivial non-nomic stability and mere truth, and so between physical necessities and accidents, ceases to have any importance; *p*'s physical necessity over and above its truth makes no difference to *p*'s role in the field. Inductive projection—projection across a certain range of subjunctive suppositions—is not set apart from mere projection across actual unexamined cases.

Although there is no work for a notion of physical necessity to do in connection with a "field" having interests confined to facts in U, I *am* prepared to recognize notions of physical necessity pertaining to many fields having very specialized interests (but encompassing some counterfactuals). In these fields, certain claims play the roles characteristic of physical necessities. For example, there may be laws of consumer behavior that sociologists and psychologists discover and that designers of stores consult. These laws concern such things as the height on shelves at which people tend first to look for items, the depth into a store that people tend to walk under various circumstances, and the directions in which people tend to turn at the end of aisles.[6]

We might be inclined to say that a law of a given field qualifies as a law *simpliciter* once it has been outfitted with provisos limiting its application to contexts in which our concerns match those of the relevant field.[7] In that case, "All human beings who drink arsenic from this bottle are poisoned" (understood to be applicable only when we are working within the field I have mentioned) is one of the laws *simpliciter*. This just goes to show that the concept of a law *simpliciter* is unimportant to scientific practice and not the concept that figures in our intuitions about the laws. Instead, we use the concept of a law of fundamental physics, the concept of a law of island biogeography, the concept of a law of human physiology, and so on. The results at which I arrived in preceding chapters regarding the root commitment associated with believing that ■*p*, the relation between physical necessity and non-trivial non-nomic stability, and so on, can be reproduced for the concept of a law of a given scientific discipline.

2. Preservation, variation, and laws concerning particular biological species

As mentioned in appendix 2 to chapter 1, two common arguments against the existence of laws concerning particular biological species are that a species is an individual thing and that a species term refers implicitly to Earth. However, these arguments presume the nonlocal-predicate requirement, against which I have already argued (in appendix 2 to chapter 1); in chapter 7, I used other means to capture the intuition that the laws are "general." Accordingly, I concentrate here on a different, more powerful argument against the existence of laws concerning particular biological species: that any such law is ruled out by the relation that laws bear to counterfactuals. (I take this argument to lie behind Beatty's remarks quoted in section 1 above.)

For example, Carl Hempel (1965, p. 175) deems "All ravens are black" to be a law. If Hempel is right and "A raven just flew by" is consistent with Λ^+ (as presumably it is), then (by Λ-preservation, or by Λ's non-nomic stability) had a raven just flown by, all ravens would still have been black, and so a black bird would have just flown by. So far, so good. But coloration typically varies within a species; as the result of random mutations, the chemicals that cause a given pigmentation can be absent from certain individuals. So if in fact all ravens are black, this is an accident; it is the case only because a certain mutation happened never to have occurred. Had a certain mutation occurred, a raven would have lacked certain chemicals, and so there would have been a nonblack raven. In certain environments, nonblack ravens might even have a selective advantage, and the mutant gene for some non-black color might well spread if it ever happens to arise. As William Kneale (1950, p. 123) points out, had a population of ravens survived for many generations in a snowy region, then it might well (have to) have included many nonblack ravens; had this fortunate mutation failed to occur, the population would likely not have survived. According to Λ-preservation (or Λ's non-nomic stability), the counterfactual "Had a certain mutation occurred in a raven, then there would have been a nonblack raven" can hold, where its antecedent is consistent with Λ^+, only if its consequent is consistent with Λ. So Λ cannot include "All ravens are black."

Similarly, consider whether it is physically necessary or accidental that all human beings deprived for x days of food containing vitamin C develop scurvy. In Peter van Inwagen's (1979) words, suppose

> that there is a certain group of biologists and bureaucrats who want to institute a program of selective breeding that is intended to produce a population of human beings who are able to get along without vitamin C. Let us further suppose that wiser council prevails, and these people are dissuaded from this idiotic and immoral undertaking; but suppose that *if* they had been allowed to have their way, they (or their descendants) *would have* succeeded: eventually there *would have been* human beings who did not satisfy ["All human beings deprived of food containing vitamin C for x days develop scurvy]. (p. 446)

Again, we have here a counterfactual antecedent that presumably is consistent with Λ^+ but under which the scurvy generalization is not preserved, showing that this generalization lacks physical necessity.

I agree with van Inwagen that our attitude toward this research project (e.g., if we were deciding whether to fund it) would be different from our attitude toward the project of designing a perpetual motion machine. But I disagree with van Inwagen's conclusion: that the scurvy generalization lacks physical necessity. Scientific reasoning in physiology, developmental biology, and the like does not seem fundamentally different from scientific reasoning in physics. If counterfactuals, inductive confirmations, and scientific explanations function in these fields similarly to the way they do in physics, why should we not expect various biological claims to play the roles in these fields that are characteristic of natural laws?

But how is the possibility of species-specific laws to be reconciled with the existence of the intraspecific variation on which natural selection operates? The most promising approach, it seems to me, is to understand remarks like "Gamma-aminobutyric acid (GABA) inhibits electrical activity of the marginal tentacles of *Cyanea capillata*" and "In *Aurelia aurita* there is a brief period of negative velocity, or backwards motion, during the refilling of the subumbrellar cavity" [both from Arai 1997 (pp. 49, 51), a text concerning the functional biology of jellyfish] as having an important proviso. In recognition of this proviso, such laws are often expressed not as "*All S*s are *T*" (where *S* is a biological species and *T* is a trait) but as "*S*s are *T*" (e.g., "Chickadees have white eggs with reddish brown dots") or as "The *S* is *T*" ("The human being has exactly 10 fingers") or as "An *S* is *T*" ("A grizzly bear is a fearsome carnivore"). It is not true that all lions are tawny, but intuitively, "The lion is tawny" is consistent with the existence of albino lions.[8] The proviso in the scurvy law is violated by the supposition that there is a program of selective breeding (or recombinant DNA research) aiming to produce human beings who do not need vitamin C, and the proviso in the raven law is violated by the supposition of a raven with a certain gene (or descended from a population that has long survived in a snowy environment).

These laws (each with its proviso) are preserved under these counterfactual suppositions, just as Λ-preservation and Λ's non-nomic stability demand; the corresponding inference rules are reliable regarding the closest possible worlds thereby posited. But it does not follow that, had the scientists carried out their program, then the resulting human beings would have needed vitamin C, or had a raven possessed a certain gene, then all ravens would have been black. The conclusions of these counterfactuals are not licensed by the inference rules applied to their counterfactual antecedents, because these antecedents violate the relevant rule's proviso and so fall outside of that rule's range of application. But what precisely would the relevant proviso be?

3. What is the proviso?

One line of thought is that "The *S* is *T*" ascribes *T*ness to all healthy *S*s, or perhaps instead to all nondefective *S*s, or to all *S*s that are the way an *S* ought to be, or to all *S*s that are functioning well or flourishing in the manner that is characteristic of *S*s. (These proposals are not synonymous. For example, in the salmon's life cycle, the salmon dies after spawning. As a salmon is dying, it obviously is not healthy, but perhaps it is doing what it ought—as a salmon—to be doing.) Advocates of this line include Gottlob Frege (1984, p. 185) and Philippa Foot (1989). Nicholas Wolterstorff (1970) is typical in holding that, when we say "The *K* is *L*" in this sense, we are sometimes claiming that most if not all *K*s are *L*, but on other occasions we are instead intending to say

> what is true of normal, properly formed K's. . . . In assertively uttering, for example, "Lions have four feet" (or "The Lion has four feet") one is simply claiming, I should think, that every properly formed lion has four feet. It might be the case that most lions did not have four feet, that most of them were maimed, and still be true that lions have four feet. (p. 170)

I do not believe that any proposal along these lines expresses the right way to understand the proviso in typical claims concerning particular biological species. To begin with, although the human being has two legs, not every healthy human being has two legs; an amputee can be perfectly healthy (or so my wife, a physician who works with amputees, tells me). If the proviso instead specifies the human being to be not only healthy but also uninjured, then this cannot do for "The human being has exactly 10 fingers," since polydactyly is compatible with being healthy and uninjured (as was Anne Boleyn—before being beheaded—despite having a sixth finger on her left hand). Furthermore, I see no reason why polydactyly should preclude a human being's qualifying as "functioning properly." By the same token, I would not regard a human being who could synthesize her own vitamin C (or who could metabolize without vitamin C) as unhealthy, defective, or improper in form or function, yet she would be an exception to the scurvy law. Admittedly, I would characterize such a human being as abnormal. But then we need an account of what it is for a member of *S* to be a "normal" *S*. Being a normal *S* is not the same as being a healthy *S*, without defect, formed and functioning properly. So what is it to be a "normal" or "normally formed" *S*?

It is also worth noting that "The *S* is *T*" may be true although no (*S* & *T*)'s are nondefective, healthy, or proper in form or function; consider "The human being with trisomy 21 has mental retardation" and "The human being with chronic rheumatoid arthritis has ulnar deviation." These laws cannot have as provisos ". . . as long as the human being is healthy, uninjured, and properly formed." (Nor does it make sense to say that any human being who has chronic rheumatoid arthritis but is *otherwise* healthy also has ulnar deviation.)

Perhaps these difficulties could be overcome, but there are others. Although the robin has greenish blue eggs, this color may not be required for the robin's health or flourishing. Perhaps this color is just a result of genetic drift; it may have nothing to do with any adaptive trait. Of course, it may be adaptively neutral in itself and nevertheless be a side effect of some mechanism that contributes to a robin's health. Yet it seems implausible to me that those naturalists who discovered the robin to have greenish blue eggs committed themselves to this character's bearing some relation, direct or indirect, to the robins' good health or proper functioning.

For that matter, "The *S* is *T*" may be true even when *T*ness is intuitively detrimental to an *S*'s flourishing. The leaf cells of various species of sorghum synthesize cyanide, bound in a safe form as a cyanoglycoside, and sequester it in vacuoles, from which it is released only when the tissue is damaged, as when insects crush it in their mouths. Upon the cyanoglycoside's release, its hydrolysis by enzymes in the cytoplasm results in toxic hydrogen cyanide gas (HCN). Insects, such as locusts, are deterred from feeding on the sorghum by the gas thereby released (Harbourne and Turner 1984, pp. 75, 93; Woodhead and Bernays 1977, p. 235). Since HCN derives its toxicity from its capacity to bind with and thereby inactivate the terminal oxidase of the mitochondrial electron-transport chain, it is just as toxic to the sorghum itself as it is to locusts (Harbourne and Turner 1984, p. 75). It can be released into the plant when the vacuole is damaged by low temperatures (Jones 1973, p. 236; Loyd, Gray, and Shipe 1971, p. 139). In species of sorghum that live in environments where it frequently turns cold, plants characteristically suffer chronic enfeeblement from slow self-poisoning by cyanide. Yet these species remain cyanogenic, since the cost of being cyanogenic in cold environments fails to outweigh the advantage of discouraging herbivory. Thus, these species of sorghum manufacture cyanide—indeed, this trait is a characteristic marker of many plant species (Harbourne and Turner 1984, p. 94; Jones 1973, p. 236)—though this trait renders these sorghum plants unhealthy, impairs their proper functioning, and is detrimental to their flourishing. Of course, to the extent that cyanide production contributes to a plant's reproductive success, the plant's *mechanism for producing cyanide* is functioning well, producing neither too much nor too little. But the poison (albeit selectively advantageous) is detrimental to a *plant's* flourishing, I would say, for the same reason as it is detrimental to a herbivore's. The plant would truly flourish only in an environment without herbivores, where cyanogenesis could be dispensed with.

To reassure ourselves that cyanogenesis does not actually contribute to a sorghum plant's flourishing, compare this case to others that appear more straightforward. It seems safe to say that a plant's possession of chlorophyll contributes to its flourishing. How does sorghum's cyanogenesis differ from its possession of chlorophyll? It might be suggested that chlorophyll is basic to the plant's means of extracting energy from its environment—to its way of "making a living"—whereas the cyanide is a defense mechanism, allowing the plant to overcome an environmental hazard. But the rose's thorns are also a defense mechanism, and yet its possession of thorns seems to bear the same relation to its flourishing as does its possession of chlorophyll. It seems to me that the reason cyanogenesis intuitively impedes sorghum's flourishing is that the cyanide is poisoning the plant, whereas the chlorophyll and the rose's thorns are not doing anything to harm the plant. This proposal might seem inconsistent with the view that a salmon's dying after spawning is intuitively part of its flourishing. But the salmon's dying after spawning is a stage in its natural "life cycle." It is the last stage, to be sure, but death (at its proper time) is intuitively part of a life that exemplifies flourishing as an animal. A sorghum plant living in an environment where it frequently turns cold is poisoning itself from time to time—is enfeebled not at some particular stage in its "life cycle," but now and again throughout its life. Intuitively, the salmon's way of life involves its dying after spawning. But part of the way of life of a sorghum species that characteristically lives in environments where it frequently turns cold cannot be being sick, *if* a sorghum plant's "being sick" just means its failing to flourish in the sense appropriate to that species and so failing in some respect to lead the way of life characteristic of that species!

This reveals the basic difficulty with trying to cash out "The *S* is *T*" by appealing to some notion of flourishing in the sense characteristic of the given species. That each species is associated with a sense of "flourishing," bound up with its characteristic "way of life," depends on a romanticized view of nature: that each species' way of life represents an elegant solution to the problems of making a living in a given environment. Selection for reproductive success acting on random variation may produce nothing of the kind. Sorghum manages to survive by poisoning itself enough to discourage herbivory but not enough to kill itself. The mechanism is inelegant; surely one can imagine a better way. (Why not thorns?) It involves no "flourishing" in any recognizable sense, and so in particular, there is no such thing as "flourishing in the manner of sorghum." Sorghum's strategy is merely what selection has happened to come up with: chronic poisoning to avert early death.

Suppose we could purge any implicit reference to "flourishing" from our intuitions regarding the "way of life" characteristic of a species. We might then try a different approach to understanding "The *S* is *T*." Perhaps "Sorghum produces cyanide" means that cyanogenesis is part of the "way of life" characteristic of this species, without any implication that this "way of life" constitutes an elegant solution to the problems of making a living or that sorghum is "flourishing" in any sense by leading this "way of life." But this analysis of "The *S* is *T*" seems unlikely to be enlightening (in just the way that reference to a "normal" *S* is unilluminating without some account of what "normal" means). What makes an individual's possession of a given property essential to its participating in the "way of life" of its species, when there is variation among the members of that species with respect to this property? This was the sort of problem that reference to health, lack of defect, or flourishing was supposed to solve; it is not clear how reference to the way of life characteristic of a species (isolated somehow from intuitions about "flourishing") will help. For instance, one intuition is that a species' way of life involves the means by which such creatures manage to carry out those activities that are necessary for them to live. For instance, the beaver requires energy to live, so the beaver's way of life involves the means by which it captures and assimilates food, from dam building to the Krebs cycle. But this characterization of a species' way of life fails to fix which claims of the form "The *S* is *T*" are true, since there is variation in the means by which *S*s satisfy these needs, not to mention variation among *S*s in these needs themselves. Moreover, some truths of the form "The *S* is *T*" do not concern traits that are necessary for survival. Some concern traits that are necessary for reproduction, and others concern traits that are not necessary for anything important. Again, the greenish blue color of the robins' egg may well play no part at all in the robins' way of life in the above sense.

It might be thought that "The human being develops scurvy when deprived for *x* days of food containing vitamin C" could be understood as a generalization covering all human beings with the proviso that the individual has not undergone any recombinant-DNA procedure, been the product of selective breeding (as in van Inwagen's example), or otherwise had his or her genome artificially manipulated. Likewise, "The robins' egg is greenish blue" might be thought to mean that every robins' egg is greenish blue as long as it was laid by a robin who had not resulted from some human intervention intended to alter its genome directly or to exert some kind of selection pressure on its lineage. But this approach does not work, and for reasons going well beyond the fact that a proviso focusing exclusively on the robins' genome must be supplemented to accommodate not only the fact that eggs may be painted, but also the fact that the genes responsible for the egg's color may yield a greenish blue egg only under certain environmental conditions. (The phenotypic expression of a genotype can be dramatically different in different circumstances; e.g., whether some aphids grow wings depends on how crowded they are at a certain period in their development.) The principal defect in this proposal is that it is simply not the case, when the *S* is *T*, that every *S*'s genome yields trait *T* (in certain environments, and in the absence of paint, injury, etc.) as long as it has not been artificially manipulated. Again, natural selection presupposes that at any moment, there is natural genetic variation among the members of a species. It is not the case that an *S*'s genome must be altered artificially in order for it to depart from a uniformity in the genes of all of the other *S*s. Typically, there is no such uniformity. Of course, we might specify the entire sequence of bases composing a given genome, along with the chemical pathways responsible for gene expression, and then say that all eggs laid by creatures having this genome are greenish blue under a stipulated range of conditions. But then we would no longer have a law concerning a particular biological species; reference to the species has been rendered superfluous. And, of course, not all robins have the specified genome (see Beatty 1995).

Again, the existence of such variation is precisely one of the fundamental reasons to deny that there are any laws concerning particular biological species. Alexander Rosenberg (1987), for instance, so argues (and then elevates the absence of such laws to a logical truth by appealing to the nonlocal-predicate requirement):

> There are no laws about particular species. . . . This fact is reflected not only in the role of specimens, but also in the decline of essentialism among biologists: variation, as Mayr has pointed out, is not viewed as a disturbance from some mean property of members of a species which provides its essence; it is viewed as the normal result of recombinations within a lineage. The generalizations about particular species on which taxonomic decisions rest are full of exceptions, and there is no background theory that will enable us to eventually eliminate, reduce, or explain these exceptions. This should be a major embarrassment for biology [unless] species are not kinds, they are individuals. As such, we can no more expect laws about particular species than we can expect *laws* about what counts as an instance of Napoleon Bonaparte or Mount Rushmore or Third French Republic. (pp. 195–96)

However, I do not agree that intraspecific variation shows that there are no species-specific laws. Consider how biologists' describe variation in natural populations. Typical is this remark of Douglas Futuyma's, from his well-regarded text *Evolutionary Biology* (1979):

> Within a single species . . . individuals sometimes have the diagnostic characteristics of related species or even genera. The form and number of teeth in mammals are important for classification; yet in a single sample of the deer mouse *Peromyscus maniculatus*, Hooper [cite] found variant tooth patterns typical of 17 other species of *Peromyscus*. Among fossils of the extinct rabbit *Nekrolagus*, Hibbard [cite] found one with the premolar pattern characteristic of modern genera of rabbits; and the *Nekrolagus* pattern is occasionally found in living species. (p. 161)

Even while emphasizing variation, Futuyma refers to the number and pattern of teeth "characteristic" or "typical" of a given species, and he refers to "the

Nekrolagus pattern." Surely these are implicit statements of laws of the form "The *S* is *T*."

4. The proviso identified

4.1. A rule for default reasoning

At the close of chapter 6, I argued that the law of multiple proportions is associated with an inference rule having the proviso "in the absence of information to the contrary." This is a rule for default reasoning. I propose that the same kind of proviso accompanies the inference rule associated with a law like "The robins' egg is greenish blue." Any counterfactual supposition positing that there exists a yellow robins' egg, or that the robin that laid a certain egg is the product of some natural-selection pressure for (or selective-breeding program intended to produce) an egg that is not greenish blue, qualifies as supplying information to the contrary. Thus, if we believe in the scurvy law (returning to van Inwagen's example), then we must believe it to be preserved under the counterfactual supposition "Had a program of selective breeding been instituted to produce human beings who do not require vitamin C." But since this supposition supplies "information to the contrary," we cannot use this inference rule's reliability regarding this possible world to conclude that the human beings who would have resulted from this program develop scurvy when deprived for *x* days of food containing vitamin C. Furthermore, any counterfactual supposition concerning an individual robin's genes or the gene frequencies among a population of robins may likewise count as "information to the contrary," since we know that genes are responsible for traits like egg color; if we have sufficient relevant genetic information, we should depend on it rather than on the individual's species affiliation in predicting biological properties. So if we believe that the raven is a black bird, then we must believe that the raven would still have been a black bird even if the relevant genes had been different in some individual raven—though we cannot use this inference rule's reliability regarding this possible world to conclude that this individual raven would then have been black, since we have "information to the contrary" regarding that raven.

How often the inference rule associated with "The *S* is *T*" leads to inaccurate predictions depends on how often we know information to the contrary regarding an *S* that is not *T*. What it takes in order for the inference rule to qualify as "reliable" depends on our purposes. Suppose that we have no information to the contrary regarding any robin, and so we use the inference rule associated with "The robins' egg is greenish blue" to conclude that *all* robins' eggs are greenish blue. This conclusion is not sufficiently accurate for the purposes of a discipline that is interested precisely in intraspecific variation, such as population genetics or evolutionary ecology. That is, biological laws of the form "The *S* is *T*" are laws of disciplines that are concerned with discovering what Ernst Mayr (1965) calls the "biological properties" of various individuals: their external morphology, internal anatomical structure, size, color, physiological properties, chemical constituents, ecological and environmental tolerances and preferences, productivity, life span, reaction to chemical stimuli, behavior, and so on. For the purposes of these disciplines, the corresponding inference rules are sufficiently accurate sufficiently often. The inference rule associated with a law of the form "The *S* is *T*" may occasionally yield an incorrect prediction, since some *S*s for which we have no "information to the contrary" may be non-*T*. But this happens insufficiently often, considering our purposes, to render the inference rule unreliable. If "The *S* is *T*" holds, then an *S* that is *T* is *normal* in the relevant respect, in that it is just the way we would, by default, expect an *S* to be.

Notice that when we make such an error, we are still correct in believing the subjunctive conditionals required by Λ-preservation (and Λ's non-nomic stability) to hold, that is, in believing the inference rule associated with a law to be reliable regarding the relevant possible world. For example, consider the counterfactual antecedent "Had Anne Boleyn never met Henry VIII." Since this antecedent is consistent with the laws, Λ-preservation says that the laws would still have held under this counterfactual supposition. This is correct; for instance, the human being would still have had exactly 10 fingers had Anne Boleyn never met Henry VIII. Of course, it is not the case that had Anne Boleyn (a human being) never met Henry VIII, Anne Boleyn would have had exactly 10 fingers. But this counterfactual conditional does not follow logically from "Had Anne Boleyn (a human being) never met Henry VIII, then the human being would still have had exactly 10 fingers," since "The human being has exactly 10 fingers" does not mean "All human beings have exactly 10 fingers." As I have shown, "The human being has exactly 10 fingers" can be true in a given possible world—that is, the associated inference rule can be reliable there for the purposes of the relevant biological disciplines—even if the rule occasionally leads us to erroneous conclusions concerning that world. For example, suppose that I had never learned that Anne Boleyn had polydactyly. Then since I take the inference rule associated with "The human being has exactly 10 fingers" to be reliable to the closest possible world in which Anne Boleyn never meets Henry VIII, I can justly apply this inference rule to make predictions regarding Anne Boleyn in this counterfactual world, and thereby conclude that Anne Boleyn would have had exactly 10 fingers had she not met Henry VIII. This counterfactual conditional is incorrect, of course. But it is not required by Λ-preservation even though "The human being has exactly 10 fingers" is a law.

There are possible worlds where very few if any *S*s are *T* and that are picked out by counterfactual suppositions that biologists in the disciplines concerned with discovering biological properties fail to recognize as supplying "information to the contrary." For example, geology and evolutionary biology have revealed that had Earth been 10% less massive, then continental drift would have proceeded vastly differently (as the internal heat source that powers it would have been considerably diminished), and so the course of evolutionary history would have been tremendously altered (as the result of changed patterns of geographic isolation). However, cardiologists, for instance, do not need to be experts in Earth science; they may well not know that the reliability of the cardiological laws of the form "The *S* is *T*" depend on Earth's mass, and so they may fail to regard this counterfactual supposition as supplying "information to the contrary." However, this is no threat to the inference rule's reliability regarding this possible world (as demanded by Λ-preservation)—its reliability for cardiological purposes, among others. Disciplines such as cardiology take no interest in such alternative evolutionary histories, so *trivially*, the inference rule associated with, say, "The human heart is four-chambered" is reliable regarding this possible world. In cardiology (as I mentioned in section 1 above), it is good enough to hold that, had Earth been 10% less massive, the human heart would still have been four-chambered. (Perhaps some of the above counterfactual suppositions, e.g., van Inwagen's, also fall outside the range of interests of these fields.) Of course, other scientific fields are interested precisely in the course that evolutionary history might have taken had Earth's mass been 10% less. In these fields, this counterfactual conditional is not close enough to the truth.

4.2. Biological species in different disciplines

That much of biology is concerned with discovering the "biological properties" of various individuals is evident from even a casual perusal of journals in such disciplines as physiology, anatomy, genetics, neurobiology, pathology, psychology, developmental biology, and so on. There one encounters scores of articles with titles like "Growth of cottontail rabbits (*Sylvilagus floridanus*) in response to ancillary sodium," "Establishment and maintenance of claw bilateral asymmetry in snapping shrimps (*Alpheus heterochelis*)," "Antibiotic activity of larval saliva of *Vespula* wasps," and "Learning to discriminate the sex of conspecifics in male Japanese quail (*Coturnix coturnix japonica*)."[9] The research discussed in these articles seeks natural laws concerning the biological properties characteristic of particular biological species (and often uses inductive strategies to arrive at them). A challenge for philosophy of biology is to explain how the existence of such laws is consistent with evolutionary theory.[10]

In these disciplines, biological species are natural kinds for the same reason as mineralogical species are natural kinds in geology, chemical species are natural kinds in chemistry, and so on.[11] As discussed in chapter 7, for each sort of natural kind, there is a law specifying the sorts of laws in which each natural kind of that sort figures. Biological species are natural kinds in certain biological disciplines because the best inductive strategies to pursue in those disciplines, considering the facts with which they are concerned, involve inductively projecting across conspecifics—using observations of one member of a given species to confirm facts about any other member of that species, as when we regard one instance of antibiotic activity on the part of larval saliva of *Vespula* wasps as confirming inductively the reliability of the salient inference rule.

In these disciplines, then, the point of grouping various organisms together as members of the same species *S* is to enable us better to predict the biological properties of unexamined organisms. This is sometimes obscured when the "biological species concept" is emphasized, according to which the capacity of two organisms to interbreed—rather than their similarity in biological properties—makes them conspecific. That two organisms can interbreed—or, to put it more accurately, belong to a common gene pool—is relevant to their belonging to the same species (in the sense of "species" used in these disciplines) only because it entails that they are likely to be similar in their biological properties. Their capacity to interbreed suggests a genetic similarity, and their belonging to a common gene pool suggests that a favorable mutation occurring somewhere in this gene pool's past may well have spread to both organisms.

That the basis of conspecificity in these disciplines is similarity in biological properties can also be obscured by the fact that classification according to common ancestry is often cited as distinguishing a natural from an artificial taxonomy. However, that certain individuals have a recent common ancestor does not *make* them conspecific in the sense relevant to these disciplines. Rather, it helps to *explain why* they have so many traits in common. That various individuals have common ancestors explains why their traits do not vary independently from one another, which would make a natural classification impossible. So although patterns of similarity in biological properties are responsible for taxonomic classification in these disciplines, the fact that all individuals are related to a greater or lesser degree by way of evolutionary descent makes such classification possible—allows these taxa to be natural kinds (Jeffrey 1982, p. 43). As Morton Beckner (1968) puts it:

> Before the advent of evolutionary theory, there was no way to state and to justify a classificatory policy that reasonably could be expected to result in . . . taxa [that] are natural kinds . . . [in virtue of the fact that they] lump together organisms with many attributes in common. [B]ut it was noticed as early as Aristotle that morphological diagnostic characters yielded a system more fruitful than classification by ecological properties. And it was discovered and accepted as an empirical fact that if a set of morphological diagnostic characters-in-common could be found for a given group, they were exceptionally reliable indices for a great many, indeed, a practically limitless number of other morphological characters. After the adoption of the Linnaean system, it was further found that the morphological diagnostic characters of the taxa of species rank were good indices for ecological properties, as well as for the important property of fertility in matings within the taxon. Evolutionary theory at once provided an explanation for these empirical facts and suggested a policy for implementing the formation of taxa that are likely to lead to the greatest number of predictions. This is to classify on the basis of phylogenetic descent, i.e., to place in the same taxa those organisms most closely related to each other. This has led to the adoption of the ideal of "monophyletic" classification, that no two distinct branches of a phylogenetic tree should lead into any one

> taxon. (pp. 52–53, see also pp. 62–63; for similar remarks, see Mayr 1969, pp. 78–80; Simpson 1961, p. 25; Hempel 1965, p. 146)

In botany, less lip service is paid than in zoology to the "biological species concept" (Stuessy 1994, p. 13) because of the relative ease of hybridization in plants as well the prevalence of apomixis (asexual reproduction. so-called "vegetative propagation"). In his *Introduction to Plant Taxonomy*, Charles Jeffrey (1982) makes clear that biological taxa are distinguished by their optimizing inductive inferences regarding biological properties:

> The most important feature of a natural classification is its very high predictive value. . . . If one knows, therefore, the properties of one member of a group of a natural classification, it is possible to predict what the properties of another member of the same group will be. This can easily be illustrated by a simple example. Suppose we are taken into a room, at the far end of which stands a screen. We can see a potted plant standing in front of the screen and we are told that behind the screen, invisible to us, stands another potted plant. We are also told that both plants belong to the genus *Primula*. On examining the plant in front of the screen, we find that it has all its leaves in a rosette at the base of the stem, that its flowers have a fairly long corolla-tube, and that its seed-capsules contain numerous seeds. . . . [S]ince the genus *Primula* is a natural group . . . , [these features] are found in all plants belonging to that genus. We can therefore say, with almost 100% certainty, that the plant behind the screen also exhibits these same features. . . . The great predictive value of the natural classification is of considerable practical importance. For example, when precursors of the drug cortisone and of the sex hormones were found to be present in certain species of the yam family (of the genus *Dioscorea*), botanists at once began to collect other species of the genus and to test them for their yield of the useful products. It was not long before other species were found which contained much higher percentages than did the original plants examined. Without a natural classification, the investigators would not have known which plants to examine next, and would have lacked a reliable guide to the furtherance of their researches. . . . [W]ith the proviso that, owing to the operation of such factors as convergent evolution, parallel evolution and different rates of evolution in different lines, the most closely similar kinds of plants may not always be the kinds that are most closely related in the course of evolution, a natural classification of plants will be to a large extent a phylogenetic one. (pp. 43–46; see also Stuessy 1994, p. 6; Stace 1980, pp. 13–15)

Although evolutionary history (or I could just as well have said genetic similarity) explains why a natural classification is possible, and although a natural classification largely reflects phylogenetic relations (and genetic similarity), what *makes* a taxonomy natural in fields concerned with "biological properties" is its role in the best inductive strategies (i.e., in the laws).

I contend, then, that disciplines such as neurology, physiology, and embryology use in practice a concept of biological species that is at base phenetic rather than phylogenetic, ahistorical rather than historical, and altogether hardly distinct from what was used by pre-Darwinian biologists. Indeed, Darwin himself contended that evolutionary theory should leave taxonomy unaffected; in *The Origin of Species* (1859/1964), he wrote:

> [S]ystematists will be able to pursue their labours as at present. . . . [They] will have only to decide whether any form be sufficiently constant and distinct from other forms, to be capable of definition and if definable, whether the differences be sufficiently important to deserve a specific name. (p. 484)[12]

Of course, accepted classifications have changed immensely since Darwin's time. But that does not mean that different taxonomic *concepts*—say, phylogenetic ones—are now in use. I believe that Clive Stace's (1980) remarks about the taxonomy of algae, bryophytes, and other nonvascular plants are broadly applicable:

> In the lower plants there have certainly been great taxonomic changes [such as splitting the algae into several putatively monophyletic groups], which might at first be seen as the result of advance in phylogenetic understanding. In fact, for the most part, these changes have had a quite different origin. An intimate knowledge of most lower plants is possible only by their study with modern biochemical and microscopic aids; their variation is at a lower order of magnitude of size than that of higher plants, and in the nineteenth century the means to study the variation adequately did not exist. Had the pre-Englerian taxonomists had access to the techniques of electron microscopy, chromatography, spectrometry and the like, there is little doubt that phenetic classifications of lower plants broadly similar to the present-day ones would have been produced. (p. 52)

On the other hand, evolutionary biology, population genetics, biogeography, ecology, and certain other biological disciplines are interested precisely in intraspecific variation. Whereas physiology, developmental biology, and so on, are interested in the *present* state of various species, evolutionary biology and the like are interested in the dynamics of species *change*. Thus, the inference rules associated with the species-specific laws of physiology, developmental biology, and so on, are not reliable for the purposes of disciplines like evolutionary biology. Accordingly, the biological species appropriate for work in physiology may not be the biological species appropriate for work in evolutionary biology, where there are laws of the form "All biological species of x kind have y property." For instance, in inductively confirming a law specifying that unisexual species are especially vulnerable to extinction, we should count two geographically isolated populations of organisms that could have interbred had they not been geographically isolated, and are very similar genetically, as two separate species; the independent extinctions of the two populations should count as distinct cases in which the law is borne out. But for the purposes of physiological investigation, these two populations should perhaps qualify as conspecific.

Here I agree with Hilary Putnam (1993):

> Imagine, for example, that someone asked the question, "Is it part of the essence of dogs that they are descended from wolves?" . . . The answer seems to be "yes" from an evolutionary biologist's point of view and "no" from a molecular biologist's point of view. From [the evolutionary biologist's point of view] species are essentially historical entities, very much like nations. . . . Evolutionary theory seeks to understand the historical origins of the populations of animals that we see in the world; and to generalize about the mechanisms of population change; from this point of view, it is very much of the essence to know that dogs are descended from wolves. From a molecular biologist's point of view, the situation is quite different. In molecular biology, animals are viewed simply as finished products; the molecular biologist is not concerned with where they came from or with what populations they belong to. From this point of view, having a certain kind of DNA is very much an essential property of a dog. It is not actually possible to give a necessary and sufficient condition for being dog-DNA (I am sure molecular biologists would be happy if it were), because, on account of the mechanisms of genetic variation that are beloved of evolutionary biologists, the DNA varies from one dog to the next. . . . These two different points of view can even lead to different decisions about counterfactual situations. . . . If we suppose that technology becomes so advanced that it is possible to synthesize a whole dog, starting from chemicals on a shelf, say, to synthesize a dog with exactly the DNA of my own dog Shlomit, then from a molecular biologist's point of view the resulting "synthetic dog" will count as a dog. . . . From the point of view of an evolutionary biologist, the situation is different. I suspect, in fact, that evolutionary biologists would not regard a "synthetic dog" as a dog at all. (pp. 130–33; see also Kitcher 1984b, esp. p. 320)

Evolutionary biology presupposes a biological species to be a historical entity in that, roughly speaking, an organism's species affiliation is determined by its belonging to a certain spatiotemporally more-or-less continuous entity (such as a certain segment of a lineage). In contrast, while it is undeniable that species arise at certain times and go extinct at other times, this fact is not central to the role played by the notion of a biological species in those fields concerned with organisms' "biological properties."

4.3. *The best inductive strategies depend on the kind of evidence available*

Biological laws of the form "The *S* is *T*" are reliable for the purposes of certain biological disciplines in virtue of accidents of evolution. They are not laws of evolutionary biology. Instead, a claim reflecting the frequency with which various traits are *currently* present in the members of a given species functions as an initial or boundary condition in evolutionary accounts of population change. On the other hand, while the inference rules associated with species-specific laws sometimes lead us astray, they represent the products of the best inductive strategies for us to carry out in fields like physiology. They can justly be used to make predictions in those fields (when we have nothing better than a given individual's species affiliation to guide us in predicting its biological properties).[13]

Evolutionary accidents are not the only kind of accidents that are responsible for the species-specific biological laws. Which set of inductive strategies is the best for the disciplines concerned with biological properties to carry out depends on what sort of evidence is typically available to workers in those fields. As long as an individual creature's complete genetic sequence is not routinely available as a basis for predictions, the best strategies have us project across conspecifics. At some future time when individuals' complete genetic sequences have become routinely available, the best inductive strategies may be different. They might aim to predict the effects of various DNA-base sequences in certain internal and external environments, no matter the species of the individual in whose genome they appear. We could then arrive justly at the scurvy generalization without needing a "fresh induction"—that is, without first having to observe any human beings deprived of vitamin C.[14]

On this view, the lawhood of "The *S* is *T*" in fields like physiology and the status of biological species as natural kinds in these fields depend on the sorts of evidence available there. This may appear counterintuitive.[15] But paleontologists, for example, have long recognized that the widespread presence of gaps in the fossil record is partly responsible for the utility of classifying fossil specimens into distinct species. There is some controversy over whether these gaps are inevitable in virtue of the nature of sedimentation and diagenesis (see Imbrie 1957, pp. 142–47), or whether they result from the fact that evolutionary novelties usually take hold in small marginal populations (hardly ever known as fossils) by evolution so rapid as to appear instantaneous in terms of detectable geological time. In any event, most paleontologists agree that no amount of future fieldwork will ever alter the fact that within typical lineages, only segments displaying little evolutionary change are represented in the fossil record. Had this been otherwise, the best inductive strategies for us to pursue in this field would have been different. As Arthur Cain (1954) puts the point:

> The imperfections of the fossil record are very useful. Because of them, the known fossils of most groups also fall into rather discrete assemblages, and the hierarchical classification, devised originally for living forms, could therefore be applied without modification to fossils. . . . But when good series are available, forms that seem to be good species at any one time may become indefinable since they are successive stages in a single evolutionary line and intergrade smoothly with each other. . . . It is interesting to reflect on what system of classification might have been adopted if for some reason good series of fossils were so well known to mankind that living animals were recognized from earliest times as the present terminators of evolutionary series. (pp. 107–08, see also pp. 112, 123; see also Ruse 1969, p. 104)

To conclude, I have argued that there are species-specific laws in certain biological fields. I acknowledge that intraspecific variation is common, so that generalizations of the form "All *S*s are *T*" typically have exceptions. I also recognize that even when they are exceptionless, their truth results from evolutionary accidents. But we can reconcile these points with the existence of species-specific biological laws as long as we recognize that laws can have exceptions (as emphasized by the "offstage miracles" in nonbacktracking contexts), that the laws of a given field are bound up with the best set of inductive strategies for workers in that field to pursue, and that a field's concerns help to fix its physical necessities. To argue that there are no species-specific laws of biology in a sense of "law" uninformed by these considerations is to employ a sense of "law" that fails to appear in the rational reconstruction of even the physical sciences. Of course, skeptics have argued that there is nothing for the concept of a natural law to do in an account of scientific reasoning. But I have shown how the traditional intuitive motivations for distinguishing laws from accidents can be captured—but only by a conception of natural law that leaves room for species-specific biological laws.

5. Micro and macro explanations

5.1. *What unique contribution do macro explanations make?*

Presumably, all entities with which science is concerned are or are ultimately composed of entities whose behavior is covered directly by the fundamental laws of microphysics. But various macroscopic states of affairs can also be explained by laws of special sciences that are not laws of microphysics. These "macro" laws are irreducible to microphysical laws, and explanations employing macro laws should not be considered convenient stand-ins for impractically complicated but more complete explanations at the micro level. In the coming sections, I contrast my account of why this is so—which involves the autonomy of macro scientific disciplines—to other proposals regarding the contribution that macro explanations make, but micro explanations cannot provide even in principle. Along the way, I at last turn to the relation that lawhood bears to explanatory power.

To explain a macro fact, we might explain why a particular microstate obtains that in this case "realizes" that fact. Alternatively, we might proceed entirely at the macro level.[16] For instance, that a certain gas introduced into one corner of a container spreads out to fill the container's entire volume can be explained by the gas's initial microstate (i.e., the positions, velocities, and other characteristics of the individual gas particles) and the laws governing the motions of and forces felt by the gas particles. These facts explain the various individual particles' trajectories, which result in a final microstate where the particles are distributed relatively evenly throughout the container's volume. Alternatively, the gas's final macrostate can be explained without explaining its particular final microstate: by the law according to which, shortly after the gas is introduced into the container, every cubic centimeter of the container has the same likelihood of containing a given gas molecule.

Similarly, that both the normal and sickle-cell hemoglobin alleles have been maintained in some human population can be explained at the micro level: by an account that explains, on a case-by-case basis, why the births and deaths of particular individuals occurred as they did, by virtue of which the population has realized the "stable polymorphism" being explained. Alternatively, this fact can be explained at the macro level: by the heterozygote's fitness exceeding that of either homozygote.

Likewise, suppose that a laboratory yeast culture is grown in a limited space and fixed environment. (I.e., the temperature and chemical composition of the medium are held constant, waste products are removed frequently, food is added in uniform amounts each day, etc.) The general shape of the population's growth curve (namely, initial exponential expansion followed by leveling off to an equilibrium size) can be explained either at the micro level—by an account explaining, on a case-by-case basis, the fates of individual organisms—or at the macro level, by the "logistic equation" of growth under limited resources: $dN/dt = rN(K - N)/K$, where rN represents the population's "intrinsic" growth rate and $(K - N)/K$ reflects how close the population has come to the environment's "carrying capacity" K.

Analogous contrasts between micro and macro explanations could be drawn in many other examples. A human action might be explained by folk-psychological laws and the agent's beliefs and desires, or at a micro level in neurological or chemical terms. An economic event might be explained at the macro level by laws of economics, or at a micro level in terms of the behavior of individual agents, or in principle even in terms of particle physics. At a macro level, the laws of island biogeography are sometimes believed to explain why more species are found on one island than on another by virtue of the islands' differences in size and distance from a source pool of potential immigrants. The species diversity of various islands could also be explained by the particular fates that befell individual potential migrants.[17]

5.2. *Are the macro explanatory generalizations analytic?*

The generalizations employed in macro explanations have sometimes been deemed to be analytic rather than logically contingent truths. For instance, Sober (1993, p. 72) so characterizes the Hardy-Weinberg equation (used in the explanation of a stable polymorphism), one version of which is that if there are exactly two alleles (A and S) at a given locus in the genome of a sexually reproducing species, and if p and q $(= 1 - p)$ are respectively the fractions of these alleles in one generation, then in the absence of disturbing factors, the likelihoods of the genotypes AA, AS, and SS in the next generation are p^2, $2pq$, and q^2, respectively. Admittedly, the Hardy-Weinberg equation is often presented in biology texts as the conclusion of a short mathematical derivation, and the proviso "in the absence of disturbing factors" *could* be interpreted so as to render the equation trivial: a "disturbing factor" is anything that would produce a departure from the above likelihoods in the next generation. But the proviso need not be understood in this way. Indeed, it is not so understood—else there would be no need for any "derivation" at all. The purpose of the "derivation," I believe, is not to demonstrate the Hardy-Weinberg equation in the manner of a geometric theorem, but to reveal the mechanism behind the Hardy-Weinberg distribution and thus the sorts of *physically possible* influences that would lead to a departure from it, such as mutations between the two alleles, migration from or to other populations, and selection pressure for one of the alleles and against the other.

In other words, the proviso in the Hardy-Weinberg equation should be understood along the same lines as the proviso in, say, Galileo's law of freely falling bodies (discussed in chapter 6). Of course, we *could* interpret that proviso as referring to any influence that leads a falling body to depart from uniform acceleration at 9.8 m/s^2, thereby trivializing Galileo's law. But in learning how to apply Galileo's law, physicists learn about or draw on their prior knowledge of some of the other sorts of influences that exist in the world as a matter of contingent fact. The reference to "disturbing factors" derives its content from this knowledge. Galileo's law is not a logical truth because there are *logically* possible influences on a falling body that are not covered by the "no disturbing factors" proviso since they are *physically* impossible (such as a kind of force that is not one of the actual kinds of force). When scientists discover the existence of a heretofore unknown non-negligible influence on falling bodies (such as being near to a pole or the equator), they reject one proposed "law of free fall" in favor of another hypothesis; none of these proposals is trivially true. The same applies to the Hardy-Weinberg equation. Scientists do not automatically learn what its proviso means simply from understanding the rest of the equation. Furthermore, just as scientists had to ascertain empirically that falls near to a pole or the equator would deviate from 9.8 m/s^2, so they had to learn empirically that "meiotic drive" is possible—a new non-negligible "disturbing factor" for the Hardy-Weinberg equation.[18]

In the case of the logistic equation of growth under limited resources, there are likewise a host of disturbing factors, such as seasonal effects on population size, crowding that fails to affect all members equally, and females who fail to find mates. As discussed in chapter 6, some provisos have the effect of limiting the inference rule's applications to cases where certain influences—such as any delay between a change in population density and its effect on birth and death rates—are negligible for the relevant purpose. Though the logistic equation purports to express (with sufficient accuracy for certain purposes) the result of many different kinds of density-dependent influences on population growth, there are different, more accurate, more complicated equations specialized for different particular kinds of influences. (I discussed the same pattern in the case of Hooke's law in chapter 1 and van der Waals's equation in chapter 7.)

5.3. *Perspicuity, hyperconcreteness, and counterfactual dependence*

One proposal regarding the unique contribution made by micro explanations is that a macro explanation is much more perspicuous than a micro explanation. The latter obscures "what is really going on" in a welter of detail concerning, for instance, the particular collisions experienced by the gas particles or the fates of individual potential migrants to the island. The micro account cannot see the forest for the trees. But thus expressed, this point appears to be merely pragmatic, possessing no greater significance than the difficulty of discovering or stating the complete micro initial conditions. A "Laplacean superphysicist" would not be bothered by any of these features of the micro explanations. Harold Kincaid (1997) puts the point well:

> [E]ven if every lower-level account were always so complex that we had difficulty in seeing the overall picture, that would not show that they are not explanations—any more than the fact that some mathematical proofs are too long and complicated for one individual to follow shows that they are not proofs. While there is no doubt a subjective sense of "explains" that equates explanation with understanding, there is also an objective sense of "explains" that I would argue is both primary and irreducible to a subjective notion. . . . Once the contrast class, relevant kind of answer, and given background information are specified, then it is a factual, objective question whether or not some proposed statement answers the question at hand. (p. 89)

The micro explanation's "hyperconcreteness," as Alan Garfinkel (1981, p. 62) calls it, might in a different way account for the macro explanation's unique contribution without turning this into a merely pragmatic point. On this proposal, the initial microstate fails to explain the final macrostate because the final macrostate would still have resulted even if the initial microstate had been different. For example, had the gas particles been differently arranged, given different velocities, and so on, then although the final microstate would have been different, the gas would still have been evenly spread in the container. To cite the gas's initial microstate as explaining its final macrostate is to suggest that the final macrostate depends upon the details of the initial microstate, which is not the case. The alleged micro explanation, Garfinkel says,

> gives us a false picture of the sensitivity of the situation to change. It suggests that, had the specific cause not been the case, the effect would not have occurred. This is false in such cases because there is redundant causality operating, the effect of which is to ensure that many *other* states, perturbations of the original microcause, would have produced the same result. (p. 62)

A similar line has sometimes been attributed (e.g., by Sober 1993, p. 75) to Putnam (1975, pp. 295–97, 301), who discusses an example involving a rigid cubical peg, each side of which is $\frac{15}{16}$ of an inch long, and a rigid board with a round hole 1 inch in diameter. Putnam argues that an account of the individual peg and board molecules cannot explain our failure to fit the peg through the hole, since the macro result would have been no different had the microstructure been different—for example, had a given board molecule been a little to the left, or had the peg been composed of glass (a supercooled liquid) or wood instead of metal. The macro result is properly explained at the macro level, in terms of the rigidity of the peg and board, and the fact that the diagonal of the peg's side exceeds the hole's diameter.[19]

As Sober (1993, p. 75) notes, this proposal presumes a counterfactual constraint on explanatory relevance: that *c* helps to explain why *e* is the case only if *e* would not have obtained had *c* not obtained (or, at least, only if it is *not* the case that had *c* failed to obtain, then *e* would still have obtained). On the basis of this counterfactual constraint, it is argued that the initial microstate cannot help to explain the final macrostate. But, of course, there are cases where *c* explains *e* even though *e* would still have occurred without *c*. For example, suppose that Jones chose to have her wedding ring made of platinum rather than gold; had her wedding ring not been fashioned from platinum, it would instead have been gold. Her ring is electrically conductive because it is platinum. But had it not been platinum, it would still have been electrically conductive, since it would have been gold. (That her ring is electrically conductive because it is platinum *or gold* does not seem correct.)

Of course, many attempts have been made to reconcile such "overdetermination" with the proposed counterfactual constraint on explanatory relevance.[20] I am not persuaded that any of these attempts works, but I do not enter into this debate, since the possibility of overdetermination is not my principal reason for rejecting the proposed counterfactual constraint. Rather, my principal worry is that many cases are unlike the ring case in that it is not at all evident *what* would have happened regarding *e* had *c* not been the case, but it is intuitively clear that *c* helps to explain *e*. The ring example was contrived so as to fix what the ring's composition would have been had it not been made of platinum. But consider a paper clip that I have just found in my desk drawer. It is electrically conductive because it is made of copper.[21] Despite my confidence in this explanation, I do not hold any confident opinion regarding what the clip would have been like had it not been made of copper. I have no belief one way or the other about whether the closest possible worlds where the clip is not made of copper include some worlds where the clip is made of hard rubber, which does not conduct electricity, along with some worlds where it is still made of metal. Given my present evidence, I am agnostic regarding whether it would still have been electrically conductive had it not been made of copper. Here, then, we have a kind of case not involving overdetermination where my belief regarding $\neg c > \neg e$ does not track my belief regarding *c*'s explanatory relevance to *e*.[22]

According to the proposal I have been critiquing, the final macrostate not only *can* be explained at the macro level, but *must* be; an account at the micro level is simply not explanatory because it fails to satisfy a counterfactual constraint on explanatory relevance. I agree with Sober (1993, p. 75) in finding this result counterintuitive, though Kitcher (1984a), Putnam, and Garfinkel apparently disagree. I ultimately argue that they are correct in emphasizing the counterfactuals in these cases. But I also contend that the macro and micro accounts are *both* explanatory and that neither renders the other superfluous, even in principle.

The role that these counterfactuals ultimately play in my argument is usefully contrasted with their role in Daniel Dennett's (1981) argument that a macro explanation involving the "intentional stance" (i.e., using folk-psychological categories, such as belief and desire) adds something that could not be supplied, even in principle, by a complete micro account:

> The power of the intentional strategy can be seen even more sharply with the aid of an objection first raised by Robert Nozick some years ago. Suppose, he suggested, some beings of vastly superior intelligence—from Mars, let us say—were to descend upon us, and suppose that we were to them as simple thermostats are to clever engineers. Suppose, that is, that they did not *need* the intentional stance to predict our behavior in all its detail. They can be supposed to be Laplacean super-physicists, capable of comprehending the activity on Wall Street, for example, at the micro-physical level. Where we see brokers and buildings and sell-orders and bids, they see vast congeries of subatomic particles milling about—and they are such good physicists that they can predict days in advance what ink marks will appear each day on the paper tape labelled "Closing Dow Jones Industrial Average." . . . Our imagined Martians might be able to predict the future of the human race by Laplacean methods, but if they did not also see us as intentional systems, they would be *missing something* perfectly objective: the *patterns* in human behavior that are describable from the intentional stance, and only from that stance, and which support generalizations and predictions. Take a particular instance in which the Martians observe a stock broker deciding to place an order for 500 shares of General Motors. They predict the exact motions of his fingers as he dials the phone, and the exact vibrations of his vocal cords as he intones his order. But if the Martians do not see that indefinitely many *different* patterns of finger motions and vocal cord vibrations . . . could have been substituted for the actual particulars without perturbing the subsequent operation of the market, then they have failed to see a real pattern in the world they are observing. (pp. 64–65; see also Dennett 1991)

I agree that we miss something if we are ignorant of these counterfactuals. But Dennett does not explain why the Laplacean superphysicists would fail to ascertain the correctness of these counterfactuals in the course of ascertaining all of the micro facts. (He says only that *if* the Martians do not see that these counterfactuals are correct, then they have missed something.) Dennett grants that all of the macro facts predicted by the intentional stance are also predicted by the Laplacean superphysicists (where by these "macro facts," Dennett obviously does not mean to include the correctness of counterfactuals). So why would the intentional stance make predictions regarding *counterfactuals* that the Laplacean superphysicists would miss when the intentional stance makes no predictions regarding *actuals* that the Laplacean superphysicists would miss?[23] I address this key question in section 6 below.

5.4. *Do micro and macro explanations answer different questions?*

Let me now turn to a different view of what macro explanations can do that micro explanations cannot: respond to certain why-questions. It has long been recognized that, in requesting an explanation of some fact, we often want to know why that fact obtained *rather than* certain alternatives. The correct answer to our question depends on the alternatives that we have cited. For instance, the following questions demand different answers, and yet all involve explaining why Jones succumbed to lung cancer: "Why did Jones *succumb* to lung cancer rather than fight off the illness and survive?" "Why did Jones succumb to *lung* cancer rather than to some other form of cancer—why *lung* cancer in particular?" "Why did *Jones* rather than *Smith* succumb to lung cancer—what set Jones and Smith apart?" The relevant alternatives to the fact being explained form the "contrast class." It is sometimes argued that macro and micro explanations answer why-questions involving different contrast classes. For example, Garfinkel (1981, p. 56) argues that to explain why a given rabbit was captured and eaten *sometime* rather than *never*, we should invoke the macro fact that the fox population was great. But to explain why the rabbit was captured and eaten *just when it was* rather than sometime earlier or later, we must discuss the situation in micro terms, explaining how the rabbit came to expose itself at that moment, why a fox was then present, and the details of the chase.

I agree that a macro account cannot address a why-question involving a contrast class at the micro level. Only a micro account can explain the micro details of the case. Return to the diffusion example: to explain why the gas ends up in one micro realization rather than another of the same macrostate, we must appeal to the gas's initial microstate. Here we have revealed something that a macro account cannot supply but a micro account can. But I do not see that an analogous argument reveals a contribution uniquely made by the macro account. To explain why the gas spreads out rather than remaining concentrated in a corner, why will the micro account not do? Why is it incorrect to say that the rabbit was eaten sometime rather than never because it exposed itself on a certain occasion (because . . .) when a fox was present, and so on? Admittedly, the micro account tells us not only why one macrostate rather than another resulted, but also why that macrostate was realized in one way rather than another. This may be more than we wanted to know, involve a cumbersome amount of detail, or be impractical to discover. But (as I argued in section 5.3 above, agreeing with Kincaid) these considerations do not show that the micro account fails to answer this why-question. It might be argued that the micro account cannot explain why the rabbit was eaten sometime rather than never because, had the micro facts been otherwise, the rabbit would probably still have been eaten sometime, given the prevalence of foxes. But this returns us to the counterfactual constraint on explanatory relevance.

5.5. *Do macro explanations unify what micro explanations treat separately?*

I now examine a different attempt to locate the macro explanation's unique contribution. I ultimately argue that this line of thought contains some truth but must be elaborated carefully. According to this line of thought, a macro explanation treats identically and thereby unifies various cases that are quite disparate from a micro perspective. For example, different instances of diffusion are explained by the same macro generalization, even though they involve different kinds of particles, exerting different forces on one another, and following different individual trajectories. Likewise, the logistic equation explains the growth of populations differing in the kinds of organism involved, their environmental requirements, and the respects in which their resources are limited. Similarly, the same macro explanation can be given for different stable polymorphisms, regardless of the precise reason for the heterozygote's greater fitness.

The point here is *not* that, by abstracting away from what Hacking (1990, p. 185) calls "the host of petty influences" responsible for a given gas particle's micro trajectory, a given yeast organism's fate, or a given peg molecule's interaction with a given board molecule, the macro explanation is more perspicuous or leaves out facts that are explanatorily irrelevant. How, then, should we cash out the intuition that the macro explanation makes a unique contribution by knitting together microscopically heterogeneous cases? It might be argued that the macro explanations allow us to diminish the number of explanatory generalizations, patterns, or (as Kitcher calls them) "schemata" that we use, just as Newton's law of gravitation did in giving a unified explanation of falling bodies, tides, and planetary motions. But although micro explanations of, say, the behavior of different gases invoke different intermolecular forces, the micro level *ultimately* achieves tremendous unity with just a few explanatory principles, uniting cases of diffusion with many other phenomena and reducing the various intermolecular forces to just a few fundamental kinds. Sober (1984) suggests that the micro perspective applied to an ecological example (such as a population's growth curve) would "probably isolate that population from other apparently similar ones" (e.g., populations whose growth is constrained by limitations on different resources); in contrast, Sober says, the macro approach is "able to treat these populations within a single framework" (p. 129). But surely the *ultimate* micro physics is able to do even better. If we operated at the level of fundamental particles, we could use the same few explanatory generalizations to understand not only any such population, but also why a certain peg fails to fit through a certain hole. That is even greater unification.

The intuition regarding macro unification could be elaborated differently. Compared to micro accounts, the macro account contributes unity that is not *greater* in degree but of a *different* kind. In other words, the macro account subsumes the case being explained not under a *broader* category than the micro account does, but under a *different* category. For example, whereas a macro account treats identically all populations growing under limited resources in a fixed environment, a micro account either subdivides this category, according different treatments to populations whose resources are limited in different respects, or instead groups the case at hand under a broader category, applying the same laws to all large collections of fundamental particles. The macro category "populations growing under limited resources in fixed environments" is not a "natural kind" at the micro level, because its micro realizations are so heterogeneous. (The same might be said for other macro categories, such as "predation," "diffusion," and "stable polymorphisms.") Any micro property common to some members of the macro category either fails to be possessed by some others, or is so general that it is also possessed by cases outside of the macro category, or is intuitively a wild disjunction that had to be gerrymandered to fit exactly the micro realizations of the macro category and so seems from a micro perspective to be drawn arbitrarily.

Fodor (1981, pp. 127–45) develops this line of thought in terms of the distinction between natural laws and physically necessary non-laws (see references to Fodor in section 2 of chapter 7). Suppose "All *F*s are *G*" to be a macro-level law. Suppose that f_1, f_2, . . . are the physically possible ways for *F*ness to be realized at the micro level, and likewise g_1, g_2, . . . for Gness. According to Fodor, there are "bridge" laws "All *F*s are f_1 or f_2 or . . ." and "All cases of g_1, g_2, . . . are *G*." Since "f_1 or f_2 or . . ." is a heterogeneous disjunction rather than a natural kind at the micro level, there are no micro laws involving this category, and likewise for "g_1 or g_2 or. . . ." However, there *are* micro laws involving the various disjuncts: "All f_1s are g_1," "All f_2s are g_2," and so on. Hence, "All cases of f_1 or f_2 or . . . are cases of g_1 or g_2 or . . ."—the micro level "image" of the macro law—is a physically necessary nonlaw. On this view, the macro laws are in a sense irreducible to micro laws (see figure 8.1). Of course, the macro laws can be deduced from micro laws and bridge laws. But the micro level image of the macro law, with its wild disjunctions, is not itself a law. The macro scientific discipline depicts a level of organization in nature that cannot be depicted as a natural level of organization from the micro perspective. On Fodor's view (1981, p. 144), there are macro explanations not for the sake of convenience, but because there are natural kinds that micro accounts cannot capture.

This is an intriguing picture. But is it correct? Fodor's argument for it is that the special sciences discover "counterfactual supporting" general truths "All *F*s are *G*" where *F* and *G* either fail to be coextensive with natural kinds in physics or are coextensive only by accident rather than as a matter of physical necessity:

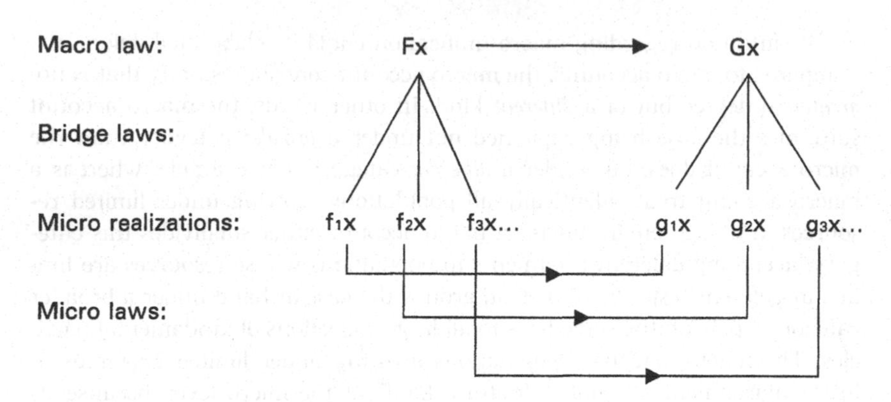

Figure 8.1. After Fodor (1981, p. 139). This is Fodor's conception of the relation between macro kinds and micro kinds.

> Suppose, for example, that Gresham's "law" really is true. (If one doesn't like Gresham's law, then any true and counterfactual supporting generalization of any conceivable future economics will probably do as well.) Gresham's law says something about what will happen in monetary exchanges under certain conditions. I am willing to believe that physics is general *in the sense that it implies that any event which consists of a monetary exchange* (hence any event which falls under Gresham's law) *has a true description in the vocabulary of physics and in virtue of which it falls under the laws of physics.* But banal considerations suggest that a physical description which covers all such events must be wildly disjunctive. Some monetary exchanges involve strings of wampum. Some involve dollar bills. And some involve signing one's name to a check. What are the chances that a disjunction of physical predicates which covers all these events (i.e., a disjunctive predicate which can form the right hand side of a bridge law of the form "x is a monetary exchange $\rightleftharpoons$ . . .") expresses a physical kind? In particular, what are the chances that such a predicate forms the antecedent or consequent of some proper law of physics? The point is that monetary exchanges have interesting things in common; Gresham's law, if true, says what one of these interesting things is. But what is interesting about monetary exchanges is surely not their commonalities under *physical* description. A kind like a monetary exchange *could* turn out to be coextensive with a physical kind; but if it did, that would be an accident on a cosmic scale. (Fodor 1981, pp. 133–34)

By this reasoning, Fodor concludes that some natural kinds (e.g., in the social and biological sciences) are not natural kinds of physics.

However, this argument presupposes that every counterfactual-supporting general truth is a law. From this premise and the fact that macro sciences have discovered *F* and *G* to figure in counterfactual-supporting general truths, Fodor concludes that the *F*s and the *G*s form natural kinds, since the natural kinds are the categories figuring in laws. Since the *F*s and the *G*s are not necessarily coextensive with natural kinds of physics, according to Fodor there are natural kinds that are not natural kinds of physics. However, Fodor must deny his own premise that every "counterfactual-supporting" general truth is a law, because physically necessary nonlaws presumably also qualify as "counterfactual supporting"; at least, they are preserved under any counterfactual supposition under which the laws are all preserved. (And, of course, Fodor cannot identify all of the physical necessities as laws, on pain of having to regard the wildly disjunctive categories as natural kinds.) It is unclear precisely what Fodor means by "counterfactual supporting" for just the reasons that led me (early in chapter 2) to eschew this term. As I have noted more than once, even an accidental generalization can apparently "support" a counterfactual, as when Trygve Haavelmo's generalization (from chapter 1) regarding the relation between a certain car's maximum speed and its gas pedal's distance from the floor supports a counterfactual regarding what the car's maximum speed would have been had the gas pedal been depressed to one-half inch from the floor.

Furthermore, that the micro-level "image" of a macro law is not a law fails to show that it could not serve as an explanatory generalization. A physical necessity may sometimes serve as an explanatory generalization even if it is not a law. For example, to explain the shape of the electric field created by a given object, we might note that the object is a uniformly charged spherical shell and that any such object must at any point outside of it create an electric field of strength kQ/r^2 directed radially (where r is the distance from that point to the shell's center and Q is the shell's total charge). This generalization is a physically necessary nonlaw; it is a special case of the laws of electrostatics. There is no separate law for spherical shells; they require no "fresh induction." Likewise, we might explain why every bracelet in the jewelry-store window is electrically conductive by appealing to the fact that every bracelet there is made from a precious metal—gold, silver, or platinum—and that an object made from any of these metals must be electrically conductive. That an explanation need not involve a natural kind is especially evident when no physical necessity appears at all in the explanation, as in Haavelmo's car example. If the micro-level "image" of a macro law *can* serve as an explanatory generalization, then it would supply precisely the same unity as an explanation employing the macro law would; it would knit together precisely the same range of cases as the corresponding macro law. The macro explanation's unique contribution would then not be some distinctive kind of unity.

6. Stability and the autonomy of macro disciplines

On what ground, then, can we argue that there are irreducible macro laws grounding explanations that contribute something that cannot be supplied even in principle at the micro level? To begin with, I suggested in section 1 above that a set of macro claims logically closed in U may possess nontrivial non-nomic stability for the purposes of a given macro discipline, and so contain exactly that discipline's physical necessities in U (of a certain grade). Likewise, the logical closure in U of the fundamental micro laws possesses nontrivial non-nomic stability for the purposes of microphysics. I also pointed out that the range of counterfactual suppositions in U that fall within the macro discipline's range of interests and are consistent with all of its physical necessities in U include some that are inconsistent with the physical necessities in U of microphysics. For instance, "Had certain birds possessed antigravity organs assisting them in takeoff" posits violations of a fundamental micro law. But under this supposition, the laws of island biogeography would still have held; for example, smaller islands would still have presented smaller targets and so (ceteris paribus) picked up fewer stray creatures as migrants. Likewise, the macro laws governing diffusion are preserved under a counterfactual supposition positing stronger or weaker intermolecular forces at work in the collisions of hydrogen gas molecules; the new forces are just another kind of "petty influence." And some organisms that are microphysically impossible (e.g., because they possess antigravity organs) would still have been governed by natural selection and the other principles responsible for stable polymorphisms. When these macro laws are projected inductively, they are projected onto these microphysically impossible cases.

This is not to say that the macro laws have a broader range of invariance than the micro laws. There are counterfactual suppositions in U that are consistent with the fundamental laws of microphysics (and so under which those laws must be preserved by virtue of their physical necessity), but under which the laws in U of a given macro discipline fail to be preserved (except perhaps trivially for the purposes of the macro discipline—because the supposition falls outside the range with which that field is concerned). For example, had worldwide sea levels fluctuated dramatically in the recent past, then the laws of island biogeography would not still have held. It is not the case that these laws have provisos that make it trivial for them to hold under this supposition; since island biogeography is interested only in possible worlds where islands form isolated patches of suitable habitat, the laws of island biogeography contain no provisos restricting them to such possible worlds (see section 1 above). The counterfactual supposition of recent worldwide fluctuations in sea levels is consistent with the fundamental laws of microphysics. So the range of invariance of the laws of island biogeography is not broader than that of the fundamental laws of microphysics. Nevertheless, the macro laws are non-nomically stable *for island biogeography*, since this counterfactual supposition is outside island biogeography's range of interests; *for the purposes of island biogeography*, the macro laws are *trivially* preserved there.

I am not contending that the macro laws would still have held no matter what the micro physics. On the contrary, certain suppositions regarding micro physics are inconsistent with the macro laws. For instance, the macro laws concerning predation require that functional morphology be governed by what Robert MacArthur (1972) calls a "continuity principle,"

> that nearby morphologies are adapted for nearby methods of harvesting food [and inversely] that nearby foods are gathered with nearby efficiencies. Thus if a flycatcher is best adapted for catching beetles 10 mm long, its efficiency at 9 mm and 11 mm should be just slightly less, and so on. (p. 59–60)

This continuity in functional morphology, though imposing no constraint on the details of the fundamental laws of microphysics, requires that they exhibit a corresponding continuity. It cannot be that $F = ma$ (Newton's second law) except when the object is 9.5–9.75 mm long, in which case some quite different relation holds. The laws of island biogeography require a similar continuity in the micro laws. They require, for instance, that the energy needed for a creature to traverse a given distance of open sea by a certain means have no sharp discontinuity at a certain distance. Otherwise, there would not be a smooth relation between an island's biodiversity (ceteris paribus) and its distance from a source pool of potential migrants. Had the law giving this relation failed to hold, it might have been because some continuity principle fails to hold, in which case other laws of island biogeography would presumably also have been violated. The range of invariance in U that a given island-biogeographical law possesses by virtue of its physical necessity in island biogeography is limited by the other biogeographical laws.

Remember that the macro laws are not analytic. Their preservation under some microphysically impossible counterfactual suppositions displays the macro discipline's autonomy: the logical closure in U of the macro discipline's laws leaves out the micro laws but nevertheless manages to possess nontrivial non-nomic stability for the purposes of the macro field. This takes a point I discussed in section 5.3 above—that the final macro state does not depend on the initial macro state's particular micro realization—and generalizes it: the final macro state (e.g., that the gas is evenly spread throughout the container's volume) would have been no different even if the initial macrostate had been realized in a microphysically impossible way. This seems to be one of the ideas floating around in Putnam's (1975) remarks concerning the peg example:

> The explanation is that the board is rigid, the peg is rigid, and as a matter of geometric fact, the round hole is smaller than the peg. . . . That is a correct explanation whether the peg consists of molecules, or continuous rigid substance, or whatever. . . . The same explanation will go in any world (whatever the microstructure) in which those *higher level structural features* are present. In this sense *this explanation is autonomous.* (p. 296).

A peg consisting of continuous rigid substance would be in violation of microphysical laws.

In this light, reconsider the macro law "All *F*s are *G*," such as the law relevant to the peg and board, and its relation to its micro-level image "All cases of f_1 or f_2 or . . . are cases of g_1 or g_2 or . . ." The micro image is supposed to be a microphysical necessity; it follows from various micro laws. For simplicity, I call these various micro laws "All f_is are g_i." The various properties f_i specify the various arrangements of molecules that would compose all of the microphysically possible realizations of a rigid cubical peg of side 15⁄16 inches aimed at a rigid board with a round hole of 1 inch diameter. But "All *F*s are f_1 or f_2 or . . . "—the "bridge law" needed to derive the macro law from its micro image—lacks physical necessity *of the sort possessed by the macro law.* From the macro viewpoint, as I have just shown, it is an accident that all actual *F*s (all rigid 15⁄16 inch cubical pegs and boards with 1 inch round holes) are made of molecules at all (rather than, e.g., continuous substances), much less that the molecular arrangements satisfy f_1 or f_2 or. . . .[24] Likewise, "All f_i are g_i" is an accident at the macro level. It says that whenever we have a peg and board of a certain specific molecular arrangement, with the peg aimed at the hole in a certain specific way, then there results a certain specific micro realization of failure to fit. This result depends on the precise forces between the peg and board molecules in this arrangement, which is a macro accident; the macrophysically nontrivially non-nomically stable set does not include the microphysical force laws. The micro-level image of the macro law, then, lacks macrophysical necessity.

Although the *truth* of the macro law follows from the micro laws and bridge laws, the macro law's possession of a certain kind of *physical necessity* does *not* so follow. The macro law's physical necessity involves its membership in a certain set that is nontrivially non-nomically stable in the macro discipline. In other words, the macro law's physical necessity involves its preservation (for the macro field's purposes) under every counterfactual supposition in a certain range. But the micro laws and bridge laws are not all preserved (for the micro field's purposes) under all of those suppositions. The microphysical necessity of the micro laws and bridge principles does not ensure that those laws are preserved under, say, the supposition of a peg and board made of continuous rigid substance. So the microphysical necessity of the micro laws and bridge laws does not entail that the macro law would still have held under the supposition of a peg and board made of continuous rigid substance. Hence, the macro law's possession of its kind of physical necessity does not follow from the micro laws' and bridge laws' possession of their kind of physical necessity. The lawhood of the macro law is thus irreducible to the lawhood of the micro laws and bridge laws. The macro discipline is autonomous. Contrary to Fodor, the macro law's micro-level image *is* counterfactual supporting (in that it is preserved under a wide range of counterfactual suppositions), since it is microphysically necessary. But the *range* of counterfactual suppositions under which the macro law's micro-level image is preserved in virtue of its microphysical necessity does not match the range of counterfactual suppositions under which the macro law is preserved in virtue of its macrophysical necessity. (Obviously, an account of this sort is possible only because the specific "counterfactual-supporting power" that is associated with physical necessity has been identified in preceding chapters.) The micro-level "image" of the macro law is not really an analog of that law, since the wild disjunction f_1 or f_2 or . . . in that image (e.g., in Putnam's example) does not cover *F*s in certain macrophysically possible worlds.

This argument helps to bear out Dennett's intuitions regarding the Laplacean superphysicists, discussed at the close of section 5.3. I asked why the intentional stance enables us to ascertain the correctness of certain counterfactual conditionals that the Laplacean superphysicists would fail to discover to be correct. This is especially curious considering that the Laplacean superphysicists would ascertain all of the micro and macro facts in U. Now we have an answer. Suppose that in the course of learning all of the facts in U, the superphysicists carry out exactly the inductive strategies that result in the microphysical laws. The counterfactual conditionals that they thereby ascertain do not include all of the counterfactuals that we would ascertain by pursuing the best set of inductive strategies for the purposes of the macro scientific discipline, that is, the inductive strategies that result in the laws of folk psychology. In particular, no counterfactuals with microphysically impossible antecedents fall within the scope of the superphysicists' inductive projections. So the superphysicists, proceeding entirely at the micro level, may indeed miss some facts in U* that can be discovered by working at the macro level.[25]

Dennett (1981) correct in pointing out that there are folk-psychological patterns that support counterfactuals—that is, that are preserved under counterfactual perturbations—and are confirmable in the manner of laws.

> Fatalists—for instance, astrologers—believe that there is a pattern in human affairs that is inexorable, that will impose itself *come what may*, that is, no matter how the victims scheme and second-guess, no matter how they twist and turn in their chains. These fatalists are wrong, but they are *almost* right. There *are* patterns in human affairs that impose themselves, not quite inexorably but with great vigour, absorbing physical perturbations and variations that might as well be considered random; these are the patterns that we characterize in terms of the beliefs, desires, and intentions of rational agents. (p. 66)

What is crucial about these folk-psychological patterns is that the *range* of "perturbations and variations" under which they are preserved is different (and in some respects goes beyond) the corresponding range for microphysical laws. In confirming the relevant folk-psychological laws, we learn, for instance, that had the stockbroker in Dennett's case been made of continuous rigid substance (to use Putnam's example), then the subsequent operation of the market would have been no different.[26] Likewise, we learn about how people would have behaved in some science-fictional situation, traveling between the stars by accelerating from rest to superluminal velocities. We must now consider how the fact that these patterns are preserved *come what may* in a certain distinctive range enables them to ground scientific explanations of a distinctive sort.

7. Explanation and possibility

Does the autonomy of a macro scientific discipline reveal what macro explanations contribute that micro explanations cannot? To find out, we must have some idea of what scientific explanations are and what laws have to do with them. As I showed in chapter 1 (section 2.3), there is an intuition that laws possess a kind of explanatory power lacking in accidental generalizations. Hempel (1965) succinctly provokes this intuition, taking as his examples the ideal gas law $PV = nRT$ and the accidental generalization "All members of the Greenbury School Board for 1964 are bald":

> [T]he two kinds of sentence differ . . . in explanatory power. The gas law, in combination with suitable particular data, such as that the oxygen in the cylinder was heated under constant pressure, can serve to explain why the volume of the gas increased; but the statement about the School Board analogously combined with a statement such as "Harry Smith is a member of the Greenbury School Board for 1964" cannot explain why Harry Smith is bald. (p. 339)

However, matters are not that simple. An accidental generalization can sometimes function just as a law does in a scientific explanation, as I showed in connection with Haavelmo's example of the car.[27] Yet Hempel is surely correct about the explanatory impotence of the school board generalization.

Furthermore, although lawhood is relevant to scientific explanation, their relation will alas shed little light on what is arguably the most deeply puzzling feature of scientific explanations: their asymmetry. The net force impressed on the particle taken together with the particle's mass physically necessitates its acceleration, and its acceleration together with its mass physically necessitates the net force impressed upon it. But whereas the force and mass can explain the acceleration, the acceleration and mass cannot explain the force. Obviously, it is not the case that the successful explanation here involves a law whereas the bogus explanation involves an accident. The relation between lawhood and explanation will apparently not account for notorious puzzles involving what makes an effect unable to explain its cause, why two effects of a common cause fail to explain each other, why an irrelevant conjunct undermines an explanation, and so on. Ideas about explanation that suffice to explicate lawhood's explanatory role may fail to take us very far toward a general account of scientific explanation.

The ideas about explanation to which I appeal fail to form either a sufficient or a necessary condition for a scientific explanation. I now set out what I take to be the "core notion" of scientific explanation, according to which the explainer shows that the fact to be explained "had to" be as it is, in an appropriate sense of "had to" (as Rescher 1970, pp. 13–14, puts it). Admittedly, this is very crude and minimal, but I believe that it contains a sufficient grain of truth for our purposes here. Certain sorts of explanation (notably statistical explanation) represent an extension of this core notion, though precisely how to extend it to these cases is not a topic on which I dwell here.

Indeed, it may be unrealistic to expect there to be a single extension that covers all statistical explanations: whereas some statistical explanations explain singular facts by explaining why those facts were likely, other statistical explanations explain singular facts by explaining why those facts had their particular likelihoods (which may not have been high), and still other statistical explanations explain singular facts by explaining simply why those facts had nonzero likelihoods.

I take the core notion of scientific explanation to be that in virtue of the explainer, the fact to be explained was bound to be, could not have been otherwise, was inevitable. In the simplest case, *Ga*'s explainer consists of an explanatory generalization "All *F*s are *G*" and an initial condition *Fa*.[28] Crucially, a generalization is "explanatory" at least partly because not only all *actual F*s, but also all *possible F*s are *G*. By "possible" *F*s here, I do not mean "logically possible" or "physically possible." Readers who have come this far will not be surprised to learn that, on my view, this notion is context sensitive, as I shortly discuss further.[29] The upshot is that since all *possible F*s are *G*, we *must* be dealing with a *G* if we have an *F*. That all *actual F*s are *G* ensures that some object *is G* given that it is *F*, but fails to ensure that it *could not* have been non-*G* given that it is *F*. For that, we need that all *possible F*s are *G*.[30] It is as if there is an urn of the possibles and every *F* in the urn is *G*. It is then understandable that every *F* that happens to be selected from the urn for actualization is *G*; no matter which of the possible *F*s had been realized, all *F*s would still have been *G*.[31] In a more complicated case, the fact to be explained is *a*'s value of some quantity *z*, the explanatory generalization is a functional relation "All *F*s satisfy $z = f(x,y)$", and the initial conditions are *Fa* and *a*'s values of the quantities *x* and *y*. Again, the generalization is explanatory partly because all *F*s would still have satisfied $z = f(x,y)$ no matter which possible *F*s had been realized. (But had different possible *F*s been realized, perhaps different values of the quantities *x* and *y* would have been instantiated by *F*s.)

On this view, an accidental generalization can sometimes serve as an explanatory generalization. Suppose I reach into an urn of marlbes and withdraw a green marble. Why did I end up with a green one? Because all of the marbles in the urn are green. This generalization is accidental. Nevertheless, any marble I *could* have ended up with (in the contextually relevant sense of "could") must have actually been in the urn to begin with. Since every *possible* withdrawal from the urn yields a green marble, it is understandable why I ended up with a green marble: I selected my marble from that urn, so a green marble is all that I could have ended up with.

In the context of this explanation, there is tacit agreement on which withdrawals from the urn were *possible*. To illustrate, consider that in this context it is correct to say "Choosing a marble from the urn, I would have gotten a green marble *no matter what*." Context supplies "no matter what" with its scope. It is understood that "no matter what" includes "no matter whether I had shaken the urn first or simply taken a marble off of the top," but does not include "no matter whether I had first added red marbles to the urn."

Likewise, Haavelmo's accidental generalization concerning the car's maximum speed can be explanatory, and in those contexts, we say that it would still have held "no matter what." Context is responsible for the fact that "no matter what" includes any realistic roadway grade but not any alteration in Earth's mass. In contrast, it is not the case that "All of the members of the Greenbury School Board for 1964 are bald" would still have held no matter what—for example, no matter who had been on the board—unless, say, on becoming a member, you are obliged to have the hair on the top of your head electrolytically removed. But then perhaps we *could* explain why Harry Smith is bald by citing his membership on the board.

A generalization with a range of invariance that is *right* (in other words, a generalization that is preserved "no matter what") can serve as an explanatory generalization. That a generalization's range of invariance is intuitively rather narrow need not keep it from playing this role, as in the marbles example. Equally, a generalization with a range of invariance that is intuitively rather broad may nevertheless typically be unable to function as an explanatory generalization. Consider the claim that agrees with Newton's gravitational force law except for dictating a different value (e.g., 5 newtons, whatever the distance) for the gravitational force between a pair of point masses of precisely 1.234 and 5.678 grams, and suppose that these mass values happen never to be instantiated. This generalization would still have held under any physically possible counterfactual supposition under which those mass values would still have been uninstantiated. Despite this generalization's relatively broad range of invariance, typically it cannot serve as an explanatory generalization since it would not have held "no matter what"; the relevant possibilities typically include masses of any values in a range that includes 1.234 and 5.678 grams. This generalization's lack of explanatory power is attributable to its lack of invariance across some *right* range.

A generalization's physical necessity is relevant to its explanatory power by virtue of bearing on its range of invariance; its physical necessity may reflect its preservation "no matter what." Macro explanations contribute something that micro explanations cannot supply because macro laws are preserved under certain microphysically impossible suppositions, and so may hold of all "possible" cases when micro laws do not. For example, the growth curve of a population in a constant environment under limited resources would still have been sigmoidal no matter what—even if the creatures had possessed antigravity organs. Considering that a given population is growing in a constant environment with limited available resources, it is understandable why its growth is sigmoidal; this kind of growth is not peculiar to the given population, but would have shown up no matter which possible population of this kind we had considered. The explanation works partly because the fact to be explained would still have obtained even if the given population had been different in some respect that is understood to fall within the scope of "no matter what." This view captures the intuition (emphasized by Garfinkel and Putnam) that the macro explanation exploits the fact that the macro result would have been no different even if the welter of micro details had been different. But this view still respects the micro account's power to explain the macro fact, since this view does not presume the counterfactual constraint on explanatory relevance that I discussed in section 5.3 above (that *c* helps to explain why *e* only if $\neg c > \neg e$).

That "All *F*s are *G*" is a law may explain why all actual *F*s are *G*: because its lawhood over and above its truth informs us of a range of counterfactual circumstances in which all *F*s would still have been *G*, and this range may include one of the right ranges to invoke in this context. That "All *F*s are *G*" is a law may thus entail that no matter which possible *F*s had been realized (in a contextually relevant sense of "possible"), they would all have been *G*. In contrast, that "All *F*s are *G*" is an accidental truth entails nothing about a range of counterfactual circumstances in which all *F*s would still have been *G*.

Suppose we want to know why all actual populations growing with limited available resources in fixed environments grow sigmoidally. We could explain this macro regularity at the micro level, dealing with each actual population individually. Certain elements are common to certain of these individual micro accounts, so it may be no coincidence that several of these populations all grow sigmoidally. The micro explanation of the macro regularity contains various micro laws, which could also be used to show that certain unrealized populations growing with limited available resources in fixed environments would have grown sigmoidally. But all of these unrealized populations must then also be governed by the actual micro laws. In contrast, the macro explanation reveals that the regularity in actual populations would still have held even if the micro laws had been different in certain respects. In this way, that the macro regularity is macrophysically necessary explains why it is true. It is not the case that the micro account supplies the very same explanation, but in a less perspicuous, unnecessarily detailed way. Rather, a micro account does not supply this explanation at all.

Afterword

I have now concluded my account of the relations that natural laws bear to counterfactual conditionals, scientific explanations, and inductive confirmations. I have tried to explain how these relations are bound up with one another, how they distinguish the laws from the accidents, and how they make it especially important for science to discover what the laws are. I have also tried to explain how the laws are related to the natural kinds, how accidents of one field can qualify as laws of another, how those laws may come in grades of physical necessity, and how the meta-laws' relation to the laws they govern reproduces the relation in which the latter laws stand to the non-nomic facts that they govern. The root commitment associated with the belief that *h* is a law—namely, the belief that *h* results from the best set of inductive strategies—captures the constraints that this account imposes on what laws of nature are.

In the philosophical literature on natural law, there has long been a tension between those who emphasize the laws' epistemic roles, especially in connection with inductive projection, and those who emphasize the laws' metaphysical significance in connection, for instance, with natural kinds, explanations, universals, and necessity. These differences in emphasis tend to be associated with different views of lawhood's ontological status. An epistemic emphasis tends to accompany the view that *h*'s lawhood is in some sense less real than *h*'s truth. On the other hand, a metaphysical emphasis tends to be accompanied by the view that *h*'s lawhood is in some sense ontologically prior to *h*'s truth.

Though I am not concerned with offering a reductive account of lawhood, I have tried throughout this essay to reconcile these contrary impulses by revealing how thoroughly interwoven are the various metaphysical and epistemological strands within the concept of natural law. I have argued that sound epistemological intuitions concerning the sort of confirmation reserved for candidate physical necessities have often been hobbled by misplaced metaphysical scruples, leading instance confirmation to be characterized without reference to the confirmation of counterfactual conditionals. We need a metaphysics that is sufficiently liberal to embrace counterfactuals if we are to explicate the range of projection distinctive of inductive confirmation. But the distinction between physical necessities and accidents that I have drawn by using this liberal metaphysics—in terms of non-nomic stability—is insufficient to account for the special scientific importance of discovering whether some reliable inference rule is associated with a physical necessity. I have appealed to an epistemology that accords special significance to inductive projection in order to recognize what is important about the range of invariance involved in non-nomic stability. Another place where I have relied on the concept of the best inductive strategies is in carving out room for physically necessary nonlaws. In short, I have used metaphysics to do the epistemology, and I have used that epistemology, in turn, to do the metaphysics.

I can well imagine some of my readers reacting unfavorably to my embrace of counterfactuals, especially considering my view that their correctness fails to supervene on the non-nomic facts. Readers who emphasize the laws' epistemic roles might cite me for violating what Earman (1984) calls the "empiricist loyalty test" (p. 195). Nevertheless, I have remained steadfast in seeking to uphold the idea that empirical evidence confirms counterfactuals for precisely the same reasons as it confirms predictions regarding actual unexamined cases.

Other readers, I feel sure, will say that my account of natural law is not metaphysically robust enough—that it depicts lawhood as an ontological lightweight. After all, I have connected lawhood to salience, and hence to our own justificatory practices. I have contended that the notion of a natural law *simpliciter* fragments into various concepts applying in different scientific disciplines. And I have argued against a tight relation between laws and regularities, allowing laws to be merely reliable for certain purposes. Of course, I introduced this reference to our purposes precisely to support the intuition that a natural law would still have held under certain counterfactual suppositions, an intuition for which nonbacktracking contexts raise serious problems. Without this appeal to our purposes, I would not have been able to capture the sense of necessity that laws possess. But this might seem to offer little consolation to those inclined toward a metaphysically robust view of laws. Still, in chapter 8, I showed how *h*'s lawhood can explain *h*'s truth (or reliability). I have thereby lent some support to what Armstrong and Heathcote (1991) call a "strong" theory of laws—one "that takes regularities, statistical distributions, etc., to be phenomena that require to be explained by an underlying law or laws rather than as constituting laws" (p. 64).

Of course, it would be nice to understand whether laws make counterfactuals correct or vice versa, or whether some other sort of fact is responsible for both of these. Such an account would presumably also shed some light on dispositions, chances, and causal relations. It have tried only to reveal some of the points at which scientific practice constrains any such account.

Philosophy of science is one of the places where metaphysics encounters epistemology. This is one source of its continuing fascination. Fortunately for those of us attached to these issues, much of the task of developing a satisfying, comprehensive account of natural law that does justice to scientific practice remains unfinished.

Notes

Chapter 1

1. Terminological note: I have just referred to a certain linguistic entity as "expressing a law." There is obviously a distinction between a fact and its linguistic expression. When it is important to attend to this distinction, I shall refer to "All gold objects are electrically conductive" and the like as "law-*statements*." (Note: The law-statement is not "*It is a law that* all gold objects are electrically conductive.") When my meaning is unambiguous, I ordinarily avoid the cumbersome term "law-statement" and refer to both facts and their linguistic expressions as "laws"; for example, I may refer to some *claim* as "Newton's first law of motion." Occasionally, I use the term "law-statement" to make plain that I am concerned with a linguistic entity, as when I discuss a ceteris paribus clause in a law-statement.

   The same considerations apply to "All gold objects are smaller than one cubic mile." In the philosophical literature, it is termed an "accidental generalization." But strictly speaking, what is accidental is not the generalization, a linguistic entity, but its truth. Sometimes, I use the term "accidental generalization" to refer to this fact.

   Some philosophers have held that when "All gold objects are electrically conductive" is used as a "law-statement," it does not function as a linguistic expression having a truth-value, that is, as a *statement* (or a *claim*). Rather, it expresses a rule of inference. (For more on the motivations offered for this view, see the appendix to chapter 6.) I do not mean to presuppose that this kind of account is incorrect when I refer to "All gold objects are electrically conductive" as a law-statement (or a claim). The term "law-statement" should be understood throughout as referring to the linguistic expression of a law, whether or not it is (properly speaking) a statement.

If a law-statement expresses an inference rule, then it is not true, but it surely must be truth-preserving or reliable or something like that. Its possession of this property, rather than its truth, would be what evidence for it confirms.

2. Lewis accepts something like this thesis once it is restricted to possible worlds that are "worlds like ours" in that "they are free of fundamental properties or relations that are alien to our world" and are similar to the actual world in various metaphysically important respects (e.g., things persist through time by consisting of distinct temporal parts). See Lewis (1994, p. 475).
3. That is not to say that *all* such regularities are laws according to Lewis. See appendix 3 to this chapter.
4. Compare Woodward (1992, p. 191) and Lewis (1983, p. 366). I discuss this more fully in section 4 of chapter 2.
5. In appendix 1 to this chapter, I give further examples of putative laws from various scientific disciplines.
6. Kneale (1950, p. 123).
7. Wigner (1967, p. 39).
8. Reichenbach (1947, p. 368; 1954, pp. 10–11); cf. Hempel (1966, p. 55).
9. Goodman (1983, pp. 18–20).
10. Reichenbach (1954, p. 9).
11. Kneale (1952, p. 75); cf. Reichenbach (1947, p. 369).
12. Chisholm (1955, p. 97).
13. Kneale (1961, p. 98).
14. Popper (1959, pp. 427f).
15. Nagel (1961, p. 52).
16. Of course, what to make of a counterfactual such as this (a "counterlegal") is a vexing issue on which philosophers have differed. Those who regard the laws as conceptual necessities (as I discuss briefly in chapter 2) will take every counterlegal as trivial (though they may disagree on whether they are all vacuously true or trivially false). In the course of this book, I work to elaborate the idea that physical necessity is a "grade" of necessity *between* logical (or conceptual) necessity and no necessity at all. Accordingly, I do not take all counterlegals as trivial. This accords with my intuitions: some counterlegals are correct and others are not. Of course, in certain contexts, it is correct to say that had copper been electrically insulating, then the wires on the table would not have been made of copper.
17. It might well be thought that had an electron once been accelerated to beyond the speed of light, then the natural laws would have been fundamentally different, and who knows what the universe would then have been like: maybe there would have been no human beings at all, much less Clinton, and no Earth at all, much less the United States and its quarter-dollar coins. I agree with this thought. But taking this view is compatible with holding that had an electron once been accelerated to beyond the speed of light *and* had Clinton still lived, and had quarters still existed, and had an accelerated electron later been in a quarter that we asked Clinton to put in his dime pocket, *then* Clinton would have refused to pocket it. In other words, "Had *p* and *r* been the case, then *q* would have been the case" plainly does not entail "Had *p* been the case, then *q* would have been the case."
18. In chapter 3, I shall argue that which non-nomic truths are physically neces-

sary does in fact supervene on which subjunctive conditionals (with nonnomic antecedents and consequents) are correct. But I do not presuppose that something like this must be so in virtue of some broad metaphysical picture of what is ontologically more basic than *p*'s physical necessity.

19. Of course, we might then wonder why all ignited copper salts burn with green flames. This is a different why-question from the previous one. The laws that explain why all ignited copper salts burn with green flames could also be used—along with some facts about the given powder—to explain why the given powder, when ignited, burns with green flames. Thus, there are at least two acceptable ways of explaining the color of the powder's flame; there is often more than one way to explain a given fact.
20. However, this is not obvious; being told that it is an accident might suffice to tell us all that we wanted to know when we asked why all of the coins in my pocket are silver. In Lewis's (1986b, p. 220) example, we ask, "Why was the CIA man there just when His Excellency dropped dead?" and we learn what we were interested in knowing simply by being told "it was a coincidence."
21. The spring constant *k* could be replaced by (F_0/x_0), where F_0 is the magnitude of the restoring force exerted by the spring when it is displaced from its equilibrium length by some small amount x_0.
22. This last point is emphasized by Cartwright (1983).
23. Here I am indebted to Teller (1983).
24. Here I agree with Fodor (1981). He invokes this idea in his account of why the explanations supplied by macro sciences are irreducible to the explanations supplied by micro sciences. In chapter 8, I disagree with that account and offer an alternative to it.
25. As Ronald Giere said in his 1994 presidential address to the Philosophy of Science Association: "[W]e need a portrait of science that captures our everyday understanding of success without invoking laws of nature *understood as true, universal generalizations*" (p. 10, emphasis added).
26. Some of the claims on this list are not stated fully. I have omitted various qualifications, ceteris paribus clauses, and so on. In chapters 6 and 8, I discuss the character of these provisos in detail. Also, note that in this book, I say relatively little about problems peculiar to understanding statistical laws. They raise issues about the nature of chance that have nothing *directly* to do with the other topics I discuss.
27. Here I agree with van Fraassen (1989, p. 37). Some claims that are customarily referred to as "laws" are not treated as laws in scientific practice, though they may once have been. These include the "Titius-Bode law" (that $d_i = a_i + 0.4$, where d_i is the distance in astronomical units of the *i*th planet from the Sun and a_i is the *i*th term in the sequence 0, 0.3, 0.6, 1.2 . . .), "Williston's law" (large numbers of similar body parts tend evolutionarily to be reduced to fewer, differently specialized units), and "Cope's law" (body size increases within evolutionary lineages). But on this last, see Stanley (1973).
28. Well, almost. Radioactive decay is an irreducibly stochastic process, so it is physically possible (although exceedingly unlikely) for a large cube of U^{235} to form without being preempted by a nuclear chain reaction.
29. After offering the above argument, Smart considers whether it can be avoided by a different conception of the meaning of "robin." Perhaps by defi-

nition, an organism is a "robin" exactly when certain nonlocal predicates apply to it (where none of these denotes the property of having greenish-blue eggs)—for example, certain predicates concerning anatomy, physiology, and so on. But, Smart says, even if all of the terrestrial creatures that qualify as robins under this definition lay exclusively greenish-blue eggs, there might well be extraterrestrial organisms that qualify as robins according to this definition but do not lay exclusively greenish-blue eggs—and if not, their absence is just an accident. So if we understand "robin" in this way, Smart concludes, we must judge that "All robins' eggs are greenish-blue" is not a law. We might reply by suggesting that the predicates in the definition of "robin" should include a specification of the robin's complete DNA sequence, which physically necessitates the greenish-blue color of robins' eggs. However, this DNA sequence alone may not suffice to physically necessitate the eggs' color; environmental factors may also play a part. Furthermore, there is genetic variation among robins, the sort of variation on which natural selection depends, and so there is no such thing as the complete DNA sequence that is shared by all and only robins. If there are—or could physically possibly be—robins whose DNA and environment result in their laying eggs that are not greenish-blue, then how can it be a law that all robins' eggs are greenish-blue? I largely agree with this argument, though I do not take it as showing that there are no species-specific biological laws. I take this up in chapter 8. For now, allow me to set this argument aside, since it does not rely on the view that local predicates are barred from statements of fundamental laws.

30. It might avoid this reference to a particular event by instead referring to the time elapsed since the latest event with such-and-such property, where the Big Bang happens to be the only such event that ever occurs. But then we would not be dealing with a law.
31. Constraints other than the nonlocal-predicate requirement have sometimes been offered as expressing the universality of laws. Nagel (1961) holds that a law's scope must be such that "the evidence for the law must not be known to coincide with its scope of predication" (p. 63); Campbell (1957, p. 69) advances the same requirement. This seems to me an inappropriate constraint if "scope of predication" refers to *actual* cases only, not counterfactual ones. Suppose we irradiate some culture of bacteria, create some very small colony of mutants that qualify as a new species, inductively confirm some generalizations regarding the members of this species by checking every present member of the culture, and (having thereby learned of the danger posed by these bacteria) eradicate the culture. (Since "inductive confirmation," as I ultimately elaborate it, involves confirming certain counterfactual conditionals, we can inductively confirm a generalization even if we believe that we have already examined each of its *actual* cases.) We know that the species (as an interbreeding population) is extinct, and so that we have observed every member of it. Nevertheless, we have established laws (I shall argue).

    Hempel (1965) offers a different requirement:

    > Surely a lawlike sentence must not be *logically* limited to a finite number of instances: it must not be logically equivalent to a finite conjunction of singular sentences. . . . Thus, the sentence "Every element of the

class consisting of the objects *a*, *b*, and *c* has the property *P*" is not lawlike . . . (p. 340)

Elsewhere, Hempel (1965, p. 267) limits this requirement to statements of fundamental laws. All of the examples he gives of candidates barred by this requirement from the class of fundamental lawlike sentences would also have been barred by this weaker requirement: "All *F*s are *G*" states a fundamental law only if no claim of the form "The number of *F*s is *n*" (for some finite number *n*) is analytic.

I can see a possible motive for this requirement. Laws are distinguished from accidental generalizations by their capacity to support counterfactuals. It would be difficult for "All *F*s are *G*" to support "Had there been *m* *F*s here, they would all have been *G*" if "The number of *F* things is *n*" is analytic and $n < m$. This motive also suggests a way around a difficulty for Hempel's requirement. Swartz (1985, p. 16) has suggested that Hempel's requirement is inconsistent with such laws as "The greatest velocity is 3×10^8 meters per second" and "The lowest temperature is −273.15 degrees Celsius," since there can be only one lowest temperature. But these statements do not take the form "All *F*s are *G*." When expressed in this form—"All material particles travel slower than 3×10^8 m/s," "All matter is warmer than −273.15 degrees C"—the exceptions vanish.

32. For useful surveys, see Armstrong (1980), Carroll (1994), and van Fraassen (1989).
33. Different regularity accounts, offering different proposals for this further condition, have been advanced by Ayer (1963), Hempel (1965, pp. 264–72), Lewis (1973, pp. 72–77; 1983, pp. 366–68; 1986b, pp. 122–24; 1994, pp. 478–80), Mackie (1962, pp. 71–73), Nagel (1961, pp. 58–68), and Reichenbach (1947, ch. 8; 1954), among many others.
34. This account is similar in some ways to proposals made by Mill and (at one stage) by Ramsey, as discussed by Lewis (1983, p. 367) and Earman (1984, p. 197). I return to Mill's view in chapter 7, and argue that it includes a valuable feature absent from Lewis's. Note also that according to Lewis (1983, p. 367), not all truths in the best system state laws; statements of particular fact are not law-statements (though it is not unclear to me precisely how Lewis intends to distinguish particular facts from regularities). Lewis (1986b, pp. 124–26; 1994, pp. 480–89) adds a further element to accommodate the fact that in an indeterministic world, some generalizations appearing in what would otherwise qualify as the best system may not be laws since they hold only by chance. For simplicity, I shall limit my attention to the deterministic case. On Lewis's view, the lawhood of a true claim is a matter of objective fact (see esp. 1994, pp. 481–82). For how objectivity is to be reconciled with the appeal to standards of simplicity, informativeness, and their best balance (a point pursued by van Fraassen 1989, pp. 42–43), see Lewis (1994, pp. 479–80). There Lewis contends that if nature "unkindly" supplies no single clearly "best" system—either because there are ties, or because which system comes out "best" varies with slight changes in the notions of simplicity and strength, or with slight changes in the exchange rate between simplicity and strength—then nothing really deserves to be called a "law"; perhaps the concept of a "law" presupposes that nature is kind.

35. See Armstrong (1983, pp. 22–23), who cites Broad (1968, pp. 174–75).
36. Lewis goes on:

> On the working hypothesis that the laws of a world are the generalizations that fit into the best deductive system true there, we can also say that the laws are generalizations which (given suitable companions) are highly informative about that world in a simple way. Such generalizations are important to us. It makes a big difference to the character of a world which generalizations enjoy the status of lawhood there. Therefore, similarity and difference of worlds in respect of their laws is an important respect of similarity and difference. (Lewis 1973, p. 74)

37. For reasons I explain in chapter 2, Lewis himself denies that the actual laws are preserved under any counterfactual antecedent *p* where *p* is physically possible. However, Lewis does accept that laws differ from accidents in their role in determining the closeness of a given possible world. According to Lewis, the fact that a possible world includes an exception to one of the actual world's laws is a more serious (albeit defeasible) impediment to that possible world's closeness to the actual world than the fact that a possible world includes an exception to one of the actual world's accidental generalizations. Lewis must explain why this is so, that is, why "miracles" (i.e., violations of actual laws) are singled out for special treatment by the metric determining a possible world's closeness to the actual world. I argue in chapter 2 that, contrary to Lewis, laws *are* preserved under all physically possible counterfactual antecedents (though laws are not invariably associated with exceptionless regularities). My present concern is with whether *any* special treatment for "miracles" (whatever the details of the treatment) could be explained by a Lewis-style account of law.
38. I return to Lewis's explanation of the special relation between laws and counterfactuals in sections 3 and 4 of chapter 2.

Chapter 2

1. Often we are in a position to know that a subjunctive antecedent is false only because it concerns some past moment. So counterfactual conditionals often appear as "Had it been the case that *p*, then it would have been the case that *q*." Typically, I shall express counterfactuals in this way, though we can know a present-tense or even a future-tense subjunctive antecedent to be false, as in "I believe that the sun will not explode in the next minute. Were it to explode, I would be surprised."
2. The counterfactuals with which we are concerned here must be distinguished from "counteridenticals." Chisholm (1955) draws this distinction neatly in connection with the accidental generalization "Every Canadian parent of quintuplets in the first half of the twentieth century is named 'Dionne'":

> We are concerned with those counterfactuals whose antecedents, "if *a* were S," may be interpreted as meaning the same as "if *a* had property S." There is, however, another possible interpretation: "if *a* were S" could be interpreted as meaning the same as "if *a* were identical with something which in fact does have property S". [The sentence "If Jones, who is Canadian, had been parent of quintuplets during the first half of the twentieth century, he would have been named 'Dionne'"] is false

according to the first interpretation, which is the interpretation with which we are concerned, but is true according to the second (for if Jones were identical with one of the Dionnes, he would be named 'Dionne'). On the other hand, the statement

> If Jones, who is Canadian, had been parent of quintuplets during the first half of the twentieth century, there would have been at least two sets of Canadian quintuplets.

is true according to the first interpretation and false according to the second. (p. 98)

I presume henceforth that the counterfactuals I am discussing are not counteridenticals.

3. I could instead have defined "Γ is preserved under Δ" as requiring merely that Γ's members all be preserved under *non-nested* subjunctive suppositions, that is, as requiring merely that for any $m \in \Gamma$ and any $p \in \Delta$, $p > m$ is correct. However, I ultimately defend theses involving preservation in the stronger—that is, nested—sense. (Later in this section, I show that all of these theses follow from a certain thesis involving preservation only in the non-nested sense, a thesis I defend in section 3.) In section 2 of chapter 3, I give the intuitive motivation for using the nested rather than the non-nested sense of "preservation" to elaborate the laws' relation to counterfactuals. That motivation is bound up with the fact that this relation is supposed to manifest the sense in which the laws are *necessary*.
4. Why is this supposition logically inconsistent with the laws? It is logically inconsistent with "All emeralds are green" and it is a law that all emeralds are green. However, in chapter 6 (as I foreshadow at the close of section 2.2 below), I drop the assumption that "It is a law that *p*" entails *p*. At that point, I suggest that *p*'s lawhood requires not *p*'s truth, but merely the "reliability" of the inference rule to which *p* corresponds. I will then redefine "counterlegal": *p* is a "counterlegal" supposition exactly when there are some claims that our belief in *p* entitles us to believe, according to the inference rules associated with the laws, where if we believe *p* and those claims, then our beliefs cannot all be reliable, that is, accurate enough. (E.g., the inference rule associated with "All emeralds are green" entitles us to infer from "There is a red emerald" to "There is a red emerald that is green.") However, it is easier now to presume that all laws are true.
5. Although many philosophers have held that every logical consequence of a law is a law, I argue otherwise in section 2 of chapter 7. I discuss the context of Fodor's remark in chapter 8.
6. Beware that though some (e.g., Carroll 1994, p. 18) define "physically necessary in world W" as I have done, others do not. Lewis (1986a, p. 20), for example, defines it to mean that *p* holds in every possible world where W's laws are *true*. (On Lewis's account, $\blacksquare p$ does not entail $\Box\blacksquare p$; if *p* is actually a law, then there are possible worlds where all of the actual laws are true but *p* is not a law—does not belong to the best system. The "closest lone-proton world," which I discuss in section 3, is such a world.) For more discussion, see van Fraassen (1989, p. 44).
7. Although I presume that "Humean regularities" (e.g., that all emeralds are green) are non-nomic facts, I am not prepared to say whether a fact concerning causal relations, explanatory relations, or objective chances is non-

nomic, since this would require analyses of these difficult concepts, which is not my aim here. Facts concerning causal relations, explanatory relations, and objective chances can perhaps be physically necessary or accidental. That each of the atoms in this vial has a 50% chance of decaying in the next 8.1 days is, presumably, an accident, whereas it is a law that each atom of iodine-131 has such a chance. Likewise, that each of my mishaps while driving a car was caused by carelessness is presumably an accidental generalization, whereas it is perhaps a law that every bolt of lightning causes a clap of thunder.

8. As in "laws are parasitic on occurrent facts" (Earman 1984, p. 195) and "[s]ingular statements are ontologically primary. Physical laws are logically derivative" (Swartz 1985, p. 81).
9. Of course, once the non-nomic facts are fixed, it is not true that the ■s and ¬■s can be distributed in any way among the non-nomic *claims*, since if *p* is a non-nomic falsehood, then ¬■*p*.
10. This leaves room for *some* constraints on the distribution of ■s and ¬■s, such as constraints of the form "If ■*p* then ■*q*." For instance, plausibly "It is a law that all *F*s are *G*" logically entails "It is a law that no *F*s are non-*G*." Likewise, perhaps "All emeralds at spatiotemporal location *l* are green" and "All emeralds at *l* are blue" cannot *both* be laws. (I see no reason why they could not both be physically necessary—when it is physically impossible for an emerald to be at *l*.) However, I do want to permit ¬■*p* and ■*q* when *p* is logically equivalent to *q*. For example, I believe that "All emeralds are green" is a law but "All nongreen things are nonemeralds" is a physically necessary nonlaw. Nongreen things do not constitute a natural kind any more than do the emeralds plus the pendulums. Our belief that "All nongreen things are nonemeralds" lacks lawhood is manifested in our beliefs about how it is supported by evidence. Suppose we discover that a given nongreen stone with specific gravity 2.2 is a nonemerald. Typically, we do not regard this evidence as confirming that a given nongreen stone with specific gravity 4.5 is a nonemerald. Obviously, this is related to Hempel's "paradox of the ravens" and to the traditional view that there is a special kind of confirmation that we may accord a hypothesis only if we believe that it may state a law. This is the subject of chapters 4 and 5. (Appendix 2 to chapter 4 is devoted to the paradox of the ravens; section 2 of chapter 7 concerns physically necessary nonlaws, such as "All nongreen things are nonemeralds," and how they can be confirmed.)

    There must be further constraints on the distribution of ■s and ¬■s among the non-nomic claims *if* the non-nomic facts include facts about objective chances (see note 7). Plausibly, "It is a law that all *F*s are *G*" logically entails "It is a law that for any *n*, no *F*s have *n*% chance of being non-*G*." (As long as probabilities *n* must be standard—i.e., noninfinitesimal—numbers, even an *F* with 0% chance of being non-*G could* be non-*G*, which cannot be if it is a law that all *F*s are *G*.) Something like this is needed for the principle I have just advanced in the main text to apply to a world with non-nomic objective chances. For if it is a law that all *F*s are *G*, then no *F* possesses (say) 20% chance of being *G*, even though an *F* having 20% chance of being *G* can logically possibly exist in a world where all *F*s are *G*—namely, when "by chance" all *F*s are *G*. This would be an example where the distribution of ■s and ¬■s among the non-nomic claims imposes a constraint on the accidental truths in U beyond that they must be consistent with the ps in U where $\blacksquare p$—it would, that is, unless "It is a law that all Fs are G" logically entails that it is physically necessary that no F has 80% chance of being non-G (i.e., 20% chance of being G). Understanding the relation between laws and objective chances would require grappling with issues concerning objective chance that are not directly relevant to the topics of this book. Accordingly, I do not develop the considerations here.

11. If no facts $\blacksquare p$ are involved, then in $\neg r$'s logically entailing $\blacksquare q$, we would have a respect in which the distribution of $\blacksquare$s over the non-nomic facts is determined by those facts.
12. So recall the above example: since the non-nomic claim p is entailed by $\blacksquare(\neg\blacksquare q \supset p)$ and $\neg\blacksquare q$, it follows that p is physically necessary. But these premises are not responsible for p's physical necessity. Rather, p is no accident because p is entailed by some truths in U that are laws.
13. We may discuss the Λ for the actual world, or we may discuss the Λ for a given possible world, which consists of the claims in U that express physical necessities *in that world.*
14. In Chapter 3, I refer to this as Λ's "non-nomic stability."
15. Of course, once the facts in U^+ are fixed, it is not true that the $\blacksquare$s and $\neg\blacksquare$s can be distributed in any way among the *claims* in U^+ but not U. That p and $\neg\blacksquare q$ determine that it is not a law that $(p \supset \blacksquare q)$, but $(p \supset \blacksquare q)$ is then not a non-nomic *fact*—it is false.
16. "Coincidental" in this sense is consistent with "physically necessary," as I elaborate in section 2 of chapter 7.
17. Recall that a law p may fail to be preserved under the counterfactual antecedent "Had $\neg\blacksquare p$." This antecedent falls outside the scope of the principle "$p \in \Lambda$ exactly when p is preserved under every counterfactual supposition in U that is consistent with Λ" because this antecedent is outside U. This antecedent falls outside the scope of the principle we have just derived because, though in U^+, it is inconsistent with Λ^+.
18. This result also follows from Λ-preservation applied solely to the actual world, if Con(Λ^+,r) and Con(Λ^+,p) hold for the actual world's Λ^+ and $\blacksquare m$ holds in the actual world. Recall that "preservation" includes nested counterfactuals.
19. Instead of using Λ^+-preservation to cash out the laws' special relation to counterfactuals, we might have tried

    Λ^+-preservation (alternate form): If $p,q \in U^+$, Con(Λ^+,p), and $p > q$, then Con(Λ^+,p & q).

    This principle says that if the antecedent of a correct counterfactual conditional (having antecedent and consequent in U^+) is consistent with Λ^+, then so is the conjunction of the antecedent and the consequent. In other words, if we start off consistent with Λ^+, so we remain. For instance, this principle says that since "Had there been an emerald under the carpet, it would have been green" is correct and Con(Λ^+, "There is an emerald under the carpet"), it must be that Con(Λ^+,"There is an emerald under the carpet and it is green"). This last does not require that "There is an emerald under the carpet and it is green" be consistent with the *truth* of the accident "There are no emeralds under the carpet," only with its nonlawhood.

    But Λ^+-preservation (alternate form) does not require separate treatment, because it is entailed by the original form of Λ^+-preservation. (Analogous remarks apply to the alternate form of Λ-preservation.) This is because, from Λ^+-preservation, it follows that, for all $p,q \in U^+$ such that Con(Λ^+,p), if q follows from p conjoined with members of Λ^+, then $p > q$. Since $(p > q) \Rightarrow \neg(p > \neg q)$, it follows that, for all $p,q \in U^+$ such that Con(Λ^+,p), if ¬Con(Λ^+,p & $\neg q$), then $\neg(p > \neg q)$. Hence, for all $p,q \in U^+$ such that Con(Λ^+,p), if $(p > \neg q)$, then Con(Λ^+,p & $\neg q$), which is equivalent to Λ^+-preservation (alternate form).

20. This "non-nested version" of Λ^+-preservation is the thesis to which I alluded in note 3.
21. As Lewis (1973) says, "The truth conditions for counterfactuals[,] . . . like the relative importances of respects of comparison that underlie the comparative similarity of worlds, . . . are a highly volatile matter, varying with every shift of context and interest" (p. 92). For example, as Bennett (1984, p. 71) remarks, our concerns influence which of these counterfactuals we would be correct in asserting: "Had I jumped from this window, I would have suffered serious injury", " . . . I would have arranged for a net to be in place below, and so I would not have suffered serious injury", " . . . the window would have to have been much closer to the ground."
22. In Lange (1993b), I also defended a view ("nomic preservation") along these lines. I said that we *preserve* those claims of the form "It is a law that . . . " and "It is not a law that . . . " that we *believe* true, under any counterfactual antecedent that is logically consistent with those claims. In contrast, Λ^+-preservation requires that we *believe* Λ^+ to be *preserved* under any counterfactual antecedent (in U^+) that is logically consistent with Λ^+. Note the difference in the order of the highlighted terms. To elucidate the difference this makes, suppose that we believe $p \in U^+$ to be logically inconsistent with the laws in U, though as yet we have not adopted any claim of the form "It is a law that . . . " or "It is not a law that . . . " with which p is inconsistent. Then we take p as falling outside the scope of Λ^+-preservation, though p falls within the scope of "nomic preservation." Thanks to Kit Fine for discussion on this point.
23. They all involve *non-nested* counterfactuals. In the following section, I consider *nested* counterfactuals.
24. I say "merely" because an additional clause might be implicit in the counterfactual antecedent and *also* obtain in the closest possible world in which the antecedent, *understood without this implicit clause*, obtains.
25. It is crucial that we believe "$p > q$, *because* r." Obviously, it is not true in general that if we accept $p > q$ and then do not accept "$p > r$," then r must have been implicit in the first counterfactual's antecedent. For instance, suppose we believe it to be a law that all emeralds are green, and so we accept "Had this stone been an emerald, then it would have been green" and do not accept "Had this stone been an emerald, it would have been nongreen." It is not then the case that the first counterfactual's antecedent was implicitly "Had this stone been an emerald and nongreen."
26. The same approach reconciles Λ-preservation with the putative counterexamples suggested by Halpin (1991, pp. 274, 288), such as "If I had become rich today, it would have taken a miracle" and "If this desk had suddenly been on Mars, that would have been a miracle." (See also Chisholm 1946, pp. 301–302.) If we asked the person who asserted the former counterfactual, "Do you believe that had you become rich today, you would still have started off the day in the humble financial position and with the meager

prospects that you did?" it seems to me that she should reply, "Of course not. My point was that to have become rich today, I would have to have begun the day with more money or better prospects than I did."

27. Of course, the laws of quantum physics are not deterministic. But even if those are the only genuine laws of nature, it is implausible to suppose that the counterfactuals we adopted long before quantum mechanics was discovered committed us to indeterminism.
28. I do not contend that this counterfactual is correct in every context. In certain contexts, we might instead assert, for example, that had this pendulum's period been half of its actual value, then the gravitational acceleration g exerted by Earth's mass would have been four times greater. But in many conversational contexts, the mass of Earth is held fixed, just as when we consider "Had the match been struck," we often do so in a context in which we hold fixed that the match is dry, well made, and surrounded by oxygen. Analogous remarks apply to other counterfactuals on this list.
29. Bennett (1984, p. 68).
30. Jackson (1977, p. 11). Again, there are surely many other means by which Kennedy could have survived the shooting, but in a given context, the bulletproof car may be the only relevant means.
31. Jackson (1977, p. 11).
32. Horwich (1987, p. 161).
33. Pollock (1976, p. 118).
34. Bennett (1984, p. 68), Horwich (1987, p. 161), and Jackson (1977, p. 11).
35. On the other hand, these at least seem compelling: "The match would have been alight only if it had first been struck," and "The match would not have been alight unless it had first been struck." Are these covert "would have to haves" or equivalent to "would haves"?
36. See note 27, above regarding quantum physics and determinism.
37. In a nonbacktracking context, we say that had Darcy asked Elizabeth for a favor, she would not have granted it, because they quarreled yesterday. Is "they quarreled yesterday" implicitly part of the counterfactual antecedent? To apply the test developed in section 2.1: we ask "Had Darcy asked Elizabeth for a favor, would they still have quarreled yesterday?" In a nonbacktracking context, the answer is "Yes," and so the test does not show that "they quarreled yesterday" is implicitly part of the original counterfactual antecedent.
38. However, if I understand him correctly, Bennett would disagree with what I am defending here; he does not countenance miracles. He contends roughly that when we assert $p > q$, we consider the p-world in which the actual laws obtain such that no other p-world's state of affairs at the time with which the counterfactual antecedent is concerned is closer to the actual world's condition at that time. So (according to Bennett 1984, p. 73) if Darcy and Elizabeth have quarreled in the actual world, then in the closest possible world in which Darcy asks Elizabeth for a favor, Elizabeth is ungenerously disposed toward Darcy, since this is the case in the actual world at the moment with which the counterfactual antecedent is concerned. Hence, had Darcy asked Elizabeth for a favor, then (running the scene forward according to the actual laws) Elizabeth would not have granted it—just as our intuitions say. But did Darcy and Elizabeth quarrel in this world? If so, then Darcy's character at the time he asks Elizabeth for a favor must be profoundly different from his actual character; he must not be so proud. If Eliza-

beth and Darcy did not quarrel in the closest *p*-world, then Elizabeth's character at the time Darcy makes his request must be profoundly different from her actual character; she must be cross with Darcy without provocation. In either event, when we include *p* in the counterfactual world and then propagate backward according to the actual laws, we must introduce changes to the past that themselves give rise (according to the actual laws) to other departures from the actual world's conditions at the time with which *p* is concerned, departures that (according to the laws) will make an important difference to Darcy's and Elizabeth's subsequent behaviors. Bennett tries to avoid this result by proposing that in any of the closest possible worlds where Darcy asks Elizabeth for a favor, either there was a quarrel but Darcy forgets it one day later or there was no quarrel but Elizabeth misremembers the next day that one occurred. But if Darcy or Elizabeth is subject to such memory lapses, then the subsequent course of their relationship will likely involve a series of strange misunderstandings. Yet we do not think (in a nonbacktracking context) that had Darcy asked Elizabeth for a favor, then Elizabeth would not have granted it but by then Darcy might well have forgotten that he had ever asked for it.

We might instead propose that the closest *p*-world involves the actual laws (understood as exceptionless regularities) and a history back to the Big Bang that differs from the actual world's in a manner sufficient to bring about *p* but insufficient to give rise, by some less direct causal pathways, to other important departures from actuality at the time with which *p* is concerned. But I, at least, am far more confident in any counterfactual conditional I assert than that there is any such logically possible world. So I do not regard my assertion of a counterfactual conditional as committing me to the existence of such a world. Recall the famous fact that, in any given breath I take, there is a good chance that there is a molecule that was exhaled in Julius Caesar's last breath. [This was first discussed by Jeans (1940, p. 32) and reflects the fact that if you used a bottle with the same volume as your lungs to measure the volume of the atmosphere, then the number of bottles counted would roughly equal the number of molecules in one breath. Jeans's remark also reflects the fact that almost all of the nitrogen in the atmosphere in Caesar's day will still be there today, and that the atmosphere mixes quickly.] This just goes to show that an event taking place now is entangled causally with a great many, seemingly unrelated past events; we can backtrack from the breath I take now to Caesar's assassination, and then forward again to who knows where!

Consider, for instance, what happens when we reason counterfactually. We might initially utter $p > q$. We might then assert $(p \& q) > r$. This process might continue repeatedly, until we reason to the conclusion in which we are interested (e.g., Had I taken the train, then I would have arrived early, and had I taken the train and arrived early, I would have had a bite to eat . . .). With each new counterfactual consequent I contemplate, I expand my range of interests regarding the relevant possible world. There are then new opportunities for an indirect causal pathway to make a difference that would be relevant to the truth of the counterfactual conditional. Yet I do not reconsider, each time, whether the laws leave room for the antecedent to come to pass in a manner that avoids leading, by some indirect causal pathway, to an effect on the new consequent that I now have in mind. That is because I did not check this matter in the first place. Rather, I posited that the antecedent came about miraculously.

39. Again, I realize that Lewis does not feel the intuition that I find so compelling. See, for example, Lewis (1986b, p. 292). In section 3 of this chapter, I defend this intuition further.

    Though Lewis holds that, had Darcy asked Elizabeth for a favor, the laws would have been different, he need not hold that there is some actual physical necessity *m* such that had Darcy asked Elizabeth for a favor, then $\neg m$ would have obtained. Remember, Lewis's view is (roughly) that $p > q$ obtains exactly when *all* of the closest *p*-worlds are *q*-worlds. It might be that in each of the closest worlds where Darcy asks Elizabeth for a favor, some law or other is broken, but different laws in different worlds.

40. Here is a similar counterfactual in actual use:

    > Living in Israel is, for me, still a spiritual adventure. . . . My children play and love and fight in a language that no one spoke for two thousand years, but for them it is full of life and is taken for granted. Were Abraham the patriarch (who would now be four thousand years old) to sit down for supper at my house, he would understand the greater part of the Hebrew spoken by my five year old daughter. What a wonder that is!

    (From "Fifty is a dangerous age" by David Grossman, in *The New Yorker* 74, no. 9 (20 April 1998), pp. 55–59. The passage appears on p. 58.) The same problem is, of course, posed by the example discussed in chapter 1: "Had Julius Caesar rather than George Bush commanded Operation Desert Storm, then he would have remembered from his campaign in Gaul that there is no substitute for total victory, and so would have persevered until he had captured or killed Saddam Hussein" (see also Halpin 1991). The same problem also arises in the following example that Alan Hajek once suggested to me as violating Λ-preservation. Suppose that, sometime ago, my friend Fred died of a seemingly minor injury—say, a scratch that became infected. I see that you have scratched yourself, and I offer you sympathy. You say, "No problem—nobody ever died of a scratch." I reply with $p > q$: "If Fred were here and heard that, he'd disagree!" Although I believe Con(Λ^{+},*p*), it seems that we are not interested in a world in which *p* and the actual laws hold, since we are interested in a world in which Fred is around to hear your remark and yet has already died of a scratch.

41. Here I disagree with Carroll (1994, p. 187).

42. Of course, *p*'s reliability in a given possible world does not *suffice* to make *p* physically necessary there; we still need to understand what sets laws apart from accidents.

43. My usage of "reliable" should not be confused with its usage in some epistemological literature to mean "leading to the truth in a high fraction of the cases." This epistemological sense of "reliable" makes no reference to our interests or purposes. A claim that never leads to the truth, but its *radically* false consequences all lie beyond our range of interests, may be "reliable" in my sense, but not in the epistemological sense I have mentioned. Perhaps some readers would find that "pragmatically satisfactory" or the like captures my meaning better than "reliable."

44. See note 4 above regarding counterlegals.

45. I do not intend this claim to be tautologous; "The glass is in working order" does not *mean* that the glass is making accurate predictions. It means that the instrument's parts and their assembly conform to certain specifications supplied by the manufacturer.
46. Obviously, the customer is using "reliable" here to mean "always accurate," not the sense of reliable that I introduced at the close of section 2.2 above.
47. Recall Λ^+-preservation (alternate form) from note 19; the above straightforwardly violates the analogous alternate form of Λ-preservation.
48. We can use the same kind of reply to account for certain other apparent counterexamples to Λ-preservation, such as the following case, which Alex Byrne suggested to me. Consider Smith, a renowned physicist. Let $p > q$ be "Had Smith proposed some alternative to special relativity that came to be accepted by the physics community, then special relativity would have been false." We have $p > q$, Con(Λ^+,p), and $\neg$Con(Λ,p & q)—apparently contrary to Λ-preservation. In the weather-glass case, we treat the glass's reliability not as physically necessary but as a hypothesis for which evidence is being posited in the counterfactual antecedent. Likewise, in the Smith case, we treat Smith's proposal and special relativity as hypotheses for which evidence is being posited in the counterfactual antecedent. Of course, there are some contexts in which we reject $p > q$, such as when we are discussing the objectivity of the natural laws, and so are emphasizing that a given law would still have held even if it had not been discovered. But in other contexts, we proceed as if we do not already believe that special relativity correctly describes the laws, as if we are considering the evidence along with the scientific community. (The scientists' conclusion suggests what their evidence must have been like.) In the same way, the weather-glass vendor in the conversation sets aside some of her beliefs about the laws, and proceeds as if she is considering the evidence from the customer's vantage point.
49. Accordingly, Goodman (1983, pp. 5–6) calls "Even if . . . still" counterfactuals "semifactuals."
50. A similar suggestion is made by Lewis (1973, p. 33n).
51. Recall that, to simplify my exposition, I presuppose here that $\blacksquare p \Rightarrow p$.
52. For more on God as legislating the natural laws in creating the world, see section 2 of chapter 7 and the references cited there. See also the frontispiece.
53. For some examples from the astrophysics literature, see Wambsganss, Cen, Ostriker, and Turner (1995); Brieu, Summers, and Ostriker (1995); and Cen and Ostriker (1996) and the references cited there. Ostriker (1996) has described this work as involving "very detailed computer simulations in which we try to take different scenarios scientists have put together on the origin of the structure of the universe, put them in the computer to find out what would happen if those scenarios were true [note the subjunctive] and compare the results to the real universe" (p. 7).
54. It might be objected that a nested counterfactual $p > (q > r)$ is logically equivalent to a non-nested counterfactual $(p \,\&\, q) > r$, and so it is misleading to describe a nested counterfactual as expressing the fact that a given non-nested counterfactual conditional $(q > r)$ obtains in a given nonactual possible world (the closest p-world). Rather, on this view, it expresses the fact that a given non-nested counterfactual conditional obtains in the actual world, that is, that the closest $(p \,\&\, q)$-world is an r-world. The problem with this objection is that $p > (q > r)$ is not logically equivalent to $(p \,\&\, q) > r$. [Here I

may disagree with Skyrms (1980, pp. 175–6).] In other words (hold on!) the *q*-world closest to the *p*-world closest to the actual world need not be the (p & q)-world closest to the actual world. In fact, the *q*-world closest to the *p*-world closest to the actual world might not even be a *p*-world. Consider this example. Suppose that you and I have just run a race, and I have won. I believe that I would always win if I really tried. Then I am willing to assert, "Suppose that you had won the race. This would have happened only if I had not been trying; had I tried, I would have won." This is $p > (q > r)$. I am not willing to assert the corresponding (p & q) > r: Had you won and I really tried, I would have won. There is no logically possible world in which you and I both win the race.

55. Perhaps this is not so. For instance, perhaps certain quantum considerations require protons to emit a constant stream of particles. But this defect in my example—rendering *p* inconsistent with the actual laws' truth—would not undermine the point I am trying to make about cases where the laws' truth *is* consistent with the counterfactual supposition, but that supposition posits a drastically impoverished universe. This defect in my example would show only that I have not, in fact, picked out a drastically impoverished universe by using a counterfactual supposition *p* that is consistent with the actual laws' truth. Our intuitions about the laws' preservation under the lone-proton supposition, even if they are based on misinformation about what the laws of nature actually are, nevertheless constrain any account of the *logical* relation between the laws in a given world and the non-nomic facts there.

    It might also be convenient to imagine that the proton is a genuinely elementary particle, so that we need not worry about the laws governing the behavior of its internal components. Perhaps, then, the lone proton should be replaced by an electron. Furthermore, in picturing the proton as alone in the universe, I am conceiving of the various forces (gravity, electromagnetism, etc.) as operating by action-at-a-distance rather than as being transmitted by fields, since in order for the proton to be genuinely alone, there can be no real fields, potentials, and so on, accompanying it. If this conception is inapplicable to contemporary physics, bear in mind again that an account of the logical relation of laws to counterfactuals should apply to any scientific theory.

56. Of course, Λ^+-preservation demands that they be laws in that world only if "There is nothing but a single proton" is consistent with the actual world's Λ^+. Whether this is so—presuming "There is nothing but a single proton" to be consistent with the actual world's Λ (see note 55)—is part of what I am examining.

57. In fact, Brieu, Summers, and Ostriker (1995) do nearly that: "it is instructive to see how the GRAPE hardware handles the least complex of astrophysical problems: the two-body problem," even though "[i]t is exactly the type of problem for which the GRAPE-3 board was *not* designed. . . . The GRAPE-3 hardware is geared toward many-body collisionless systems . . . " (p. 568).

58. It is a law that all particles are electrons, or muons, or so on. Hypermesons are not on the list. Hence, though the laws would have been different had there been hypermesons, this counterfactual's correctness does not violate Λ-preservation because its antecedent is inconsistent with Λ^+.

59. Ewing (1935, 1943, 1962) and Kneale (1952, 1961) deem the laws to be logically or conceptually necessary. Swoyer (1982) and Bigelow, Ellis, and

Lierse (1992) take the laws to be essential to the properties and kinds that figure in them. See also my discussion of Sellars in the appendix to chapter 6.

60. This view has been suggested by Armstrong (1983, p. 71) and is encouraged by Jackson's (1977, esp. p. 4) talk of the "Hume world," a logically possible world in which all of the non-nomic facts agree with those in the actual world but every regularity is accidental. Carroll (1994) also provides an extended argument that the laws fail to supervene on the Humean facts.
61. According to Earman (1984), by espousing nonsupervenience I disqualify myself from numbering among the "empiricists" regarding natural law. Readers should judge for themselves after considering later chapters.
62. Armstrong and Earman reject Λ^{+}-preservation in connection with an example concocted by Tooley (1977, p. 669). Armstrong (1983, pp. 117–18) describes the case as follows:

> Tooley imagines a world containing ten, and only ten, types of fundamental particle. Allowing that a particle may interact with a particle of its own type as well as with particles of other types, this allows for 55 interaction laws governing the interaction of pairs of particles. Suppose that 54 of these laws are known. They prove in each case to be so idiosyncratic that, given *any* 53 of them, the nature of the 54th could not be known, or even rationally conjectured, antecedently to experience. But suppose that the 55th law, the law of the interaction of B-type with J-type particles, is not known. This occurs because, although this type of interaction is physically possible, yet boundary conditions in the universe are such that, throughout all time, no B-particle is ever close enough to a J-particle to interact with it.

Tooley holds that in this world, there is intuitively a B–J interaction law.

Tooley offers this case as a challenge for regularity accounts of law. For instance, it is difficult to see how Lewis's account (extended—beyond the range intended by Lewis—to cover all possible worlds, not just those without "alien" properties) could accommodate the existence of such a law: Any generalization about B–J interactions adds neither strength (because it is uninstantiated) nor simplicity (because each of the laws is so idiosyncratic). Tooley also offers this case as a challenge for accounts of laws as relations of necessitation among universals. Presumably, one of the universals that would have to participate in the relation that constitutes the B–J law—namely, the property of being a B–J interaction—is uninstantiated. But on many accounts, a universal must sometime be instantiated in order to exist. Otherwise, as Armstrong (1983, p. 82) says, the existence of universals is contrary to "naturalism"; uninstantiated universals cannot be regarded as simply the repeatable features of the spatiotemporal world studied by science.

Armstrong (1983, p. 124) and Earman (1984, p. 210) say that, in Tooley's possible world, there is no B–J interaction law but had there been a B–J interaction, then there would have been a B–J interaction law. This counterfactual violates Λ^{+}-preservation because it denies that all of the B–J regularities that lack physical necessity in Tooley's world also lack physical necessity in the closest world to Tooley's in which there is a B–J interaction. It seems to me that this move, intended to gratify somewhat Tooley's intuition that there is a B–J interaction law in his world, is the result of misplaced metaphysical scruples: in Earman's case, his insistence that the laws supervene on the Humean facts, and in Armstrong's case, his insistence that laws are relations among universals coupled with his devotion to "naturalism."

The move suggested by Armstrong and Earman would not violate Λ^+-preservation if the supposition of a B–J interaction conflicted with the Λ^+ of Tooley's world. But as Tooley and Armstrong describe Tooley's world, it is merely the "boundary conditions" that prevent the B and J particles from interacting. This suggests that there is a law covering their interaction but, as a matter of sheer happenstance, B and J particles never have an opportunity to interact. This exploits the distinction between laws and initial conditions to which I referred above: that the initial conditions are able to vary tremendously under the same laws. Tooley's example thus exploits the same intuition as the lone-proton case; nonsupervenience follows from the thought that many B–J interaction laws are consistent with the non-nomic facts in Tooley's world.

Perhaps Earman and Armstrong would say that in the closest lone-proton world, there is no electron–proton interaction law, but there would have been some such law had there been an electron. But then why would it have been the same law as in the actual world if there is no interaction law in the closest lone-proton world?

63. Appendix 3 of chapter 1 discusses Lewis's explanation of why miracles possess special significance in determining a possible world's closeness.
64. In section 1 of chapter 1, I directed a similar point against Armstrong's account. Lewis (1983, p. 366) makes what I take to be a similar objection.
65. This is my reply to Armstrong's (1983) "second objection" (p. 51).
66. This is erroneous. In chapter 1, I gave examples in which hypotheses believed physically unnecessary were confirmed by their instances to hold in unexamined cases. These examples must somehow be reconciled with the view that hypotheses believed physically unnecessary cannot be confirmed "inductively." I set this problem aside for the moment; it is the principal topic of chapter 4.

Chapter 3

1. It might be suggested that unless we believe it to be a *law* that all *F*s are *G*, we cannot be justified in believing that all *F*s are *G* while believing that there remain *F*s that we have not yet examined for their *G*ness. Such a view has been defended by many philosophers (see my discussion of Mackie at the close of chapter 2) and is the subject of chapter 4. For now, let me say only that this view, as I have just expressed it, is mistaken (as was foreshadowed in section 2.4 of chapter 1). But I ultimately argue that this view contains a key grain of truth: there is an important *kind* of confirmation (which I call "inductive" confirmation) that is available to a hypothesis only if it is believed that the hypothesis may express a physical necessity. To explain what distinguishes "inductive" confirmation, why it is important, and why its availability is related to whether the hypothesis is believed to express a candidate physical necessity, I must first discuss the relation of laws to counterfactuals, as I do in this chapter.
2. I anticipated this point in section 2.2 of chapter 1.

3. Actually, these two remarks turn out to be a bit oversimplified, as I discuss in section 3 below.
4. Actually, (iv) renders (ii) superfluous: let p be a logical truth in U^+. Then (iv) requires that for any $q \in \Gamma$, $p > q$. Since p is a logical truth, $(p > q) \Leftrightarrow q$. Note also that $\Gamma \Rightarrow p$ means that p is logically entailed by some members of Γ.
5. Let us take this last step more slowly. On a typical possible-worlds account of counterfactuals, $p > q$ is correct (when it is not the case that $\neg p$ is logically necessary) if and only if there is a possible world in which p & q is true that is more like the actual world (in the special sense of "similarity" relevant to counterfactuals) than is any possible world in which p & $\neg q$ is true. [It is not the case that $(r \,\&\, r')$ is logically necessary, since r' does not belong to Γ, but Γ is stable and so contains all of the logical truths in U^+.] So if $(\neg r \vee \neg r') > r$, then there must be a world in which $(\neg r \vee \neg r') \,\&\, r$ is true that is more like the actual world than is any world in which $(\neg r \vee \neg r') \,\&\, \neg r$ is true. But any world in which $(\neg r \vee \neg r') \,\&\, r$ is true is a world in which $\neg r'$ is true. So there must be a world in which $(\neg r \vee \neg r') \,\&\, \neg r'$ is true that is more like the actual world than is any world in which $(\neg r \vee \neg r') \,\&\, \neg r$ is true—and so, in particular, than is any world in which $(\neg r \vee \neg r') \,\&\, \neg r \,\&\, r'$, that is, any world in which $(\neg r \vee \neg r') \,\&\, r'$. It follows that $(\neg r \vee \neg r') > \neg r'$. Thus, $(\neg r \vee \neg r') > r \Rightarrow (\neg r \vee \neg r') > \neg r'$.

   On a possible-worlds account of counterfactuals, $(r \vee s) > t$ if $r > t$ and there is a world in which r is true that is more like the actual world than is any world in which s is true—whether or not $s > t$. It has sometimes been suggested that $(r \vee s) > t$ requires $r > t$ *and* $s > t$, considering certain English sentences (e.g., "Were you to stay home or to go out, you would still complain," "Had he done the deed personally or merely given the order, he would have been responsible."). But Loewer (1976) and McKay and van Inwagen (1977) argue persuasively that, in such cases, the English "or" does not function as the logical "$\vee$."
6. To recognize the influence that context can exert here, recall Bennett's example of counterfactuals' context sensitivity (from chapter 2, note 21), involving "Had I jumped from this window, I would have suffered serious injury" (call this $p > q$). This counterfactual can be recast in the form I am discussing here, since $(p > q) \Leftrightarrow ((p \,\&\, (q \vee \neg q)) > q) \Leftrightarrow (((p \,\&\, q) \vee (p \,\&\, \neg q)) > q)$. Let $a = \neg(p \,\&\, q)$ and $b = \neg(p \,\&\, \neg q)$. Context influences whether we would be correct in asserting $(\neg a \vee \neg b) > \neg a$, $(\neg a \vee \neg b) > \neg b$, or neither.
7. We could put this assertion to the test by selecting as our a an accidental truth that is preserved under a tremendously broad range of counterfactual antecedents consistent with the logical closure of it together with the members of Λ^+. For instance, let a be "Sometime in the history of the universe, there exists some matter." It is perhaps initially difficult to find any counterfactual antecedent consistent with the logical closure of a together with the members of Λ^+ under which a might not be preserved. But let b be "The universe's energy is insufficient to return the universe to a Big Crunch in much less than 15 billion years [the current age of the universe]." I see no reason to say that $(\neg a$ or $\neg b) > (a \,\&\, \neg b)$—that is, "Had there either been no matter ever or else enough energy to close the universe in much less than 15 billion years, then there would still have been matter sometime and sufficient energy to close the universe so quickly."

8. In chapter 7, I draw a distinction between the natural laws and the physical necessities; all laws are physically necessary, but some physical necessities are not laws. Accordingly, suppose now that Γ is a special proper subset of Λ^+: the logical closure in U^+ of the truths of the form □*p* in U^+. Again, consider the counterfactual antecedent "Had it not been a law that every material particle accelerated from rest fails to be accelerated beyond the speed of light." It can be *physically necessary* that no material particle is accelerated from rest to beyond the speed of light even if it is not a *law*, and so this counterfactual antecedent is consistent with every member of Γ. But consider the following counterfactual conditional, which must be correct for the given set to be stable: "Had it not been a law that every material particle accelerated from rest travels slower than the speed of light, it would still have been physically necessary."
9. I reconsider this argument in section 3 below, in connection with "grades of necessity." I also reconsider this argument at the close of chapter 7, in connection with laws governing laws. But to simplify my exposition, I defer these reasons why Λ^+ may not, in fact, be the unique nontrivially stable set. Also notice that if we remove from Λ^+ some physically necessary, logically contingent $p \in U$, or if we remove □*p*, then for the resulting set to be logically closed, we must also remove some truth of the form ■*q*. But if we remove ¬□*p*, we need not also remove ¬■*p*, since (I argue in chapter 7) ¬(¬■*p* ⇒ ¬□*p*); not every physical necessity expresses a law. However, the logically closed set Γ of truths that results from removing ¬□*p* while retaining ¬■*p* is unstable; Con(Γ,□*p*), but there is no guarantee that □*p* > ¬■*p*, though it is required by Γ's stability. Perhaps *p* would have been a law had *p* been physically necessary.
10. Actually, (iv) again renders (ii) superfluous; see note 4.
11. If Γ is a stable set, then Γ's members in U must form a non-nomically stable set. For if Γ is stable, then (in particular) Γ's members *in U* are preserved under (in particular) every *p in U* that is consistent with Γ. But these include every *p* in U that is consistent with Γ's members in U (and so Γ's members in U form a non-nomically stable set). That is, if *p* in U is consistent with Γ's members in U, then *p* is consistent with Γ. That follows from its contraposition: if *p* in U is not consistent with Γ, then Γ must entail ¬*p*, so ¬*p* must belong to Γ, so *p* is not consistent with Γ's members in U.
12. Because I just showed that Λ lacks stability *simpliciter*, the "non-nomic" qualification is needed.
13. Notice that [Λ ∪ {*a*}]-preservation (second revision) from section 1 of chapter 2 entails that [Λ ∪ {*a*}]'s closure in U has non-nomic stability. So since [Λ ∪ {*a*}]'s closure in U lacks non-nomic stability (except trivially), [Λ ∪ {*a*}]-preservation (second revision) fails (except when *a* is so full as to make [Λ ∪ {*a*}]'s closure in U contain all truths in U).
14. Later I argue that whether or not ■*p fails* to supervene on the truths in U*.
15. Remember that the laws in these sets do not conflict in any way; all are true, and the counterfactual conditionals supported by one set form a proper subset of the counterfactual conditionals supported by another.
16. In chapter 8, I argue that different sets possess non-nomic stability *for the purposes of different scientific disciplines*. That the sort of necessity associated with laws of, say, island biogeography is distinct from the sort of necessity associated with laws of particle physics is ultimately responsible for the fact

that the scientific explanations given by a macro discipline contribute something that cannot be supplied by the scientific explanations given by fundamental physics. The different senses of necessity associated with the laws of different disciplines are not the different "grades" of necessity I have discussed here. Rather, there might be different grades of physical necessity in the sense used in island biogeography, different grades of physical necessity in the fundamental microphysics sense, and so on.

17. If it is logically possible for there to be more than one grade of physical necessity, then it is worth considering how this multiplicity could be accommodated by various proposed accounts of natural law. Consider, for instance, the Armstrong-Dretske-Tooley account of laws as relations of nomic necessitation holding among universals. Presumably there would have to be several different kinds of nomic necessitation. I do not regard this as much of an argument against the account; if we are willing to countenance one such primitive relation among universals, then I suppose we would not mind positing a multiplicity of them. But it is a complication. It is not clear to me how various grades of physical necessity would be accommodated by a Lewis-style regularity account.
18. The same argument I have just given for $p > \Box m$ also entails $p > (q > \Box m)$, and so on, that is, *nested* preservation of the actual physical necessities' physical necessity.
19. Λ^+ (and Λ) preservation also concern certain *nomic* counterfactual antecedents: those in U^+ that are logically consistent with Λ^+. I return to these in chapter 7.
20. Recall that early in chapter 2, I appealed to Λ^+-preservation in order to show that for any accident *a*, there is some counterfactual antecedent for which $[\Lambda \cup \{a\}]$-preservation (first revision) fails. In particular, I used Λ^+-preservation (and Popper's moa example) to show that when *a* is "All *F*s are *G*," one such counterfactual antecedent is "Had there been an *x* such that *Fx* and *Px*" where *Px* means $Gx \supset \Box a$. But now we have seen that Λ^+-preservation may be violated; if there is more than one grade of physical necessity in the closest *p*-world, then it may be that had *p* been the case, something would have been physically necessary that is not physically necessary in the actual world. I must therefore amend the counterfactual antecedent in the preceding argument to be "Had there been an *x* such that the following are all the case: *Fx*, *Px*, and there is a unique grade of physical necessity." When I originally discussed the "ideal test moa", I found no intuitive need for this amendment, but I have just suggested that we commonly disregard the possibility of multiple grades of necessity "between" logical or conceptual necessity and no necessity.
21. A further reason for some unease with this proposal regarding the root commitment is that it applies only to our believing $\blacksquare p$ *where p* is in U. I discuss meta-laws at the close of chapter 7. To extend an account to meta-laws, we need resources for distinguishing laws from physically necessary nonlaws, since "All laws of nature are Lorentz-invariant" and other such laws governing laws are logically entailed by the distribution of $\blacksquare$s and $\neg\blacksquare$s among the members of U—and so are physically necessary—but not all physical necessities in U^+ are meta-laws.

Chapter 4

1. As in chapters 2 and 3, I assume here that $\Box h \Rightarrow h$.
2. Here I follow standard practice in representing our degrees of belief as subjective probabilities, with a probability of 1 as full belief and 0 as denial, where pr(*h*) is our degree of confidence in *h* before discovering *e*, and pr(*h* | *e*) is our degree of confidence in *h* after discovering *e*. (Though I presuppose Bayesian updating throughout, nothing important will turn on this.) For a brief introduction to this approach, see Skyrms (1986).
3. To "confirm" *h* is to confirm *h*'s truth. Confirmation of *h* may but need not be accompanied by confirmation of $\Box h$.
4. By a "case" of "All *F*s are *G*," I mean an *F*; by an "unexamined case," I mean an *F* that has not yet been examined for whether it is *G*; by an "instance," I mean that some thing, which we already believe to be an *F*, is *G*. I use this terminology to make better contact with the intuitions that are driving Goodman and others. Later in this chapter, I discuss Hempel's raven paradox, which is also provoked by this terminology. Ultimately (in chapter 5), we can do without this terminology, since our attention will no longer be restricted to hypotheses of the form "All *F*s are *G*."
5. I discuss Mill's view further in section 1 of chapter 7. In the passage I have just cited, he writes: "[E]very well-grounded inductive generalisation is either a law of nature or a result of laws of nature, capable, if those laws are known, of being predicted from them." For more on this distinction between laws and physically necessary nonlaws, see section 2 of chapter 7. Mill (1893, bk. 3, ch. 3, at the close of sec. 3) discusses inductive confirmation in Goodmanlike terms.
6. A little care is needed to see just how Reichenbach presupposes that only claims that we believe may be physically necessary can receive inductive support. In distinguishing laws from accidental generalizations, Reichenbach first presents some formal conditions (e.g., being an "all statement," containing no proper names or local predicates) such that if a claim satisfies them, then we cannot be justified in believing that we have checked all of its instances. It follows, according to Reichenbach, that to justify accepting such a claim, we need to use inductive confirmation. He then specifies that the "original nomological statements" are the claims satisfying these conditions that are "demonstrable," that is, whose acceptance can be justified. Reichenbach (1954) summarizes the point thus: "It is the inductive verification, not mere truth, which makes an all-statement a law of nature" (pp. 11–12).
7. Perhaps Lewis (1973, p. 74; 1994, p. 479) should be added to this list.
8. This "standard history" is told in many books and articles on confirmation, such as van Fraassen (1983), Rosenkrantz (1981, pp. 1.4-1–1.4-4), Lambert and Brittan (1987, pp. 67–110), and Suppes (1966, pp. 202–204).
9. The example of ships lost at sea is due essentially to Good (1967). He imagines a ship, lost in a storm at night, that at dawn finds itself near an unknown island. The captain, consulting his charts, concludes that this island must be either island X or island Y, which can be distinguished by their bird populations. Both islands have a great many birds. On island X, all of the crows are black, but only 5% of the birds are crows. On island Y, most of the birds are crows, but not all crows are black; 95% of the birds are black crows, 1% are white crows, and 4% are noncrows. The captain dispatches a

shore party to examine the bird population. A few minutes later, the captain spies the shore party standing on the beach, waving a black crow. The captain would be incorrect to regard this discovery, an instance of "All crows on the island are black," as confirming this hypothesis, and thereby confirming that the ship is located at island X. He would be ignoring the background information that most crows are black on island Y and that the fraction of crows among the birds on island Y is much greater than the fraction of crows among the birds on island X. He would also be ignoring the information in the evidence beyond the mere instantiation of "All crows on the island are black"—namely, that a crow was found quickly.

The example of philosophers checking their hats, due to Rosenkrantz (1977), is essentially similar. Three philosophers—A, B, and C—check their hats with the bell boy. They are the only ones who do. When they leave, the bell boy mistakenly returns B's hat to A. This is an instance of "Each philosopher receives someone else's hat." Likewise, the bell boy returns A's hat to B—another instance. But these instances, far from confirming the hypothesis, demonstrate that it is false. We must not ignore the background knowledge that there were three philosophers and three hats. We must also not ignore the evidence beyond the fact that A and B each received someone else's hat; the evidence includes whose hats they received. It follows that C receives her own hat, contrary to the hypothesis.

The example of grasshoppers outside of Pitcairn Island is from Swinburne (1971). Suppose we begin with a background belief that animals located on Pitcairn Island are much like animals located elsewhere, especially on islands of similar climate. Suppose, to embroider Swinburne's example, that we are viewing a grasshopper exhibit in a museum. We see many preserved specimens of grasshoppers from various islands, though none from Pitcairn. We thereby discover instances of the hypothesis "All grasshoppers are located in parts of the world other than Pitcairn Island." Yet these instances might well disconfirm the hypothesis.

10. By referring to a "special kind" of confirmation, I do not mean a kind of confirmation that operates outside of the probabilistic framework, or that it is special simply because it involves a hypothesis that we believe may be physically necessary. How "inductive" confirmation is special is the subject of this chapter and the next.

11. It is often remarked that the probabilistic machinery of recent confirmation theory *does* save what is right about instance confirmation, by the following argument: according to Bayesian confirmation theory, *e*'s discovery confirms h if and only if $\mathrm{pr}(h \mid e) > \mathrm{pr}(h)$, and according to Bayes's theorem, this inequality holds if and only if $\mathrm{pr}(e \mid h) > \mathrm{pr}(e)$, but if e is an instance of h, then h (together with background beliefs) entails e, so $\mathrm{pr}(e \mid h) = 1$, and so [since $\mathrm{pr}(e) < 1$ because e is new evidence] $\mathrm{pr}(e \mid h) > \mathrm{pr}(e)$, and so *e*'s discovery confirms h. But this does not capture the intuition motivating the "classical conception" of induction as "projection" from instances to unexamined cases, since this argument does not show anything about whether e confirms that h holds *of unexamined cases*. The analogous argument concerning *e*'s confirmation of *h*'s holding in unexamined cases does not lead anywhere, since even if e is an instance of h, so that e is entailed by h (together with background), e is not in general entailed by *h*'s *holding of unexamined cases* (to-

gether with background), and so pr($e \mid h$ holds of unexamined cases) is not equal to 1.

12. Thus, we do not need to have already justly arrived at such background beliefs in order to confirm a given hypothesis inductively. This anticipates a point I emphasize in chapter 5, that an "inductive strategy" is a "free move" precisely because inductive confirmation is not sustained by a certain sort of background opinion—an opinion that we could be warranted in holding only by virtue of some prior empirical investigation.
13. By Bayes's theorem, $\mathrm{pr}(h_1 \mid b_1) = \mathrm{pr}(h_1)\ \mathrm{pr}(b_1 \mid h_1)/\mathrm{pr}(b_1) = (.25)(1)/(.5) = .5$. Similarly, $\mathrm{pr}(h_2 \mid b_1) = (.25)(.5)/(.5) = .25$.
14. By Bayes's theorem, $\mathrm{pr}(h \mid b_1) > \mathrm{pr}(h)$ exactly when $\mathrm{pr}(b_1 \mid h) > \mathrm{pr}(b_1)$. Since b_1 is new evidence, $\mathrm{pr}(b_1) < 1$, whereas $\mathrm{pr}(b_1 \mid h) = 1$.
15. By the theorem of total probability, $\mathrm{pr}(b_2 \mid b_1) = \mathrm{pr}(b_2 \mid h_1\&b_1)\ \mathrm{pr}(h_1 \mid b_1) + \mathrm{pr}(b_2 \mid h_2\&b_1)\ \mathrm{pr}(h_2 \mid b_1) + \mathrm{pr}(b_2 \mid h_3\&b_1)\ \mathrm{pr}(h_3 \mid b_1) + \mathrm{pr}(b_2 \mid h_4\&b_1)\ \mathrm{pr}(h_4 \mid b_1) = (1)(.5) + (.5)(.25) + (.5)(.25) + 0 = .75 > .5 = \mathrm{pr}(b_2)$.
16. Let d be that the marble yielded by the second withdrawal is distinct from the marble yielded by the first withdrawal. By the theorem of total probability, $\mathrm{pr}(b_2 \mid d\&b_1) = \mathrm{pr}(b_2 \mid d\&h_1\&b_1)\ \mathrm{pr}(h_1 \mid d\&b_1) + \mathrm{pr}(b_2 \mid d\&h_2\&b_1)\ \mathrm{pr}(h_2 \mid d\&b_1) + \mathrm{pr}(b_2 \mid d\&h_3\&b_1)\ \mathrm{pr}(h_3 \mid d\&b_1) + \mathrm{pr}(b_2 \mid d\&h_4\&b_1)\ \mathrm{pr}(h_4 \mid d\&b_1) = (1)(.5) + 0 + 0 + 0 = .5 = \mathrm{pr}(b_2 \mid d)$. Note, for instance, that $\mathrm{pr}(b_2 \mid d\&h_2\&b_1) = 0$ because if the first marble withdrawn is black and the second marble withdrawn is distinct from the first and there is one black marble and one white marble, then the second marble withdrawn must be white.

    Here's another way to bring out that the instance's power to bear upon an unexamined case depends on our belief that $\neg d$ may obtain—that the unexamined case may involve the same marble as the instance. Notice how $\mathrm{pr}(b_2 \mid b_1)$ comes to exceed $\mathrm{pr}(b_1)$. Apply the theorem of total probability: $\mathrm{pr}(b_2 \mid b_1) = \mathrm{pr}(b_2 \mid d\&b_1)\mathrm{pr}(d \mid b_1) + \mathrm{pr}(b_2 \mid \neg d\&b_1)\ \mathrm{pr}(\neg d \mid b_1) = (.5)(.5) + (1)(.5) = .75$. [Notice $\mathrm{pr}(b_2 \mid \neg d\&b_1) = 1$ because if both withdrawals yield the same marble (i.e., $\neg d$), then if the first withdrawal yields a black marble, the second must do so too.] Compare $\mathrm{pr}(b_2) = \mathrm{pr}(b_2 \mid d)\ \mathrm{pr}(d) + \mathrm{pr}(b_2 \mid \neg d)\ \mathrm{pr}(\neg d) = (.5)(.5) + (.5)(.5) = .5$. So all of the increase in b_2's likelihood as a result of b_1's discovery is given $\neg d$; the instance's confirming power is insulated by d.
17. This is easily shown by reinterpreting b_1 to represent h's many past instances and b_2 to be that the next withdrawal will yield a black marble. As in note 16, $\mathrm{pr}(b_2 \mid d\&b_1) = \mathrm{pr}(b_2 \mid d\&h_1\&b_1)\ \mathrm{pr}(h_1 \mid d\&b_1) + \mathrm{pr}(b_2 \mid d\&h_2\&b_1)\ \mathrm{pr}(h_2 \mid d\&b_1) + \mathrm{pr}(b_2 \mid d\&h_3\&b_1)\ \mathrm{pr}(h_3 \mid d\&b_1) + \mathrm{pr}(b_2 \mid d\&h_4\&b_1)\ \mathrm{pr}(h_4 \mid d\&b_1) = (1)(.5) + 0 + 0 + 0 = .5 = \mathrm{pr}(b_2 \mid d)$. Notice that $\mathrm{pr}(h_1 \mid d\&b_1)$ remains .5, just as it was when we took b_1 to represent only a single instance of h, because d tells us that the instances represented by b_1 all involve the same marble.
18. Of course, a given instance does not confirm, of a case posited as involving the withdrawal of a white marble, that this case would accord with h. (The same goes for a case posited as involving the withdrawal of a black marble.) But a case posited as possessing this property is not an *unexamined* case in the relevant sense. In section 3 I refine this point considerably.
19. My means of distinguishing "inductive" confirmation—so far, by the fact that the instance's bearing on all actual unexamined cases is not sustained by our beliefs about which hypothetical unexamined cases, of those we do not believe physically impossible, happen to be actual—is usefully compared to Nagel's (1961) approach in this passage:

> S [that all of the screws in Smith's car are rusty—which is presumed to lack physical necessity] might be accepted on the ground that we have examined only a presumably "fair sample" of screws in Smith's car, and have inferred the character of the unexamined screws from the observed character of the screws in the sample. But . . . the presumption of the inference is that the screws in the sample come from a class of screws that is complete and will not be augmented. For example, we assume that no one will remove a screw from the car and replace it by another, or that no one will drill a fresh hole in the car in order to insert a new screw. . . . On the other hand, no analogous assumption appears to be made concerning the evidence on which statements called laws are accepted. . . . If we accept S as true on the basis of what we find in the sample, we do so in part because we assume that the sample has been obtained from a population of screws that will neither increase nor be altered during the period mentioned in S. . . . In calling a statement a law, we are apparently asserting at least tacitly that as far as we know the examined instances of the statement do not form the exhaustive class of its instances. Accordingly, for an unrestricted universal to be called a law it is a plausible requirement that the evidence for it is not known to coincide with its scope of predication and that, moreover, its scope is not known to be closed to any further augmentation. (pp. 62–63)

I join Nagel in emphasizing that the instance's confirming power in this example depends on background beliefs about which physically unnecessary possibilities may happen to be realized. For this reason, I say that the confirmation is not "inductive." But Nagel focuses on our belief that the set of screws in Smith's car will not be augmented. As discussed in appendix 2 of chapter 1 (especially note 31), and emphasized in section 3 of chapter 2 (in connection with the lone-proton world), I do not agree with Nagel's view about what we are tacitly asserting in calling something a law. The correct moral, as I see it, is drawn by focusing on the "freshness" of the additional screws (that they were just bought from the store) rather than on the mere fact that they are additional. Instances of S fail to confirm that, had there been fresh screws in Smith's car, they would have accorded with S. Hence, instances fail to confirm S inductively. The mere fact that the screws are posited to be additional does not suffice to insulate them from instances' confirming power. The Native-American case bears this out: the instances confirming that all Native Americans are blood type O or A confirms that, had an additional Native American been born, then he or she would have had blood of type O or A.

20. This does not mean that those background beliefs are entirely idle. Perhaps we discover the instance by making a certain observation, and it is only in virtue of a background belief about a physically unnecessary matter of fact (e.g., that some measuring device happens to be well calibrated) that this observation persuades us that the instance obtains. Nevertheless, our beliefs about physically unnecessary facts are not responsible for the breadth of the instance's confirming power. See also note 28.
21. The two "structure-dependent" accidental generalizations to which Feigl (1961) alludes are as follows:

Any eruption of Old Faithful is followed, roughly 64 minutes later, by another.

Any year of maximum solar activity is followed, about 11 years later, by another.

Both were highly confirmed by instances to hold in unexamined cases. Nevertheless, neither was confirmed *inductively*. Although we do not know precisely what "structure" is responsible, we can posit a case that we believe lies outside of whatever structure is common to all actual cases. This posited case is not confirmed by the instance to accord with the hypothesis. In the solar-maximum example, the past instances fail to confirm that, were the Sun to become a red giant star in the 10 years following the next solar maximum, then the next solar maximum after that would occur about 11 years later. Likewise, we may not yet understand precisely how the size and shape of the cavern beneath Old Faithful combines with the rest of the neighborhood's geology to determine the geyser's period. Nevertheless, an instance of the Old Faithful generalization fails to confirm it inductively, since the instance does not confirm that, were there an eruption of the geyser followed by an earthquake that closed the underground cavern, then the next eruption would occur in about 64 minutes.

Suppose, on the other hand, that we have a precise theory regarding the structure responsible for Old Faithful's period. Suppose we then modified Feigl's generalization to read "Any eruption of Old Faithful after which the underground cavern continues to have *A* structure, the water table continues to have *B* structure, and the underground heat source continues to have *C* structure is followed in 64 minutes by another eruption." Any additional property *P* that, when attributed to a posited case of the original generalization, insulated that case from an instance's confirming power cannot consistently be ascribed to a case of the modified generalization. Accordingly, we regard the modified generalization as confirmed inductively. Of course, we believe that the modified generalization does not express a natural law, since we believe that no natural law concerns Old Faithful in particular (see appendix 2 of chapter 1). Nevertheless, we believe the modified generalization to be physically necessary—a logical consequence of various laws (that fill out *A*, *B*, and *C*). For more on physically necessary nonlaws, see section 2 of chapter 7.

22. For more on the intuitive motivation for my conception of "inductive" confirmation, and on its differences from the ways that other philosophers have cashed out the intuitions motivating Goodman and others, see appendix 1 to this chapter.
23. A property *P* such that Fx & Px is logically inconsistent—and so *trivially* logically necessitates Gx—is unsuitable by virtue of the constraint on suitability that I discuss next.
24. As is customary in accounts of confirmation, I am assuming that we know all of the relevant logical truths.
25. I also motivated this intuition in connection with the marbles example in section 2.2 above.
26. A further restriction would be needed to deal with *P*s that refer rigidly to the actual world (see section 2.6 of chapter 2). For simplicity, I disregard these *P*s here.

27. Obviously, this definition treats the inductive confirmation of "All *F*s are *G*" as requiring a different kind of evidence, and as requiring the confirmation of different subjunctive conditionals, from the inductive confirmation of "All ¬*G*s are ¬*F*." This is an artifact of talking about projecting a *generalization* from observed instances onto unexamined cases (which I am doing to make better contact with the intuitions motivating Goodman and others to distinguish "inductive" confirmation). This feature disappears in chapter 5.
28. Nevertheless, when the evidence confirms the hypothesis inductively, background beliefs about which suitable kinds of unexamined cases there happen to be might influence the *degree* to which the evidence confirms that all of the actual unexamined cases accord with *h*. This raises an important point: the discussion in this chapter has depicted the continuum running from mere content cutting to inductive confirmation as solely concerning the *range* of unexamined cases on which the evidence bears. This is an oversimplification, since it neglects *quantitative* features of confirmation. For instance, a piece of evidence might confirm *h* inductively but bear to different degrees upon different suitable unexamined cases. Two pieces of evidence might each confirm *h* inductively but bear to different degrees upon a given suitable kind of unexamined case. I lack the space here to discuss these matters in more detail, but that is not needed to develop a concept of "inductive" confirmation that does the job I am asking it to perform. See also appendix 1 to this chapter.
29. Actually, at the end of chapter 3, I amended this *P* by adding "There is a single grade of physical necessity." For the sake of simplicity, I disregard this complication here.
30. If this argument goes through, can we likewise show that we cannot confirm *h* inductively if we believe that *h* may be physically necessary—by the analogous procedure of allowing *Px* to mean "$Gx \supset$ (*h* is an accidental generalization)"? No. This *P* was one of my earlier examples of a property that is unsuitable when we believe that *h* may be physically necessary. We might then worry that the definition of "suitability" has been gerrymandered to produce this result. See the discussion in section 3 of the intuitive motivations for the various elements of the definition. See also chapter 5, where my definition of "inductive" confirmation is generalized and thus simplified.
31. For any *j* believed physically unnecessary, the property *P* where *Px* means "$Gx \supset \Box j$ and there is a single grade of physical necessity" is suitable when we believe $\neg\Box h$ and violates (4.1), and so shows that *h* cannot be confirmed inductively. In my ideal "test moa" argument of chapter 2 section 1.5, I let *j* be *h*, making it easier to see that *h* is not preserved under *p*. But I could have chosen any such *P*: if we believe *h* to be accidental, then *h* is not preserved under a counterfactual antecedent positing an *F* whose *G*ness requires the physical necessity of something (whether *h* or something else) that actually is physically unnecessary (where there is a single grade of physical necessity). For instance, suppose that, as a matter of physically unnecessary fact, all creatures who breathe a certain oxygen molecule die before taking their next breath (*j*). Consider the counterfactual antecedent having the intuitive import: "Had there been a moa, 20 years old, ideally healthy and apparently ideally situated for future health, who now breathes the given oxygen molecule (and so will live beyond age 50 if it does not die before its next breath)." We believe that if even this moa would have died before age 50, it must be *physically necessary* that any creature who breathes the given oxy-

gen molecule dies before taking its next breath. So this counterfactual antecedent posits a moa with property P: if it dies before age 50, then $\Box j$. It would not die before age 50. Once again, the Ps that keep h from being confirmed inductively are exactly the Ps that keep h from being preserved in the manner of a physical necessity.

32. That is, the limit in the weak future-moving sense is the limit of the sequence $pr(G_2 \mid G_1)$, $pr(G_3 \mid G_2,G_1)$, $pr(G_4 \mid G_3,G_2,G_1)$. . . . In contrast, one limit in the weak past-reaching sense is the limit of the sequence $pr(G_2 \mid G_1)$, $pr(G_2 \mid G_1,G_0)$, $pr(G_2 \mid G_1,G_0,G_{-1})$. . . .
33. My sense of "projection" does not quite satisfy Earman's demand, but I am not sure that the demand is appropriate. Suppose our background beliefs include that *a* (the object under examination) is an emerald slice that precedes 3000. Let Px mean that x is at a given spatiotemporal location during or after the year 3000, so that the subjunctive antecedent "Were there an emerald slice possessing P" posits an emerald slice on the opposite side of the year 3000 from the emerald slice *a* actually under examination. Suppose that we hold there could be at most one emerald slice possessing P, and that, as yet, we have adopted neither the green hypothesis nor the grue hypothesis. Suppose we believe that the green hypothesis may express a physical necessity, and likewise for the grue hypothesis, so that each is a candidate for inductive confirmation. To confirm the green hypothesis inductively, *Ga* must confirm the conditional C_P: "Were there an emerald slice possessing property P, then it would be green." (Presumably, P is "suitable.") To confirm the grue hypothesis inductively, *Ga* must confirm the conditional C'_P: "Were there an emerald slice possessing property P, then it would be blue." Suppose we believe there to be laws entailing an object's color from the fact that it is an emerald slice and perhaps its spatiotemporal location. Suppose also that we believe the entailed color to be "green" or to be "blue." That is, we believe that "All emerald slices are green or blue" is physically necessary. Therefore, we believe that the laws conjoined with the antecedent of C_P and C'_P either logically entail the consequent of C_P or logically entail the consequent of C'_P. By Λ-preservation, then, we believe that either C_P or C'_P. Therefore, *Ga* cannot confirm them both. Consequently, *Ga* cannot inductively confirm both the green hypothesis and the grue hypothesis. Under these conditions, then, my concept of "projection" satisfies Earman's criterion of adequacy. But suppose the assumption that we believe "It is physically necessary that all emerald slices are green or blue" is relaxed, and instead we believe "It is physically necessary that all emerald slices are green or blue or yellow or . . . ." Then we cannot regard *Ga* as inductively confirming each of the hypotheses "All emerald slices are green," "All emerald slices are grue," "All emerald slices are grellow," and so on, though we *can* regard *Ga* as inductively confirming all but one of these hypotheses. This seems right; Earman's criterion of adequacy seems in this respect stricter than it ought to be. In another respect, however, it is not strict enough. I believe that one of the valid lessons to be drawn from Goodman's examples is the relation between projection and physical necessity. Indeed, this relation is one of Goodman's central concerns. But Earman does not discuss it; he appears to deny that a faithful elaboration of Goodman's sense of "projection" must allow this relation to be expressed.

34. Here is how this can happen. Let our evidence e be Ga, where we already believed Fa but did not already believe Ga. Suppose we believe that h ("All Fs are G") may be physically necessary. So Ga confirms $\Box h$, as I have just shown. Ga can nevertheless fail to confirm j: that a given unexamined case accords with h (i.e., that a given object—which we already believe to be F and do not already believe to be G—is G). By the theorem of total probability,

$$\text{pr}(j) = \text{pr}(j \mid \Box h)\text{pr}(\Box h) + \text{pr}(j \mid \neg\Box h)\text{pr}(\neg\Box h).$$

Since $\Box h$ entails j, $\text{pr}(j \mid \Box h) = 1$, and so

$$\text{pr}(j) = \text{pr}(\Box h) + \text{pr}(j \mid \neg\Box h)\text{pr}(\neg\Box h).$$

Likewise,

$$\text{pr}(j \mid e) = \text{pr}(\Box h \mid e) + \text{pr}(j \mid \neg\Box h, e)\text{pr}(\neg\Box h \mid e).$$

If $\text{pr}(j \mid \neg\Box h, e) = \text{pr}(j \mid \neg\Box h)$, then since $\text{pr}(j \mid \neg\Box h) < 1$ [because otherwise $\text{pr}(j) = 1$, contrary to our assumption that we do not already believe j], $\text{pr}(\Box h \mid e) > \text{pr}(\Box h)$ entails $\text{pr}(j \mid e) > \text{pr}(j)$. That is, an instance confirming $\Box h$ thereby confirms j. But if $\text{pr}(j \mid \neg\Box h, e) < \text{pr}(j \mid \neg\Box h)$, then e can confirm $\Box h$ without confirming j. Intuitively, e can bear upon j either through bearing upon the possibility that j and $\Box h$ hold or through bearing upon the possibility that j holds but $\Box h$ does not. The point of using the theorem of total probability is to separate e's supporting $\Box h$ from e's supporting j. Intuitively, the support of j deriving from the confirmation of $\Box h$ can be balanced or even outweighed by our decreased confidence in j given $\neg\Box h$. To illustrate, let me put in definite values:

$$\text{pr}(\Box h \mid e) = .8 > \text{pr}(\Box h) = .2$$
$$\text{pr}(j \mid \neg\Box h) = .8 > \text{pr}(j \mid \neg\Box h, e) = .2$$

So

$$\text{pr}(j) = (.2) + (.8)(.8) = .84$$
$$\text{pr}(j \mid e) = (.8) + (.2)(.2) = .84.$$

If each unexamined case is like j, then h is merely content-cut by e, even though we believe that h may be physically necessary.

Why did these particular numbers yield $\text{pr}(j \mid e) = \text{pr}(j)$? Let

$$\Delta = \text{pr}(\Box h \mid e) - \text{pr}(\Box h) = \text{pr}(j \mid \neg\Box h) - \text{pr}(j \mid \neg\Box h, e).$$

Then

$$\begin{aligned}\text{pr}(j \mid e) &= \text{pr}(\Box h \mid e) + \text{pr}(j \mid \neg\Box h, e)\text{pr}(\neg\Box h \mid e) \\ &= \text{pr}(\Box h) + \Delta + [\text{pr}(j \mid \neg\Box h) - \Delta]\,[\text{pr}(\neg\Box h) - \Delta] \\ &= \text{pr}(\Box h) + \Delta + \text{pr}(j \mid \neg\Box h)\,\text{pr}(\neg\Box h) + \Delta^2 - \Delta\,\text{pr}(\neg\Box h) \\ &\quad - \Delta\,\text{pr}(j \mid \neg\Box h).\end{aligned}$$

Let $\alpha = \text{pr}(\neg\Box h) = \text{pr}(j \mid \neg\Box h)$. So

$$\begin{aligned}\text{pr}(j \mid e) &= \text{pr}(\Box h) + \text{pr}(j \mid \neg\Box h)\,\text{pr}(\neg\Box h) + \Delta^2 + \Delta - 2\Delta\alpha \\ &= \text{pr}(\Box h) + \text{pr}(j \mid \neg\Box h)\,\text{pr}(\neg\Box h) + \Delta\,(\Delta + 1 - 2\alpha) \\ &= \text{pr}(j) + \Delta\,(\Delta + 1 - 2\alpha).\end{aligned}$$

So under these simplifying assumptions, $\text{pr}(j \mid e) = \text{pr}(j)$ iff $\Delta\,(\Delta + 1 - 2\alpha) = 0$, which occurs when $\Delta = 0$ (the trivial case in which e does not bear on $\Box h$) or $\Delta = 2\alpha - 1$. (In my example, $\Delta = .6$ and $\alpha = .8$.)

35. For versions of this approach, see Hosiasson-Lindenbaum (1940), Pears (1950), Alexander (1958b), Good (1960), Mackie (1963), Suppes (1966), and Horwich (1982, pp. 54–63).
36. Some who offer this response to the raven paradox make this assumption explicitly. For instance, Pears (1950) says that when we discover a case of a generalization, the "general hypothetical does not merely escape falsification; but is confirmed *to the extent that it ran the risk of being falsified*" (p. 5; emphasis added).
37. I have in mind here examples in which we proceed in connection with the raven hypothesis much as Newton did in connection with his hypothesis of the gravitational force law. But there are certain difficulties involved in regarding a claim of the form "All members of biological species *S* have property *T*" as expressing a law, which I discuss in chapter 8. For now, to make matters less problematic, we might replace the hypothesis about ravens in this example with, say, "All emeralds are green."
38. Again, many years ago now, Ken Gemes made a suggestion to me along roughly these lines: we regard a black raven but not a nonblack nonraven as confirming *h*'s predictive accuracy. He did not employ my concept of "inductive" confirmation to elaborate this suggestion. It must be cashed out so as to accommodate the fact that we regard the discovery of *certain* nonblack things (e.g., white birds with certain characteristics) to be nonravens as confirming *certain* other nonblack things (white birds with those same characteristics) to be nonravens. We can do this by appealing to "inductive" confirmation: that a nonblack thing is a nonraven typically bears on *some* suitable kinds of nonblack things, but not on *each* of them.
39. That (roughly speaking) we regard ravens as relevant to all other ravens, but we usually do not regard nonblack things as relevant (even as far as nonravenhood is concerned) to all other nonblack things, I take to result most immediately not from the nature of our cognitive makeup, but from the sorts of "inductive strategies" that we choose to pursue, in view of our beliefs (or hunches) about what the natural kinds are (see chapters 5 and 7). Of course, to resolve the raven paradox, I do not need to say where these beliefs about natural kinds come from. It suffices to derive our attitude toward indoor ornithology from our prior opinion that ravens may well form a natural kind whereas nonblack things (or even nonblack birds) do not. In like manner, the common Bayesian response that I discuss purports to derive our attitude toward indoor ornithology from our prior opinion that there are many more nonblack things (or nonblack birds) than ravens. This response does not have to say how we came justly to hold such an opinion; it need

show only that this prior opinion can be responsible for our intuitions regarding how to confirm "All ravens are black."

40. Again, I borrow this example from Ken Gemes.

Chapter 5

1. To simplify, in this case I omit those predictions that contain a rigid reference to the actual world, as I did in chapter 3.
2. There is no need for probability (pr) here, because (i) we know which hypotheses we are prepared to confirm inductively, that is, which hypotheses are in Θ, and (ii) logical omniscience—complete knowledge of the (relevant) logical truths—is being presupposed, so Con(Θ's closure in U,*p*) holds if and only if pr(Con(Θ's closure in U,*p*)) = 1. I have referred here to Θ's *logical closure* in U in order to rule out *p*s that are consistent merely with each of the hypotheses in Θ *taken individually*.
3. Here again I presume logical omniscience.
4. To demonstrate this equivalence, first show that if Θ's closure in U is non-nomically stable, then every prediction $p > q$ made by Θ's members, regarding each *p* that is consistent with Θ's closure in U, is true. (The predictions involving nested conditionals work analogously.) Consider any such prediction; let it be made by $h \in \Theta$. By the definition of "prediction," $p \in$ U and p & h logically entails q. So $p > q$ holds if $p > h$ holds. If Θ's closure in U is non-nomically stable, then by (iv) in the definition of "non-nomic stability" (chapter 3, section 2.3), $p > h$ holds.

   Now I show the reverse direction. The set Θ's closure in U is stipulated as satisfying (i) and (iii) in the definition of "non-nomic stability." Suppose that every prediction $p > q$ holds that is made by Θ's members where *p* is consistent with Θ's closure in U. Then in particular, for any such *p* and any $h \in \Theta$, $p > h$ holds (and analogously for the nested case). Hence, Θ's closure in U satisfies (ii) and (iv) in the definition of "non-nomic stability."

   It is *trivially* the case that every prediction $p > q$ holds that is made by Θ's members, where *p* is consistent with Θ's closure in U, exactly when (1) Θ's closure in U is the set of logical truths in U, since then every prediction is a logical truth, or (2) Θ's closure in U is the set of all truths in U, since then *p* must be true (since no falsehood in U is consistent with Θ's closure in U) and $p > q$ is trivially true if *p* and *q* are true. This is exactly when Θ's closure in U *trivially* possesses non-nomic stability.
5. Actually, this is a slight oversimplification: I have ignored the fact that we can confirm a prediction *f* only if, prior to confirming it, pr(*f*) ≠ 1. Notice, then, that even if some of the predictions $p > m$, $p > (q > m)$, and so on, made by $h \in \Theta$—where Con(Θ's closure in U,*p*), Con(Θ's closure in U,*q*), and so on—are false, it can happen that each of *h*'s predictions *that we confirm*, in confirming *h* inductively, is true—as long as we (mistakenly) already assign the maximal degree of confidence to each of the (unbeknownst to us) false predictions. When we are willing to confirm each of the h_i inductively (if the right evidence comes along), then each of the predictions that we are thereby willing to confirm is true, *no matter which (if any) of the predictions* p > q, p > (q > m), *and so on, where Con(Θ's closure in U,* p*), Con(Θ's closure in U,* q*), and so on, we already assign the maximal degree of confidence*, if and only if Θ's closure in U possesses non-nomic stability.

6. Well, I almost explain: in this chapter, I do not consider what would make the h_i belonging to one set having Λ as its closure in U better to confirm inductively than the h_i in another such set. This distinction ultimately corresponds, I believe, to the distinction between the natural laws and the physical necessities that are not natural laws. I discuss this distinction in chapter 7. Here I am explaining only why science should especially care about discovering which non-nomic truths are physically necessary; I am not yet in a position to explain why it is so scientifically important to discover which physical necessities in U are laws. Notice that, for each of the sets having Λ as its closure in U, the logical closure in U* of the predictions that are confirmed, in confirming every member of that set inductively, is the same. (In chapter 7 section 4, I discuss laws outside U.)

As I defined "inductive" confirmation in chapter 4, the inductive confirmation of "All ravens are black" requires different evidence (the discovery of a raven to be black) and involves the confirmation of different subjunctive conditionals C_P from the inductive confirmation of "All nonblack things are nonravens." However, on the definition of "inductive" confirmation given here, *h*'s inductive confirmation does not require a different kind of evidence, and does not involve the confirmation of different conditionals, from *k*'s inductive confirmation, where *h* and *k* are logically equivalent (see chapter 4, note 27). This might appear to undermine my remarks concerning the raven paradox (in appendix 2 of chapter 4), which emphasized that ravens bear widely on unexamined ravens whereas nonblack things do not bear nearly so broadly on unexamined nonblack things. One way to try to maintain that line under this new definition of "inductive" confirmation would be to contend that, typically, only *certain* of the predictions made by "All *R*s are *B*" (or, equivalently, made by "All ¬*B*s are ¬*R*") are regarded as capable of confirming it (or any hypothesis equivalent to it) inductively—namely, predictions involving the discovery of a raven to be black. But below I introduce the notion of an "inductive strategy," in the course of which some hypothesis is confirmed inductively by *any* of its predictions, whether that a given *R* is *B* or that a given ¬*B* is ¬*R*. If we carry out an inductive strategy in the course of which "All *R*s are *B*" is confirmed inductively, then the above move to maintain my original response to the raven paradox, that only certain predictions can confirm the hypothesis inductively, is unavailable, since discovery that a given nonblack thing is a nonraven must also confirm the hypothesis inductively. I ultimately take a different approach, which preserves the idea behind my original remarks concerning the raven paradox: in chapter 6, I argue that what we are really dealing with here is confirmation of the "reliability" of *inference rules*. The inference rule corresponding to "All *R*s are *B*" mediates inferences from *Rx* to *Bx* and is distinct from the inference rule corresponding to "All ¬*B*s are ¬*R*s," which mediates inferences from ¬*Bx* to ¬*Rx*. Ultimately, I say that we inductively confirm an inference rule's reliability only when we discover one of its "predictions" to be borne out and thereby confirm some of its other "predictions," where a rule's "predictions" are all conditionals $p > m$, $q > (p > m)$, and so on, where that rule can mediate the inference from *p* to *m*. So again, to confirm inductively the reliability of the inference rule corresponding to "All ravens are black," we must discover a given raven to be black and this discovery must bear on the blackness of various unexamined ravens (actual and counterfactual), whereas to confirm inductively the reliability of the inference rule corresponding to "All nonblack things are nonravens," we must discover a given nonblack thing to be a nonraven and this discovery must bear on the nonravenhood of various unexamined nonblack things. I argue that we carry out an inductive strategy in the course of which "All ravens are black" is confirmed inductively, but not one on which "All nonblack things are nonravens" is confirmed inductively. This returns me to my initial point: that ravens bear widely on unexamined ravens whereas nonblack things do not bear nearly so broadly on unexamined nonblack things. In virtue of our choice of inductive strategies, we regard "All ravens are black" as a law but "All nonblack things are nonravens" as a physically necessary nonlaw. This is closely related to the fact that we regard ravens as forming a natural kind of a certain sort (a biological species) but do not so regard nonblack things (or even nonblack birds). (Actually, the raven law contains implicit provisos; as I discuss in chapter 8, the relevant law is "The raven is black" rather than "All ravens are black." But, of course, the raven paradox could be posed with a nonbiological example, such as "All emeralds are green.")

7. It might be suggested that such confirmation is sometimes virtually automatic according to Bayesian conditionalization, since if $h \Rightarrow e$ (as when h is "All of the pears on the tree are ripe" and e is "The first pear selected from the tree is ripe"), then $pr(e \mid h) = 1$, and so when $0 < pr(h) < 1$ and $0 < pr(e) < 1$, we have $pr(h \mid e) > pr(h)$. But even presupposing Bayesian confirmation theory, this Bayesian argument avails us not, as I explained in chapter 4. The real question concerns how we justify taking e as confirming some of h's *predictions* (e.g., "Were we to select a second pear from the tree, then it would be ripe"). No such prediction f entails e, and so it remains unclear on what ground we can justify $pr(e \mid f) > pr(e)$, which is necessary (according to Bayesian conditionalization) for e to confirm f.
8. By saying that psychiatrists are "permitted to" (i.e., "entitled" to, "justified" in deciding to) pursue this hunch, I obviously do not mean that they make this decision on the basis of having good reason to believe that this gamble in particular will pay off, since their predicament is precisely that they lack any such reason. Rather, I mean that in pursuing their hunch, psychiatrists are not acting in an epistemically irresponsible fashion, but rather are faithfully carrying out their epistemic duties, acting within their rights, subject to no epistemic reproach.
9. In terms of Bayesian confirmation theory, I am suggesting that scientists are sometimes permitted to decide simply to adopt some new probability distribution, rather than arriving at it by updating their prior distribution by Bayesian conditionalization upon the receipt of new evidence. Their new probability distribution then supplies them with the conditional probabilities that express their willingness to regard some class of evidence as relevant confirmationwise to some class of predictions. But I am not suggesting that scientists are permitted to decide to adopt just any new probability distribution at all. So which non-Bayesian shifts of opinion are permissible under these circumstances? Which opinions would be arbitrary or biased in a sense that would render a shift to them unjustified? That is the question I am asking. Obviously, non-Bayesian shifts of opinion raise a great many issues (e.g., regarding Dutch book arguments that Bayesian conditionalization is the only

rational way to change opinions) that are beyond the scope of this book (but see Lange 1999).

10. Which hypothesis we believe this is depends on our background beliefs. Inductive strategies cannot be employed without *some* background beliefs; the observation that a given behavior by unexamined cases would represent a departure from the way that examined cases have been found to behave depends (for its status as an *observation*) on the putative observer justly holding certain prior opinions, just as any other observation does (I discuss the notion of a "salient" hypothesis much further in chapter 7). But a psychiatrist's justified opinions, prior to engaging in an inductive strategy, are insufficient to justify her taking one autistic patient's response to a given drug as relevant confirmationwise to any other's.
11. Though inductive strategies are common in science, a scientist may discover a law other than by carrying out an inductive strategy (see chapter 1, section 2.4). But even if our belief in a given law was not originally justified through an inductive strategy, we can later use an inductive strategy to justify that belief. In fact, though an inductive projection may be offered only after the law has been arrived at on the basis of significantly more equivocal evidence, an inductive strategy often constitutes the most effective of all the sorts of arguments that could be available for it. See Kuhn (1977, p. 327).
12. Someone carrying out these inductive strategies could nevertheless confirm some of these predictions, but not in virtue of carrying out these inductive strategies. That is, someone may regard a discovery as confirming a certain range of predictions in virtue of carrying out some inductive strategy and, in addition, as confirming various other predictions for other reasons.
13. Or, at least, the physical necessities in U. See note 6 above.
14. The principle I am implicitly employing here is not that if I now believe that I may later believe *h*, then I must now believe that *h* may be true. For this principle is incorrect: I might now believe that $\neg h$, and yet also believe I may later become misled, confused, inebriated, deranged, or what have you, and in this state believe that *h*. Rather, the principle to which I am appealing here is somewhat narrower: if I have adopted, as my rule (policy, strategy, etc.) for belief revision, a rule that I believe might later—on accumulation of further evidence—require my believing *h*, then I must now believe that *h* may be true, for otherwise, I would not now adopt this principle as my rule for belief revision; I would instead adopt a principle that I believe cannot later require that I believe *h*. (Relevant to this point is the current controversy over van Fraassen's "Reflection" principle; see van Fraassen 1997.)
15. The principle cited in note 14 above is also being applied here; the various inductive strategies that we are now pursuing can be regarded as a single policy for belief revision. Furthermore, this result constrains the inductive strategies that we can jointly be pursuing. For instance, we cannot simultaneously pursue an inductive strategy on which *h* is currently salient and an inductive strategy on which *k* is currently salient, where *h* is logically inconsistent with *k*.
16. I have portrayed the concept of "inductive" confirmation developed in this chapter as a generalization of the concept elaborated in chapter 4: that earlier concept applies only when the hypothesis being confirmed takes the "All *F*s are *G*" form, and the predictions it makes that are confirmed when it is confirmed inductively all take the C_P form. This earlier concept specifies that

C_P is confirmed in the course of h's inductive confirmation if and only if pr(C_P) $\neq 1$ and we believe that Con(Λ^+,p) holds where p is the antecedent of C_P—which is to say (for non-nomic p) that we believe that Con(Λ,p) holds. Since we believe that Θ's closure in U may be Λ, we believe that Con(Λ,p) holds only if Con(Θ's closure in U,p) holds—which [along with pr(f) $\neq 1$] is what it takes for a prediction f made by h to be confirmed when h is confirmed "inductively" in the sense elaborated in the present chapter. So if we think of C_P as a prediction f made by h (where p is non-nomic), then if C_P is confirmed when h is confirmed "inductively" in the sense given in chapter 4, it follows that C_P is confirmed when h is confirmed "inductively" in the sense given in this chapter.

However, in one important respect, the chapter 4 concept of "inductive" confirmation is *more* general than the concept given here: it allows p in a confirmed prediction $p > m$ to fall outside U (as long as it belongs to U^+). In contrast, here I have limited "predictions" to non-nomic claims (in the broad sense of U*) because I am trying to explain how the lawhood of a truth in U derives special scientific importance by virtue of lawhood's relation to *non-nomic matters* (broadly construed as U*). Having here told this story, in chapter 7 I expand my concept of h's "predictions" to include claims having antecedents and consequents outside U.

17. Of course, we may not pursue the right inductive strategies; we cannot know in advance which inductive strategies would be best for us to pursue. But neither can we know the laws of nature a priori. Furthermore, our inductive strategies need not have rendered salient *all* of the laws in order for all the predictions they yield that are relevant to our project to be accurate; it is sufficient that the salient hypotheses entail all of the *relevant* laws. For instance, we do not need to have rendered salient all of the laws of elementary particle physics (or the laws of psychiatry) in order for our inductive strategies to result in the confirmation of "Were a Cepheid-type variable star to exist in the Andromeda nebula, then its period–luminosity relation would be . . ." but not in the confirmation of "Were a Cepheid-type variable star to be accelerated from rest to beyond the speed of light, then its period–luminosity relation would be. . . . "
18. If an account of law construes the fact that there is a given natural law as going well beyond the facts in U, then the account threatens to make it difficult to see how we could be justified in believing that there is a given law. As Armstrong (1983, pp. 107–108) notes, this point has been offered by Braithwaite (1953, pp. 11–12), Popper (1959, p. 437), and Mellor (1980, p. 108) as favoring more modest conceptions of lawhood.
19. Foster (1982–83) appears to be making roughly the same objection to Ayer. But he contends that "if extrapolative induction [i.e., an extension to all or some of the unexamined cases—e.g., to all "nomologically possible" cases—of what we have found to hold for the examined cases] is the only form of inference [to ■h], then Ayer is clearly right" (p. 88). I have argued in chapter 5 that this is mistaken; inductive projection (a kind of "extrapolative induction") can justify acceptance of ■h but not of h without ■h. Foster instead defends inference to the best explanation as playing this role. As mentioned in section 1 of chapter 3, I have no problem with using science's interest in explanations rather than its interest in counterfactuals in order to account for its interest in the laws. But this approach requires an argument that laws

bear a special relation to explanations, just as I argued that laws bear a special relation to counterfactuals. Foster argues that accidental generalizations are not explanatory because "[i]n subsuming the past regularity under a universal regularity [that we do not believe to be physically necessary] we would not be diminishing its coincidental character, but merely extending the scope of the coincidence to cover a larger domain" (p. 91). But as I showed in chapter 4, some accidental generalizations are not utter "coincidences." In the pear example, I explain why all of the examined pears from the given tree are ripe by subsuming this fact under the regularity that all of the pears on the tree are ripe. This regularity is (modestly) explanatory despite its lack of physical necessity; it shows that it was no coincidence that all of the pears picked for examination were ripe. That is, (as I argue in chapter 8), this accidental regularity's explanatory power derives partly from its preservation under a certain relevant range of counterfactual suppositions. In particular, it would still have held had we checked different pears from the tree. On this view, a law's preservation under counterfactual suppositions partly accounts for its explanatory power. (For more discussion of explanatory nonlaws and their invariance under counterfactuals, with particular attention to explanatory generalizations in the social sciences, see Woodward 1995.) Foster's approach, I think, fails to explain why laws as distinct from explanatory nonlaws are so important to science.

Foster favors inference to the best explanation over any sort of extrapolation because he believes that "[w]hen rational, an extrapolative inference can be justified by being recast as the product of two further steps of inference, neither of which is, as such, extrapolative. The first step is an inference to the best explanation—an explanation of the past regularity whose extrapolation is at issue. The second is a deduction from this explanation that the regularity will continue or that it will do so subject to the continued obtaining of certain conditions" (p. 90). I am unsure of this. Why couldn't we extrapolate some past regularity (to some unexamined cases or to all "nomologically possible" cases) without adopting any explanation of that regularity, as long as the past regularity (and our background beliefs) give us good reason to believe that the explanatory factors (whatever they are) will continue to hold? For instance, scientists expect the roughly 22-year sunspot cycle to persist even though they do not understand why it holds, only that it has held steady for a long time (see chapter 4, note 21).

Chapter 6

1. A law's relations to counterfactuals, inductions, and explanations can be explicated whether the law-statement is construed as expressing the "regularity" that a certain inference rule is reliable or construed as expressing that rule of inference itself. (On the former construal, the law-statement has a truth-value; on the latter, not. See note 1 of chapter 1 regarding my use of "law-statement.") I am not offering an account of the meaning of "All *F*s are *G*" when it functions as a law-statement, so I have no need to decide between these alternatives. Nor am I contending that laws *are* inference rules. As I state in chapter 1, I am not offering a reductive account of natural law.
2. Of course, belief in a given inference rule's reliability is only *part* of the root commitment undertaken in believing in a certain law, since accidents are

truths and hence are also associated with reliable inference rules. In chapter 7, I resume my discussion the rest of the root commitment.

3. Thanks to John Earman and John Roberts, who kindly showed me a draft paper in response to an earlier paper of mine on provisos. Their reactions helped me to see where I needed to express myself more clearly.
4. This use of "provisos" is similar to but slightly different from Hempel's (1988) use, as I shortly explain.
5. Similar remarks often apply to the kinds of approximations, models, and idealizations used in science; a model can be used successfully long before it is understood what that model misdescribes or omits. See, for instance, Woolley (1978).
6. It is not relevant here that Newton's gravitational force law is not the basis of the best current theory of gravity. Newton's law was once taken to be a fundamental law, and so can be used to illustrate my point.
7. See Cartwright (1983, pp. 60–73) and Woodward (1992, pp. 193–96).
8. Any mechanics textbook will include a discussion of this example; see, for instance, Kibble (1985, p. 106). The Coriolis force law is $F = -2m\ \omega \times v$, where m is the mass of the body on which the Coriolis force F acts, ω is the constant angular velocity of the rotating frame of reference in which we are working, and v is the velocity of the body relative to that frame.
9. In section 6 of this chapter, I return to issues surrounding the regularities associated with the laws governing the fundamental forces. In chapter 8, I take up certain other issues arising in connection with "derivative" laws. For instance, does Galileo's law qualify as a "law" at all, since it depends on certain physically unnecessary facts about Earth? This is an interesting question, yet it is irrelevant to the present discussion; if Galileo's "law" is not a law, even a "derivative" one, then a different example could be chosen—for example, Snell's law.
10. Elsewhere, Hempel (1965) acknowledges that some law-statements have such completeness clauses. But, he is quick to point out, they are not empty: "all of the factors considered relevant . . . are clearly understood (as in the familiar formulation of Galileo's law, which is understood to refer to free fall in a vacuum near the surface of the earth)" (p. 167). I could not agree more, as section 3 should make evident.
11. Along with the references I have already given, versions of this problem have been discussed by Canfield and Lehrer (1961), Coffa (1968), and Molnar (1967), in many places by Cartwright (1983 and after), by Giere (1988), Fodor (1991), and Schiffer (1991), among many others, as well as by Hempel (1965, p. 167). The issue, in general terms, was anticipated by Scriven (1961), and ultimately, my response to this issue agrees with a central point in Scriven (1959). Difficulties similar to the problem of provisos have also been noted in ethics—for example, with regard to Ross's (1930) definition of a "prima facie duty." See also Armstrong (1983, pp. 7, 28).

    It might be thought that Armstrong's account avoids the problem of provisos, since he distinguishes between "iron" laws and "oaken" laws, where an oaken law does not entail the existence of a corresponding regularity. But this would be to misread Armstrong (1983), who believes that certain states of affairs intrinsically constitute lacks or absences, and that such "negative" properties are not universals (p. 147). Consider, then, Newton's first law of motion: a material body acted on by no forces continues in its present state of motion or rest. Armstrong cannot take the universals standing in the relation of nomic necessitation that constitutes this law to include the universal "not being acted on by a force," since this "negative" property is not a universal. The law is, rather, the relation of nomic necessitation between being a material body and continuing in the present state of motion or rest. This is one of Armstrong's examples of an "oaken" law, since it does not entail that all material bodies continue in their present state of motion or rest, since otherwise, it would entail something false. But Armstrong notes (p. 149) that even an oaken law is associated with a regularity, just not the regularity corresponding directly to the universals figuring in the relation of nomic necessitation that constitutes the law. For instance, there *is* a regularity that is associated with Newton's first law: all material bodies *feeling no forces* continue in their present state of motion or rest. Armstrong's notion of "oaken" laws, then, arises from his view of what can constitute a universal; here he is not concerned with provisos, and the examples he gives of "oaken" laws (e.g., Newton's first law of motion) are laws that (Hempel would say) involve no provisos at all, unlike, say, the law of thermal expansion. (It is perhaps worth noting that Armstrong is careful to express Newton's first law in terms of a body's "continuing in its present state of motion or rest" rather than in terms of its "undergoing no acceleration." I wonder if the property of undergoing no acceleration is a lack or absence, like the property of being acted on by no forces. Isn't the property of undergoing no acceleration the same as the property of continuing in the present state of motion or rest? What makes a property "negative"?)

12. Even Swartz (1985), who appears willing to distinguish the genuine or "physical" laws from the merely approximate or "scientific" laws, and who explicitly commits himself therefore to discussing metaphysics rather than scientific practice (pp. 4–12), cannot bring himself wholly to divorce the genuine law-statements from actual science. For example, he attacks rival accounts of "physical" law as unable to explain how science can discover these laws (pp. 67–72), and he goes on to endorse a very scientific "experimental basis," involving "laboratory data," for justifying the adoption of "physical" laws.
13. The provisos that pose interesting problems are those that are difficult to cash out without appealing to some such expression as "when . . . is the only relevant factor" or "in the absence of factors like . . . and so on." Although these provisos are the focus of my attention, it is worth noting that there are other, more innocuous sorts of clauses implicit in certain law-statements. Consider, for instance, the law of definite proportions (Proust's law), which is typically expressed as the following claim: "Any pure sample of a given chemical compound is made up of various elements in the same proportions by mass as any other pure sample of that compound." Although chemists may once have accepted a "law of definite proportions" lacking any provisos, they have since discovered that the genuine law has further conditions. But a complete list of these provisos contains no problematic term such as "relevant factors." The provisos accommodate the "nonstoichiometric" (i.e., Berthollide) compounds, and each of the several, very specific types of nonstoichiometric compounds can be picked out without resorting to "relevant factors" or the like. For instance, one kind of nonstoichiometric compound consists of polymers. Both

$$CH_3\text{—}CH_2\text{—}O\text{—}CH_2\text{—}CH_2\text{—}O\text{—}CH_2\text{—}CH_2OH$$

(which is $C_6H_{14}O_3$) and

$$CH_3\text{—}CH_2\text{—}O\text{—}CH_2\text{—}CH_2\text{—}O\text{—}CH_2\text{—}CH_2\text{—}O\text{—}CH_2\text{—}CH_2OH$$

(which is $C_8H_{18}O_4$) are molecules of polyoxyethylene, and so the proportions of the various elements are not the same in each molecule of polyoxyethylene. Another type of nonstoichiometric compound consists of network solids, that is, crystals in which atoms of one element are replaced at randomly scattered sites in the lattice by atoms of another element, where the two elements' atoms are of similar size and bonding propensities. In a ruby, for example, chromium atoms appear in place of some of the aluminum atoms in the lattice. Thus, the law of definite proportions, as commonly stated, has various provisos referring to the various kinds of nonstoichiometric compounds. But these provisos can be expressed in unproblematic terms. (A similar interpretation of the law of definite proportions is offered by Christie 1994.)

14. After reading a preliminary version of this chapter, George Molnar kindly referred me to his 1967 "Defeasible Propositions," in which he advances views similar to mine. In particular, he argues that "no logically complete specification of what constitutes standardness [i.e., the absence of disturbing factors] can be given without invoking the notion of relevance" (p. 189) and that "[t]he non-tautologous character of defeasible propositions is safeguarded by the fact that we can confirm an individual's FG-standardness independently of confirming that it is G" (p. 190). Scriven (1959) also sets out a position like mine when he argues that many laws are expressed by "normic statements":

> The normic statement says that *everything* falls into a certain category *except* those of which *certain special conditions* apply. And although the normic statement itself does not explicitly list what count as exceptional conditions, it employs a vocabulary which reminds us of our knowledge of this, our trained judgment of exceptions. . . . Now if the exceptions were few in number and readily described, one could convert a normic statement into an exact generalization by listing them. Normic statements are useful where the system of exceptions, although perfectly comprehensible in the sense that one can learn how to judge their relevance, is exceedingly complex. . . . The physicist's *training* makes him aware of the system of exceptions and inaccuracies, which, if simpler, could be put explicitly in the statement of scope. (p. 466)

15. Cartwright (1983) has emphasized this point.
16. However, this revision applies only to conditionals having consequents in U. Any reference to the correctness of a conditional with a *nomic* consequent must be treated more carefully: insofar as its consequent goes beyond some claim in U, the conditional must be correct. For example, Λ^+ is preserved under p only if the actual laws would still have been laws had p been the case. That a given inference rule is good enough for certain purposes is itself not *merely* good enough for certain purposes.

17. Unless we so redefine "Γ's logical closure in U," we will run into a problem if Γ contains, say, statements of the ideal gas law and of van der Waals's gas law. The corresponding inference rules are both reliable for certain purposes; the relevant predictions they make nearly agree. But these predictions are different. So Γ contains logically inconsistent claims about gases, and therefore as ordinarily construed, Γ's logical closure in U would have to include "There are no gases."
18. Or I might simply drop this requirement altogether, since it is already implicit in requirement (iv) in the definition of non-nomic stability (chapter 3, section 2.3). See chapter 3, notes 4 and 10.
19. If we *are* interested in the causal antecedents of Darcy's request, then we are in a backtracking context, and a different possible world—where no quarrel takes place—is the closest possible world in which Darcy makes his request. Again, concerning those events that we care about, the inference rules associated with the actual laws are reliable.
20. Toulmin (1953) also distinguishes a law-statement from a claim that specifies the law's applicability to a given case. He alleges that this distinction reveals "the distinguishing mark of a law" (p. 78; see also p. 86) because it shows that law-statements specify rules of inference, not regularities. It is unclear to me whether part of Toulmin's motivation is some worry about provisos. I discuss Toulmin's position in the appendix to this chapter.
21. For my purposes, it will not matter whether this "law" is *currently* taken to be a law. I am concerned with how it was understood when it was so regarded.
22. The same considerations apply to the notion of "simple ratios" in Gay-Lussac's law "that gases always combine in the simplest proportions when they act on one another; and we have seen in reality in all the preceding examples that the ratio of combination is 1 to 1, 1 to 2, or 1 to 3" [from Gay-Lussac's *Memoir on the Combination of Gaseous Substances With Each Other* (1809), quoted in Nash 1956, p. 54]. This law allowed Gay-Lussac to arrive justly at the combining volumes of various gases in reactions in which their direct measurement would have been experimentally difficult.
23. Dalton apparently regards the rule of greatest simplicity as associated with laws (or, at least, with physical necessities), since he seems to regard its reliability as having been confirmed inductively and to use it to support counterfactuals in the manner of a law. He also refers to it as "the law of chemical synthesis" (Dalton 1953, vol. 1, p. 223).
24. The law of Dulong and Petit is roughly that an element's atomic weight, multiplied by the quantity of heat needed to raise a given mass of that element by a given temperature, gives the same constant for all solid elements. Gay-Lussac's law is given in note 22.
25. Other laws function in the same way as the law of multiple proportions, and they are accompanied by the same kind of proviso. For example:

    > [T]he 'principle of superposition' [that upper rock formations in a sequence are younger] is one of the first things that an undergraduate learns in a geology class. . . . Werner confidently asserted that the observable character of position in the rock succession was a reliable indicator of the essential, but unobservable, character of time of formation. His confidence turned out to be somewhat premature. In the succeeding

generation the recognition of the igneous origin of a wide range of rocks, the recognition of metamorphism, and the recognition of the scale of displacement to which strata were subject once they had been deposited made the sorting out of the order of succession much more difficult than Werner had anticipated. Nonetheless, the principle of superposition offered a key to at least a preliminary sorting of the formations. With this as a base, historical geologists were able to integrate trickier cases into the succession." (Laudan 1987, pp. 96–97)

26. These policies of judging might even be different for different people. Ramsey (1931, p. 241) expresses the rules that are natural laws in the indicative mood with "I" ("If I meet an *F*, I shall regard it as *G*") rather than as commands ("Expect an *F* to be *G*") or as universal norms ("Whenever you find an *F*, you ought to expect it to be *G*"), and Ryle (1963) often makes a point of doing likewise (e.g., p. 309). This has understandably suggested to some (e.g., McPherson 1950) that Ramsey takes such a rule to be equivalent to a description of one's own policy ("I make it a practice to expect *F*s to be *G*"). Ramsey (1931) is unclear on this point: he refers to a natural law both as "my rule" (p. 253) and as "a formula from which we derive propositions" (p. 251). Yet the notion that such a rule governs a game or social practice suggests that the rule is everyone's, not just mine.
27. See especially Sellars (1948a, pp. 297, 300–301; 1948b, p. 439; 1953, pp. 318–19; 1963a, pp. 292–93, 317, 331, 355).
28. Moreover, Sellars (1953, pp. 334–35) says that even if the term figures in observation-reports, its use must be governed by material rules of inference. Two claims with different contents, in virtue of having different inferential roles, may be appropriately uttered as observation-reports in exactly the same possible circumstances. For the difference in their inferential roles to constitute a difference in their contents, it must be manifested in differences in the material rules of inference governing the use of these claims.
29. In addition, Sellars's (1964) account of why induction is justified depends on his conception of induction "as establishing principles *in accordance with which* we reason, rather than major premises *from which* we reason" (1958, p. 286). But induction as Sellars purports to justify it (namely, as the "straight rule": if *f* is the fraction of *F*s we have found to be *G*, then believe "*f F*s are *G*" to be a law) would seemingly lead us to adopt Reichenbach's (1947, p. 368; 1954, pp. 10–11) generalization about gold cubes as a law. Sellars (1968) says that to believe "All *F*s are *G*" to state a law is to respect the corresponding inference rule as expressing "how we ought to think about the world" and "what ought or ought not to be the case with respect to *our beliefs about* the world" (p. 117)—to use that rule in a manner analogous to the way in which we use rules of inference from deductive logic (1958, p. 270). But Sellars does not adequately explain what, if anything, would make it the case that we ought to use certain rules but not others in this way, beyond his discussion of inductive inference, and this does not suffice to distinguish the laws from the accidental uniformities.
30. (It does not affect my point that this "moral law" about promise keeping possesses some provisos that I have omitted.) Likewise, an accidental generalization in science bears to counterfactuals the same relation as does an acci-

dental generalization in ethics. In a different connection, Cohen (1962) has offered an example:

> [I]f we consider some presumably rare vice, like necrophily perhaps, we can produce an example partly analogous to [Popper's] case of the moas. It may well turn out as a matter of pure accident, so rare are the kinds of people concerned, that all and only hunchback albinos are necrophiles (though so as not to give offense, I must add that I know of no evidence for this and can equally well imagine that all hunchback albinos are models of virtue). But this accidentally true identity-statement does not license substitution in "It is a moral principle deserving of universal respect that anyone who indulges in necrophily does something that he ought not to do." (p. 309)

Chapter 7

1. Shortly, I discuss why "All copper objects are electrically conductive" is not merely physically necessary in the closest lone-proton world but, moreover, expresses a natural law there, according to the proposed root commitment.
2. There are other important points of difference as well: Lewis's account purports to reduce the facts in Λ^+ to facts in U, and likewise regards counterfactual conditionals in U* as made true by facts in U, and of course does not regard Λ as preserved in the way that I contend it is.
3. I noticed this resemblance to Mill's view after developing mine, and I am not sure how deep it goes beyond the particular respects I discuss here. For instance, are counterfactual conditionals confirmed in the course of Mill's "inductions"?
4. That only *generalizations* can state laws seems to be an ad hoc restriction in certain accounts. For example, Lewis recognizes that the deductive system of truths with the best combination of simplicity and informativeness may include some claims that are not generalizations (in some unspecified sense). Lewis stipulates that only the generalizations in that system state laws. I don't know of a place where Lewis explicitly considers the status of a generalization that is a theorem in the best deductive system and follows from axioms one of which is not a generalization. Would this axiom describe initial or boundary conditions, or would it be a physically necessary nonlaw? Moreover, if the laws' distinctive roles in connection with inductions, explanations, and counterfactuals are supposed to follow from the laws' inclusion in the best deductive system, then are the nongeneralizations in that system also capable of playing these roles, even though they aren't laws?
5. For instance, see Hempel (1965, pp. 346–47), Lewis (1983, pp. 367–68), Braithwaite (1953, p. 305), Nagel (1961, p. 60), Davidson (1990, p. 218), and Smart (1985, p. 276).
6. The electron, which possesses a charge of this magnitude, is the only negatively charged particle having this rest mass.
7. I first suggested this point in appendix 2 to chapter 4, in connection with the paradox of the ravens.
8. Instead of saying that scientists have treated these examples as physically necessary nonlaws, I suppose I might just as well have said that scientists treat them as "derivative" rather than as "fundamental" laws. It is not terri-

bly important to me which terminology is used to mark the distinction I am drawing. My point is that this distinction (whatever it is called) is highly intuitive and (for reasons I discuss in this chapter) must figure in our account of scientific reasoning. I prefer not to express this distinction in terms of "fundamental" versus "derivative" laws because the paradigm examples of "derivative" laws in the philosophical literature are Kepler's three laws of planetary motion and Galileo's law of free fall. These are thought to qualify as "derivative" because they refer to particular entities: the Sun and Earth, respectively. Yet it is not at all clear to me what is intended by calling them derivative *laws* since they derive from Newton's laws of motion and gravity and various facts traditionally considered to be *physically unnecessary*, such as that the Sun is far more massive than any planet and that Earth is 5.98 $\times 10^{24}$ kg. Philosophers who term these "derivative laws" apparently do not believe them to be physically necessary at all; they are more like applications of the laws to certain initial conditions. I therefore find the term "derivative law" puzzling as it is traditionally used. Since the claims with which I am concerned have been considered physically necessary, I prefer to refer to them as "physically necessary nonlaws."

9. We might be tempted to put this point by saying that the force between two Old English Sheepdogs is not *explained* by the fact that they are Old English Sheepdogs, but merely by the fact that they are material. I resist this way of capturing the intuition I am after. It can be a law that all *F*s are *G* even if an object's *F*ness does not explain its *G*ness. For instance, it is a law that all pendulums of period *T* have length $gT^2/4\pi^2$, and yet (at least typically) a pendulum's period does not explain its length. I defer until chapter 8 any further discussion of the laws' relation to explanations of facts in U. (Thanks to John Carroll for the pendulum example and for urging me not to put my point here in terms of explanation.)

   My contention that the laws fail to form a logically closed set offers a useful context in which to view the notorious difficulty, raised by Hempel and Oppenheim (1948; repr. Hempel 1965, p. 273 n.33), that the conjunction of Kepler's laws and Boyle's law does not help to explain either of its conjuncts. Perhaps the reason is that laws must be explained by other laws, and the conjunction of Kepler's and Boyle's laws is not a law, though it is physically necessary.

10. Perhaps by saying "It is a law that no signal exceeds twice the speed of light" we conversationally imply that it is not a law that all signals travel slower than or equal to the speed of light. But if we cancel that implication—by saying "It is a law that all signals travel slower than twice the speed of light, but I don't mean to imply that there is no law specifying an even lower maximum velocity for all signals"— then it seems to me that the intuitive reply is: "Hold on—if the laws dictate a lower maximum velocity, then it is a physical necessity that all signals travel slower than twice the speed of light, but the law here gives the limiting velocity."

11. In section 5 of chapter 8, I discuss the argument in the service of which Fodor makes this remark.

12. There is one more: a remark by Goodman, which he later disavowed. See note 18 below.

13. See Ayer (1963, pp. 210–11), Rescher (1970, pp. 111–21), van Fraassen (1989, pp. 1–10), and Carroll (1994, pp. 17–18). Ayer finds the notion of

a command that it is impossible to disobey incoherent. He might have been interested to know that animals and plants have been tried and sentenced for (supposedly) violating the laws of nature (and so manifesting diabolical possession). For example, in 1474, the magistrates of Bale sentenced a cock to be burned at the stake "for the heinous and unnatural crime of laying an egg" (Evans 1906, pp. 10–11, 162–63), and similar cases have been recorded as late as 1730. (Apparently, the plumage of a hen can match the typical plumage of a cock so closely as to mislead those ignorant of the internal anatomical differences, according to Cole 1927, p. 97.) Even the traditional notion of a natural law permitted exceptions, whether by Divine or diabolical action.

14. For many similar passages, see Winsor 1991 (pp. 8, 24–26) and Ridley (1986, pp. 107–08). Of course, Agassiz is referring here to what makes the various taxonomic categories in biology constitute natural kinds. Agassiz's discussion is pre-Darwinian (from 1857), but whether it actually applies to biological species is beside the point. I am discussing a traditional intuition about natural law according to which a logical consequence of laws need not be a law itself, since the categories it employs need not be natural. If this intuition does not apply to biological species because (as many natural philosophers believe today) they are not natural kinds and there are no biological laws that refer to particular biological species, then my point still applies to scientific disciplines that are directed toward the discovery of natural laws. (In chapter 8, however, I argue that some fields of biology today do treat biological species as natural kinds and do purport to discover biological laws that refer to particular biological species.)
15. For more on von Pettenkofer, see van Spronsen (1969, pp. 73–74).
16. For each of h's predictions $p > q$ that we must confirm in order to confirm h inductively, there is some conjunction k of salient hypotheses, not including j, such that j & k makes this prediction. So j predicts $(p \,\&\, k) > q$, and this prediction is confirmed in confirming j inductively. Now

$$\begin{aligned}\mathrm{pr}((p\&k) > q) &= \mathrm{pr}(((p\&k) > q) \mid p > k)\,\mathrm{pr}(p > k) + \mathrm{pr}(((p\&k) > q) \mid \neg(p > k))\\ &\quad \mathrm{pr}(\neg(p > k))\\ &= \mathrm{pr}(p > q \mid p > k)\mathrm{pr}(p > k) + \mathrm{pr}(((p\&k) > q) \mid \neg(p > k))\\ &\quad \mathrm{pr}(\neg(p > k)), \text{ and}\end{aligned}$$

$$\mathrm{pr}(p > q) = \mathrm{pr}(p > q \mid p > k)\,\mathrm{pr}(p > k) + \mathrm{pr}(p > q \mid \neg(p > k))\,\mathrm{pr}(\neg(p > k)).$$

Substituting and rearranging,

$$\begin{aligned}\mathrm{pr}((p\&k) > q) = \mathrm{pr}(p > q) + \mathrm{pr}(((p\&k) > q)\&\neg(p > k)) - \mathrm{pr}((p > q)\\ \&\neg(p > k)).\end{aligned}$$

If e confirms j inductively, e confirms $(p \,\&\, k) > q$. So if e has no effect on the difference between the second and third terms above, then e confirms $p > q$, and so (if this holds generally) e confirms h inductively.

17. See chapter 2, note 10, and the close of chapter 4, appendix 2, regarding the confirmation of the physically necessary non-law "All nonblack things are nonravens." This point also ultimately bears upon the close of chapter 4,

appendix 1, section 2, regarding the grue hypothesis. I return to the grue hypothesis in section 3 of this chapter.

18. Goodman once considered a similar example. Say that a temporal slice of an "emerose" is either a temporal slice of an emerald and precedes a certain future time *t* or a temporal slice of a rose and not before *t*. Say that a temporal slice of an object is "gred" if and only if it is either green and before *t* or red and not before *t*. Davidson offered the hypothesis "All temporal slices of emeroses are gred" as a counterexample to Goodman's use of entrenchment to deal with his "new riddle of induction," the grue problem. Davidson (1966) claimed that this hypothesis is "projectible" even though "emerose" and "gred" are not well entrenched. Goodman's (1966) initial response was that we believe that this hypothesis is a nonlaw, and yet that if it is true, then it follows entirely from laws. Goodman explained that the hypothesis "is unprojectible in that positive instances do not in general increase its credibility; emeralds found before *t* to be green do not confirm it" (p. 328). Goodman's remark is problematic in two respects. (1) He identifies an emerald slice found before *t* to be green as constituting an *instance* of the hypothesis. But by Goodman's own lights (as noted by Davidson 1966), to discover an instance of the hypothesis is actually instead to discover simply that something, already known to be a temporal slice of an emerose, is gred. Therefore, even if we find this emerose slice to be a green emerald slice before *t*, this fact is (somehow) excluded from the "instance." So Goodman's remark about the confirmatory power of discovering an emerald slice before *t* to be green is not obviously relevant. (2) Contrary to Goodman's remark, we might well increase our confidence in the hypothesis by discovering an instance. Of course, Goodman (1983) says that the sort of confirmation with which he is concerned is not merely increased confidence in the hypothesis, but "occurs only when an instance imparts to the hypothesis some credibility that is conveyed to other instances" (p. 69). However, as I noted in Chapter 4, this is a poor characterization of "inductive" confirmation. After all, *some* as yet unexamined emerose slices—those preceding *t* where the emerose slice under examination is already known to precede *t*—*are* confirmed to be gred by the gredness of the examined emerose slice. What Goodman must have had in mind is that the gredness of a *certain* (perhaps merely hypothetical) emerose slice is not confirmed by discovering that the examined emerose slice is gred. In particular, we do not confirm an emerose slice *after t* to be gred (presuming that we already knew the examined slice to precede *t*). So we do not regard the hypothesis as confirmed inductively. Accordingly, since Goodman (1966) thinks that lawlike hypotheses can be confirmed inductively, he says that we can believe that a given hypothesis is non-lawlike even though we believe that it follows logically from hypotheses each of which is lawlike (p. 328).

    Later, Goodman (1983, pp. 104–108; see Schwartz, Scheffler, and Goodman 1970) abandoned this response to Davidson in favor of an alternative that embraces "All emerubies are gred" as projectible. But I regard this retreat, made in response to objection (1) above, as a departure from the intuitions that Goodman originally set out to capture.

19. In fact, the inductive strategy that Boyle actually pursued may have aimed to find a reliable rule covering only atmospheric air, not a reliable rule covering gases of all kinds (see Shapin 1994, pp. 323–24, 345–47). Nevertheless,

we believe that the broader inductive strategy is what Boyle *should* have pursued. In justifying Boyle's law (e.g., during a lecture-demonstration or in a chemistry text), we invite our audience to regard certain observations as bearing inductively on the behavior of *any* sample of gas. It is in this retrospective sense that I refer to the inductive strategy covering all gases as "Boyle's inductive strategy." See chapter 2, note 65, and chapter 5, note 11.

20. Notice that if we pursue the broader inductive strategy, then also to pursue one of the narrower strategies would add nothing. Just at the point when some rule of the sought-after kind that belongs to a nontrivially non-nomically stable set becomes salient on one of the narrower strategies, the corresponding broader rule becomes salient on the broader strategy. Once we have come inductively to believe in the broader inference rule's reliability, there is nothing left to be confirmed by pursuing one of the narrower inductive strategies.
21. It might be that if we have observed no *B* cases, then observations of *A* cases alone are insufficient to suggest a rule that is reliable to *B* cases, but if we have observed some *B* cases, then observations of *A* cases do help to make salient a rule that is reliable to *B* cases. For instance, Newton's second law ($F = ma$), covering net forces of all magnitudes, would presumably not be rendered salient by observations of net forces measuring exactly 1 newton (*A* cases), no matter which or how many. However, if we have already observed several net forces of various other magnitudes (*B* cases) and Newton's law is not yet rendered salient, then the addition of several observations of net forces measuring exactly 1 newton could suffice to suggest it.
22. Salmon (1963) suggests that the grue hypothesis is unjustified because "grue" cannot be defined ostensively. Salmon apparently thinks that grue cannot be defined ostensively because if we are shown a collection of grue objects (various green things before the year 3000) and told that they are all "grue," we fail to latch onto "grue." In particular, we then fail to recognize a blue thing after 3000 as "grue", that is, as relevantly the same as the grue things that we have been shown. Salmon, then, seems to be gesturing toward the fact that I exploit: that instances preceding 3000 do not suffice to render the grue hypothesis salient (see section 3.2 of this chapter). Of course, Salmon does not put matters this way. While I make salience relevant to inductive confirmation by relating both to lawhood, it is unclear to me why Salmon takes our inability to define "grue" ostensively as bearing some relation to the capacity of the grue hypothesis to be confirmed.
23. It might be objected that such a hunch is rational only if the estimated risk of being led to false beliefs, coupled with the disutility of so being, is outweighed by the estimated prospect of being led to increased knowledge, and the value of so being—and surely, since we must at the outset regard the likelihood of any particular strategy's success as rather small, our gamble cannot be rational. However, part of being free to elect a strategy is being free to believe, in carrying out this strategy, that the likelihood of its success is favorable. Adopting a strategy represents changing various opinions (conditional probabilities, etc.), including opinions regarding its likelihood of success.
24. To do this, van der Waals does not need to have identified *all* of the influences that Boyle's law neglects. As I argued in chapter 6, the complete list of responsible factors is often unknown when such a law is discovered. To

derive his equation, van der Waals does not pick out the non-negligibly large factors from the complete list of neglected influences. He simply posits certain influences as being too great to ignore.

25. See van der Waals (1890, p. 407) and Loeb (1934, p. 172).
26. Here I am deliberately paraphrasing Sellars (1965, p. 194). That observations often cannot justify belief in a law (even one that exclusively concerns observable matters), if those observations are considered from a nontheoretical outlook, is the basis of Sellars's arguments against phenomenalism (1963a) and for scientific realism (1965, 1977). (See Lange 1998.) Sellars argues, for instance, that even if our observations were made in strictly phenomenal terms, the phenomenal "counterparts" of the kinds figuring in material-object laws would be "salient" (as I would put it) if and only if we thought about our observations in material-object terms. The concepts in many actual natural laws are not salient on "autonomous inductive reasoning in the observation framework" (Sellars 1977, p. 319). For the idea of using van der Waals's law to illustrate Sellars's point, I am indebted to van Fraassen (1977, p. 337), though my elaboration of this example differs considerably from his.
27. Wigner (1967, p. 43) also regards principles of symmetry and invariance as meta-laws.
28. In chapter 2, note 58, I said that "All leptons are electrons, muons, tauons, or neutrinos," for example, is a law. It belongs to Λ; it does not require that electrons form a natural kind. Campbell (1957, p. 54), Caldin (1960, p. 211), and Nagel (1961, pp. 75–76) assert that claims like "Silver exists" are laws. It is hard to see "Silver exists" as expressing a matter of natural law if it says merely that "silver" is instantiated; surely, it is merely an accidental matter of fact that there are silver atoms, since there might have been nothing but a lone proton, or (to choose a less extreme possibility) insufficient matter in the universe to form stars massive enough to create supernovae, and thereby to synthesize silver. Perhaps these philosophers intend that "Silver exists" be understood as asserting that silver is a natural kind of a certain sort. Then it is physically necessary, since it is bound up with the existence of various sorts of laws.
29. In chapter 3, I showed that, for any stable set, its members belonging to U form a non-nomically stable set. If Ψ consists entirely of claims like "All laws of motion are Lorentz-invariant" and "For each property *F*ness covering only particles, if there are laws "All *F*s are *G*" of at least *n* out of some *m* specified sorts, then there are laws of all *m* sorts", then there are no non-trivial members of Ψ belonging to U. The set of logical truths in U trivially possesses non-nomic stability. In addition, I showed that, for any two stable sets, one must be a proper subset of the other, and I argued that Λ^+ may be stable. I noted that Ψ is a proper subset of Λ^+ since any meta-law is true—though perhaps not a law—in any possible world with exactly the same laws in U as the actual world, and since any meta-law belongs to U^+ and Λ^+ is logically closed in U^+. Notice also that perhaps it is not the case that all laws of motion would still have been Lorentz-invariant had it not been a *law* that all laws of motion are Lorentz-invariant. Yet the failure of "All laws of motion are Lorentz-invariant" to be preserved under this counterfactual supposition does not undermine Ψ's stability, since the counterfactual antecedent does not belong to U^+. (In U^+, the only expressions in the scope of ■ are expressions

in U.) Likewise, let h be a physically necessary nonlaw about laws. Perhaps Ψ is not preserved under $\blacksquare h$. But again, this is consistent with Ψ's stability because $\blacksquare h$ does not belong to U^+ since h does not belong to U.

30. For that matter, let h be "All particles are protons or electrons or . . . or muons." Then $\blacksquare h$ where $h \in U$, and hence $\Box \blacksquare h$, but not $\blacksquare \blacksquare h$. So $\blacksquare h$ is a physically necessary nonlaw about laws, and it need not be preserved under $p \in U^+$ such that Con(Ψ,p), as when p posits hypermesons.
31. I call these "cross-inductions" because they are akin to what Reichenbach (1938, pp. 365–73) so terms. He has in mind our inferring, from the fact there are laws for iron, copper, tin, aluminum, and so on, that specify their melting points, that there is a law specifying a melting point for carbon (under certain conditions), though we have not yet identified that temperature or even observed carbon to melt. Although this is similar to the kind of confirmation I have in mind, Reichenbach does not consider how we confirm that carbon belongs in the same category as iron, copper, and so on; his inference presupposes our justified belief that carbon figures in the same kinds of laws—whatever they are—as these other substances do. The kind of confirmation I have in mind here also recalls Goodman's (1983, pp. 109–14) remarks on "overhypotheses."
32. To take advantage of the potential for cross-inductions, the best set of inductive strategies includes, for instance, several inductive strategies seeking reliable rules concerning all copper objects and several other inductive strategies seeking reliable rules concerning all objects made of nickel. So even though all objects made either entirely of copper or entirely of nickel are electrically conductive, the best set contains no single inductive strategy seeking a reliable rule concerning the electrical conductivity of all such objects. Intuitively, that all such objects are electrically conductive is a physically necessary nonlaw since it fails to carve nature at its joints. See section 2 of this chapter.
33. Any p in U that is consistent with the salient h's in U falls within the scope of inductive projection as it has just been delimited (see chapter 2, section 1). So the results in U of the best set of inductive strategies make accurate enough predictions $p > m$, $q > (p > m)$, and so on, in U*, where $p, q, \ldots$ are consistent with the closure in U of those results. So those results' closure in U forms a non-nomically stable set.

    The definition of "inductive" confirmation as just amended was not given in chapter 5 because my purpose there was to show how the laws derive scientific importance from their special relation to *non-nomic facts* (broadly construed as facts in U*). That argument could not presume our interest in any truths $p > m$ with p, m outside U, since this presupposition would have begged the question. So "inductive" confirmation was then defined as projection over certain predictions having antecedents and consequents entirely within U (see chapter 5, note 16).

Chapter 8

1. I used arguments like this in chapter 3 to show that a proper subset Γ of Λ typically lacks non-nomic stability, since Γ typically fails to be preserved under counterfactual suppositions that are inconsistent with the members of Λ

that are omitted from Γ. (I say "typically" to leave room for grades of physical necessity.)

2. A fact that does not concern health, disease, anatomy, injury, or anything like that is not automatically offstage for medical purposes. For instance, the fact that I am now wearing a yellow shirt is *not* offstage for medicine. If it were offstage, then it would trivially be accurate enough for medicine with regard to any possible world. So consider the subjunctive supposition that I am wearing a yellow shirt ⊃ the human heart is not four-chambered. (This supposition falls within medicine's range of concern, just as it is medically relevant to point out that Jones would still have suffered a heart attack even if Jones had been wearing a differently colored shirt that day.) With regard to the closest possible world in which this supposition holds, it is *not* accurate enough to say that I am wearing a yellow shirt (and so the human heart is not four-chambered) there. In a medical context, it is correct to say that, had I been wearing a differently colored shirt or the human heart not been four-chambered, then I would have been wearing a differently colored shirt. In contrast, a fact *p* about human evolutionary history *is* offstage. It is accurate enough for medical purposes to say that had *p* ⊃ the human heart is not four-chambered, then *p* and the human heart would not have been four-chambered. That is because the counterfactual supposition concerns human evolutionary history and so falls outside of medicine's range of concern.
3. The term "physically necessary" is potentially misleading here, since it incorrectly suggests that we must be talking about physics. Though "nomically necessary" might be better, the term "physically necessary" is so well entrenched that I retain it here.
4. Of course, ecologists may be mistaken concerning the area law and other putative laws of island biogeography; they remain controversial. I presuppose these theories to be correct only in order to explain what it would be for there to be laws of island biogeography.
5. Had there existed creatures with such antigravity organs, the values of c and z for birds in a given island group might have been different. But species distribution would still have approximately conformed (ceteris paribus) to *some* curve of the form $S = cA^z$, which is all that the area law demands.
6. For an entertaining look at such laws, see Hitt (1996) and Underhill (1999).
7. No such proviso accompanies the law as it figures in the set Γ possessing non-nomic stability *for that field's purposes*.
8. Sellars calls "The *S*" a "distributive singular term" (1963b, pp. 631–35; 1968, p. 79; 1989, pp. 234–46, 245–49) and analyzes "The *S* is *T*" as "All *S*s are *T**," where * is "indicating that the right-hand side is a 'non-accidental' truth about" *S*s (1963b, p. 632). I agree that some claims of the form "The *S*s are *T*" are physically necessary, but clearly, some of them are not, though perhaps these are not generalizations either (e.g., "The Metasequoia was unknown to Western botanists before it was brought out of China in 1948," "The Apple Blossom is the state flower of Michigan," "The Snowy Egret, which was near extinction, has made great gains in recent years"). Sellars (1989, pp. 235–38) acknowledges that ceteris paribus clauses are often implicit in claims of the form "The *S* is *T*." Sellars also notes that not every distributive singular term takes the form "The *S*"; he offers "Man is rational" as an example (1963b, p. 633). Anscombe (1958, p. 14) agrees that "The *S* is *T*" cannot be understood as "All *S*s are *T*" without adding

provisos: "[M]an has so many teeth, which is certainly not the average number of teeth men have, but is the number of teeth for the species. . . ." Below I review others who have discussed claims of the form "The *S* is *T*."

9. Titles are from, respectively, Clark D. McCreedy and Harman P. Weeks, Jr., *Journal of Mammalogy*, 74, 1 (19 February 1993): 217–24; R. E. Young, J. Pearce, C. K. Govind, *The Journal of Experimental Zoology*, 269, 4 (15 July 1994): 319–26; Parker Gambino, *Journal of Invertebrate Pathology*, 61, 1 (January 1993): 110–11; Susan Nash and Michael Domjan, *Journal of Experimental Psychology: Animal Behavior Processes*, 17, 3 (July 1991): 342–53.
10. This challenge cannot be recognized, let alone met, as long as philosophy of biology attends primarily to evolutionary biology rather than to these disciplines. This is a theme that Philip Kitcher has sounded. See, for instance, Kitcher (1984b).
11. I am unsure of the extent to which my remarks concerning biological *species* apply to higher taxonomic categories in biology. (See the upcoming quotation from Jeffrey, where he takes the genus *Primula* as his example.) It is often thought that species are real in a way that higher taxa are not. Sometimes these arguments are based on the "biological species concept," which I shortly discuss.
12. For more on Darwin's view that the groups within groups into which biologists classify organisms are natural kinds, that a classification that best predicts biological properties is apt to reflect phylogeny, and that the taxonomic rank (species, genus, family, etc.) of a given natural group is arbitrary, see Winsor (1991, pp. 23–24).
13. This proviso accounts for the fact that "The *S* is *T*" does not logically entail "The *S* with property *P* is *T*."
14. An analogous argument could be made regarding meteorology (or "atmospheric science"—a change that itself is partly a reflection of my point). We now have a great deal of data regarding large-scale fluid motions in the atmospheres of other planets. These planets differ from Earth and from one another in many relevant respects: mass, rotation rate, solar heating, axial tilt, atmospheric surface pressure, internal heating, radiative time constant, surface boundary conditions, and so on. The best inductive strategies for meteorology now to pursue project from those other planets to Earth. For example, observations of the spread of Martian dust storms and the effect of dust on the Martian atmosphere were among the factors that led to widespread discussion of nuclear winter by planetary scientists. They made predictions regarding what *would* happen on Earth *were* a nuclear exchange to occur. That dust and smoke from a meteorite impact (as may have occurred at the time of the dinosaurs' extinction) or a nuclear exchange would spread globally was first well confirmed by applying to Earth dynamical modeling developed to account for data from Mars (Haberle, Ackerman, Toon, and Hollingsworth 1985). Prior to the availability of such extraterrestrial data, the best inductive strategies for meteorology to pursue yielded inference rules involving parameters that, although the result of accidents of Earth's development, functioned as "universal constants" in meteorology. These inference rules are reliable with regard to Earth.

    For example, consider the logarithmic wind profile: that in the surface boundary layer under conditions of neutral stability (e.g., no large vertical temperature gradients), the average wind speed v varies with height z according to $\partial v/\partial z = u/kz$, where u is the friction velocity (the turbulent horizontal wind stress in the surface boundary layer divided by the air density there) and k is von Karman's constant (usually taken to be 0.4). For many purposes, this equation gives a generally satisfactory fit to observed wind profiles in the surface layer (see McIlveen 1992, p. 307).

With the availability of extraterrestrial data, however, there are better inductive strategies for arriving at a reliable inference rule regarding the wind profile. Inference rules like the above are not associated with laws of atmospheric science (though perhaps they then become considered "derivative laws"), and the constants appearing in them are treated merely as accidental features of our planet. The log wind profile in meteorology may be completely recast in atmospheric science; at the very least, von Karman's "constant" will become a function of several factors that differ on different planets, such as the average mass of a parcel of atmosphere. It is believed that the best inductive strategies for atmospheric science use extraterrestrial data to yield atmospheric scaling laws that relate physical parameters to features of the global circulation of planetary atmospheres. These will permit atmospheric scientists to anticipate terrestrial climate variations as external parameters continue to change (see Kahn 1989).

15. On this point, see the Afterword. Of course, this does not mean that the laws of nature change when the range of information available to us changes. Rather, as I emphasized in chapter 7, the concept of "natural law" refers essentially to *us* (the speakers), here in the *actual* world (in a rigid sense), with our justificatory practices, the range of information to which we typically have access, and so on. The laws reflect the best inductive strategies for us (in a token-reflexive sense) to carry out. Of course, Λ-preservation demands that had the range of information accessible to us been different, the natural laws would still have been the same.
16. It is widely accepted that the same fact can be explained in many different ways—*even in* the same conversational context, so that the different explanations must respond to the same interests as well as the same "contrast class" (a term I explain in section 5.4 below). Salmon (1989, p. 183) offers a nice example. A helium-filled balloon moves toward the front of an airplane upon acceleration for take-off. This fact can be explained by the increase in air pressure from the front to the rear of the plane, resulting from the inertia of the air molecules when the airplane accelerates. Because of the pressure gradient, the air molecules in the plane behind the balloon push it forward more strongly than the air molecules to the front of the balloon push it back. The balloon's forward motion can also be explained by Einstein's principle of equivalence, according to which an acceleration is physically equivalent to a gravitational field. The balloon floats toward the front of the plane just as it floats toward the top of Earth's atmosphere.
17. Compare Sober (1983, p. 207), Kitcher (1989, pp. 426–27), and Hacking (1990, pp. 180–88).
18. These are cases of "segregation distortion": a successful gamete produced by a heterozygote is more likely to contain one allele than the other, though neither allele confers any fitness advantage to the organism harboring it. The excess recovery of the "driven" allele in an individual's gametes may result from the dysfunction of some fraction of the gametes carrying the

other allele, or their relatively lower likelihood of fertilizing an egg when competing with sperm carrying the "driven" allele (see Lyttle 1991).

19. Again, the generalization in this explanation—say, that a rigid peg fails to fit through a hole in a rigid board when that hole is smaller than the peg's cross-section—is not an analytic truth. "Rigid," a term from classical mechanics belonging with "moment of inertia" and "center of mass," means that the object's parts do not change their relative distances (or, at least, do so only negligibly) when a force (of a contextually understood sort) is applied. What it takes for some object to qualify as "rigid" depends on how much change in its parts' relative distances can safely be ignored for the relevant purposes. The appeal to rigidity does not trivialize the explanatory generalization. It is logically possible (though physically impossible) for the utterance of a magic word to make a rigid peg that is approaching the board gradually become longer and thinner (though remaining rigid and otherwise retaining roughly its original shape, as objects sometimes appear to do in a funhouse mirror) so as to fit the nearest hole. (Perhaps the hole must have a certain minimum cross-sectional area.) Because the peg returns to its original proportions immediately after passing through the hole and in some respects retained its original shape throughout, we can justly say that the peg, and not merely its matter, fits through the hole. So again, the explanatory generalization is not an analytic truth. Since there is actually no such magic word, there is no need for a "no disturbing factors" clause to rule it out.
20. See, for instance, Lewis (1986b, pp. 193–212), though Lewis's direct concern is causation rather than explanation.
21. I do not contend that this is the only way to explain its electrical conductivity; we might instead describe its electron-band structure. See note 16.
22. It does not help to weaken the counterfactual constraint to read that c helps to explain e only if $\neg(\neg c > e)$. In the clip example, I do not know whether or not $\neg c > e$. That I do not believe $\neg c > e$ does not mean that I believe $\neg(\neg c > e)$.
23. Of course, left to themselves, the Laplacean superphysicists might never ask themselves whether the intentional strategy would yield accurate predictions of macro facts. They might never formulate the concept of an intentional strategy. But if they became acquainted with the intentional strategy and considered its accuracy, their knowledge of micro laws would lead them to conclude that its predictions are fairly accurate—that is, fairly often coincide with the predictions made by the micro laws.
24. By the same token, the macro state of being an F cannot be *defined* as the state of satisfying this disjunction, since a microphysically impossible F fails to satisfy the disjunction.
25. I have not ruled out the possibility of the superphysicists' ascertaining these counterfactuals by employing some kind of ampliative reasoning other than inductive projection.

    It might be objected that, since by stipulation the superphysicists know all of the facts in U, they must know the correctness of any counterfactual conditional in U* as well. But the counterfactuals fail to supervene on the truths in U (recall chapter 2).

    Also notice a departure here from Dennett (1991, pp. 37–42), who makes much of the macro regularities in J. H. Conway's "game of life" as an analogy to the macro patterns in folk psychology that the intentional stance would uncover but a micro perspective would miss. This "game" (really, an exercise for one player) is played on a grid, each square of which is either occupied or unoccupied by a piece. The single micro law of the game is that, at any given time step, a square will be occupied if and only if, at the preceding time step, (1) it and exactly two of its eight neighboring squares were occupied, or (2) exactly three of its eight neighbors were occupied. Complicated patterns are thereby generated, involving kinds of shapes such as "glider guns" and "pi-heptiminos." Dennett regards these macro patterns as important facts that would be missed by working at the micro level alone, in terms of the single micro law. But whereas, in the examples I have been discussing, the macro laws are preserved under various microphysically impossible suppositions, the macro patterns in the game of life are not preserved under small changes in the micro law, so the arguments I have given for the autonomy of the macro level do not apply. Unlike "rigid cubical peg," we *could* define "glider gun" in terms of microphysically possible states because "glider guns" are not realizable apart from the micro law. (Of course, different physical media could serve as the grid and pieces of Conway's game, but the micro level in this example is supposed to be the level of pieces and grids, not molecules.) It does not follow that a macro regularity, such as "A block is indigestible and will damage its eater beyond its ability to repair itself," is not explanatory. Rather, my contention is that there fail to be two sorts of physical necessity here.

26. Insofar as Dennett contends that the Laplacean superphysicists fail to discover that the market would have behaved no differently had the stockbroker said "I want to buy 500 shares of General Motors" instead of "I want to buy 500 shares of GM," I disagree with Dennett. The Laplacean superphysicists would surely discover *some* counterfactuals, since their reasons for taking the evidence as bearing upon certain actual unexamined cases also render the evidence relevant to certain counterfactual cases.
27. For further discussion, see Woodward (1995). One of his examples involves the explanatory power of the (physically unnecessary) connection between endorsement by the Manchester, New Hampshire, *Union-Leader* newspaper and the outcome of New Hampshire elections.
28. A case requiring no initial condition is perhaps even simpler. For example, that all observed *F*s are *G* might be explained by the fact that all *F*s are *G*. See the example below of the urn of green marbles.
29. I discussed would-have-to-have counterfactuals in chapter 2, section 2.
30. That its *G*hood is inevitable given its *F*hood does not ensure that its *F*hood *made* it *G*; this "core idea" is too minimal to account for explanatory asymmetries. But it does account for the modal element in a scientific explanation, and hence for the difference that *p*'s lawhood makes to scientific explanations.
31. This approach can be extended to explanations involving contrast classes. To explain why *Ga rather than Hb*, the initial conditions and explanatory generalizations must entail *Ga* and ¬*Hb*. When *Ga* entails ¬*Hb*, this requires nothing more than explaining *Ga*, as in "Why did Jones *succumb* to lung cancer rather than fight off the illness and survive?" But to answer "Why did Jones *rather than Smith* succumb to lung cancer?" we must explain why Smith failed to succumb to lung cancer. (Of course, in answering this question in a conversation, it may suffice merely to mention the points at which

the initial conditions in the explanation of Jones's death differ from the initial conditions in the explanation of Smith's survival, just as in Lewis's (1986b, p. 220) example, to answer "Why did the dictator die just when the CIA agent was in the country?" it may suffice to reply "Just coincidence.") Likewise, the prevalence of foxes cannot explain why the rabbit was captured and eaten just when it was rather than earlier or later, whereas, I hold, the micro account *can* explain why the rabbit was eaten sometime rather than never.

References

Achinstein, Peter (1971) *Laws and Explanation* (Oxford: Clarendon Press).
Adams, Ernest (1970) "Subjunctive and Indicative Conditionals," *Foundations of Language* 6: 89–94.
Agassiz, Louis (1962) *Essay on Classification* (Cambridge, Massachusetts: Harvard University Press).
Alexander, H.G. (1958a) "General Statements as Rules of Inference?" *Minnesota Studies in Philosophy of Science, vol. 2*, ed. H. Feigl, M. Scriven, and G. Maxwell (Minneapolis: University of Minnesota Press), pp. 309–29.
——— (1958b) "The Paradoxes of Confirmation," *British Journal for the Philosophy of Science* 9: 227–33.
Anscombe, G.E.M. (1958) "Modern Moral Philosophy," *Philosophy* 33: 1–19.
Arai, Mary N. (1997) *A Functional Biology of Scyphozoa* (London: Chapman and Hall).
Armstrong, David M. (1978) *Universals and Scientific Realism* (Cambridge: Cambridge University Press).
——— (1983) *What Is a Law of Nature?* (Cambridge: Cambridge University Press).
Armstrong, David M., and Heathcote, Adrian (1991) "Causes and Laws," *Nous* 25: 63–73.
Ayer, A.J. (1963) "What Is a Law of Nature?" in *The Concept of a Person and Other Essays* (London: Macmillan), pp. 209–34.
——— (1976) *The Central Questions in Philosophy* (London: Pelican).
Beatty, John (1981) "What's Wrong with the Received View of Evolutionary Theory?" *PSA 1980*, vol. 2, ed. Peter D. Asquith and Ronald N. Giere (East Lansing, Michigan: Philosophy of Science Association), pp. 397–426.

——— (1995) "The Evolutionary Contingency Thesis," in *Concepts, Theories, and Rationality in the Biological Sciences*, ed. Gereon Wolters and James Lennox (Pittsburgh: University of Pittsburgh Press), pp. 45–81.

Beckner, Morton (1968) *The Biological Way of Thought* (Berkeley: University of California Press).

Bennett, Jonathan (1974) "Counterfactuals and Possible Worlds," *Canadian Journal of Philosophy* 4: 381–402.

——— (1984) "Counterfactuals and Temporal Direction," *Philosophical Review* 93: 57–91.

Bigelow, John; Ellis, Brian; and Lierse, Caroline (1992) "The World as One of a Kind: Natural Necessity and Laws of Nature," *British Journal for the Philosophy of Science* 43: 371–88.

Birch, Thomas (1772) *The Works of the Honourable Robert Boyle*, vol.1 (London: J. and F. Rivington).

Black, Max (1967) "Induction," *Encyclopedia of Philosophy*, vol. 4, ed. P. Edwards (New York: Macmillan), pp. 169–81.

Braithwaite, R.B. (1927) "The Idea of Necessary Connexion (I)," *Mind* 36: 467–77.

——— (1953) *Scientific Explanation* (Cambridge: Cambridge University Press).

Brieu, P.P.; Summers, F.J.; and Ostriker, J.P. (1995) "Cosmological Simulations Using Special-Purpose Computers," *Astrophysical Journal* 453: 566–73.

Broad, C.D. (1968) "Mechanical and Teleological Causation," *Induction, Probability, and Causation: Selected Papers by C.D. Broad* (Dordrecht: Reidel), pp. 159–83.

Burnham, Robert (1977) *Burnham's Celestial Handbook*, vol. 1 (New York: Dover).

Cain, Arthur J. (1954) *Animal Species and Their Evolution* (London: Hutchinson).

Caldin, E.F. (1960) "Theories and the Development of Chemistry," *British Journal for the Philosophy of Science* 10: 209–22.

Campbell, Norman Robert (1957) *Foundations of Science* (New York: Dover).

Canfield, John, and Lehrer, Keith (1961) "A Note on Prediction and Deduction," *Philosophy of Science* 28: 204–208.

Carnap, Rudolf (1937) *The Logical Syntax of Language*, trans. A. Smeaton (London: Routledge and Kegan Paul).

——— (1950) *Logical Foundations of Probability* (Chicago: University of Chicago Press).

——— (1966) *Philosophical Foundations of Physics* (New York: Basic Books).

Carroll, John (1994) *Laws of Nature* (Cambridge: Cambridge University Press).

Cartwright, Nancy (1983) *How the Laws of Physics Lie* (Oxford: Clarendon Press).

——— (1989) *Nature's Capacities and Their Measurement* (Oxford: Clarendon Press).

Cen, R.Y., and Ostriker, J.P. (1996) "Hydrodynamic Simulations of the Growth of Cosmological Structure—Summary and Comparisons among Scenarios," *Astrophysical Journal* 464: 27–43.

Chisholm, Roderick (1946) "The Contrary-to-fact Conditional," *Mind* 55: 289–307.

——— (1955) "Law Statements and Counterfactual Inference," *Analysis* 15: 97–105.

Christie, Maureen (1994) "Philosophers versus Chemists Concerning 'Laws of Nature,'" *Studies in the History and Philosophy of Science* 25: 613–29.

Coffa, J. Alberto (1968) "Discussion: Deductive Predictions," *Philosophy of Science* 35: 279–83.
Cohen, L. Jonathan (1962) *The Diversity of Meaning* (London: Methuen).
Cole, L.J. (1927) "The Lay of the Rooster," *Journal of Heredity* 18: 97.
Dalton, John (1953) *A New System of Chemical Philosophy* (London: William Dawson and Sons). (Originally published 1808)
Darwin, Charles (1964) *On the Origin of Species*, facsimile reprint (Cambridge, Massachusetts: Harvard University Press). (Originally published 1859)
Davidson, Donald (1966) "Emeroses by Other Names," *Journal of Philosophy* 63: 778–80.
——— (1990) *Essays on Actions and Events* (Oxford: Clarendon).
Dennett, Daniel (1981) "True Believers: The Intentional Strategy and Why It Works," in *Scientific Explanation: Papers Based on Herbert Spencer Lectures Given in the University of Oxford*, ed. A.F. Heath (Oxford: Clarendon Press), pp. 53–75.
——— (1991) "Real Patterns," *Journal of Philosophy* 88: 27–51.
Dirac, P.A.M. (1938) "A New Basis for Cosmology," *Proceedings of the Royal Society (London) Series A* 165: 199–208.
Downing, P.B. (1959) "Subjunctive Conditionals, Time Order, and Causation," *Proceedings of the Aristotelian Society* 59: 125–40.
Dretske, Fred I. (1977) "Laws of Nature," *Philosophy of Science* 44: 248–68.
Earman, John (1984) "Laws of Nature: The Empiricist Challenge," in *D.M. Armstrong*, ed. R.J. Bogdan (Dordrecht: Reidel), pp. 191–223.
——— (1992) *Bayes or Bust?* (Cambridge, Massachusetts: Bradford Books).
Evans, E.P. (1906) *The Criminal Prosecution and Capital Punishment of Animals* (London: Heinemann).
Ewing, A.C. (1935) "Mechanical and Teleological Causation," *Aristotelian Society Supplement*, 14: 66–82.
——— (1943) *Idealism* (London: Methuen).
——— (1962) *The Fundamental Questions of Philosophy* (New York: Collier Books).
Feigl, Herbert (1961) "On the Vindication of Induction," *Philosophy of Science* 28: 212–16.
Fine, Kit (1975) "Critical Notice of *Counterfactuals*, by D. Lewis," *Mind* 84: 451–58.
Fodor, Jerry A. (1981) *Representations* (Cambridge, Massachusetts: MIT Press).
——— (1991) "You Can Fool Some of the People All of the Time, Everything Else Being Equal; Hedged Laws and Psychological Explanations," *Mind* 100: 19–34.
Foot, Philippa (1989) "Naturalism," Romanell Lecture to the Pacific Division of the American Philosophical Association, 27 March.
Foster, John (1982–83) "Induction, Explanation and Natural Necessity," *Proceedings of the Aristotelian Society* 83: 87–102.
Frege, Gottlob (1984) "On Concept and Object," trans. Peter Geach, in *Collected Papers*, ed. Brian McGuinness (Oxford: Blackwell), pp. 182–94. (Originally published 1892)
Futuyma, Douglas J. (1979) *Evolutionary Biology* (Sunderland, Massachusetts: Sinauer).
Garfinkel, Alan (1981) *Forms of Explanation* (New Haven, Connecticut: Yale University Press).

Giere, Ronald (1988) "Laws, Theories, and Generalizations," in *The Limits of Deductivism*, ed. A. Grunbaum and W. Salmon (Berkeley: University of California Press), pp. 37–46.

——— (1994) "Presidential Address: Viewing Science," *PSA 1994*, ed. D. Hull, M. Forbes, and R.M. Burian (East Lansing, Michigan: Philosophy of Science Association), pp. 3–18.

Good, I.J. (1960) "The Paradoxes of Confirmation," *British Journal for the Philosophy of Science* 11: 145–49; 12: 63–64.

——— (1967) "The White Shoe Is a Red Herring," *British Journal for the Philosophy of Science* 17: 322.

Goodman, Nelson (1947a) "On the Infirmities of Confirmation Theory," *Philosophy and Phenomenological Research* 8: 149–51.

——— (1947b) "The Problem of Counterfactual Conditionals," *Journal of Philosophy* 44: 113–28.

——— (1966) "Comments," *Journal of Philosophy* 63: 328–31.

——— (1983) *Fact, Fiction, and Forecast*, 4th ed. (Cambridge, Massachusetts: Harvard University Press).

Gray, Alonzo (1848) *Elements of Chemistry* (New York: Newman).

Haavelmo, Trygve (1944) "The Probability Approach to Econometrics," *Econometrica*, 12 (Suppl.): 1–117.

Haberle, Robert; Ackerman, Thomas; Toon, Owen; and Hollingsworth, Jeffrey (1985) "Global Transport of Atmospheric Smoke Following a Major Nuclear Exchange," *Geophysical Research Letters* 12: 405–408.

Hacking, Ian (1990) *The Taming of Chance* (Cambridge: Cambridge University Press).

Halpin, John (1991) "The Miraculous Conception of Counterfactuals," *Philosophical Studies* 63: 271–90.

Hanson, Norwood Russell (1969) *Perception and Discovery* (San Francisco: Freeman, Cooper).

Harbourne, J.B., and Turner, B.L. (1984) *Plant Chemosystematics* (London: Academic Press).

Harre, Rom (1960) *An Introduction to the Logic of the Sciences* (London: Macmillan).

Hempel, C.G. (1962) "Deductive-Nomological vs. Statistical Explanation," *Minnesota Studies in the Philosophy of Science*, vol. 3, ed. H. Feigl and G. Maxwell (Minneapolis: University of Minnesota Press), pp. 98–169.

——— (1965) *Aspects of Scientific Explanation* (New York: Free Press).

——— (1966) *Philosophy of Natural Science* (Englewood Cliffs, New Jersey: Prentice-Hall).

——— (1988) "Provisos," in *The Limits of Deductivism*, ed. A. Grunbaum and W. Salmon (Berkeley: University of California Press), pp. 19–36. (Reprinted in *Erkenntnis* 28 (1988): 147–64)

Hempel, C.G., and Oppenheim, P. (1948) "Studies in the Logic of Explanation," *Philosophy of Science* 15: 135–75.

Hitt, Jack (1996) "The Theory of Supermarkets," *New York Times Magazine*, 10 March, p. 56.

Horwich, Paul (1982) *Probability and Evidence* (Cambridge: Cambridge University Press).

——— (1987) *Asymmetries in Time* (Cambridge, Massachusetts: MIT Press).

Hosiasson-Lindenbaum, Janina (1940) "On Confirmation," *Journal of Symbolic Logic* 5: 133–68.

Hull, David L. (1987) "Genealogical Actors and Ecological Roles," *Biology and Philosophy* 2: 168–84.

Imbrie, John (1957) "The Species Problem with Fossil Animals," *The Species Problem*, ed. Ernst Mayr, American Association for the Advancement of Science Publication 40: 125–53.

Jackson, Frank (1977) "A Causal Theory of Counterfactuals," *Australasian Journal of Philosophy* 55: 3–21.

Jeans, James (1940) *An Introduction to the Kinetic Theory of Gases* (Cambridge: Cambridge University Press).

Jeffrey, Charles (1982) *An Introduction to Plant Taxonomy*, 2nd ed. (Cambridge: Cambridge University Press).

Jones, David A. (1973) "Coevolution and Cyanogenesis," in *Taxonomy and Ecology*, ed. V.H. Heywood (London: Academic Press), pp. 213–42.

Kahn, Ralph (1989) *Comparative Planetology and the Atmosphere of Earth*, a report to the Solar System Exploration Division, NASA.

Kibble, T.W.B. (1985) *Classical Mechanics*, 3rd ed. (London: Longman).

Kim, Jaegwon (1992) "Multiple Realization and the Metaphysics of Reduction," *Philosophy and Phenomenological Research* 52: 1–26.

Kincaid, Harold (1997) *Individualism and the Unity of Science* (Lanham, Maryland: Rowman and Littlefield).

Kirsten, G., and Korber, H.-G., eds. (1975) *Physiker uber Physiker* (Berlin: Akademie-Verlag).

Kitcher, Philip (1984a) "1953 and All That: A Tale of Two Sciences," *Philosophical Review* 93: 335–73.

——— (1984b) "Species," *Philosophy of Science* 51: 308–33.

——— (1989) "Explanatory Unification and the Causal Structure of the World," *Minnesota Studies in the Philosophy of Science*, vol. 13, ed. Philip Kitcher and Wesley Salmon (Minneapolis: University of Minnesota Press), pp. 410–505.

Kneale, William (1950) "Natural Laws and Contrary-to-Fact Conditionals," *Analysis* 10: 121–25.

——— (1952) *Probability and Induction* (Oxford: Oxford University Press).

——— (1961) "Universality and Necessity," *British Journal for the Philosophy of Science* 12: 89–102.

Kuhn, Thomas S. (1977) *The Essential Tension* (Chicago: University of Chicago Press).

Lambert, Karel, and Brittan, Gordon G. (1987) *An Introduction to the Philosophy of Science*, 3rd ed. (Atascadero, California: Ridgeview).

Lange, Marc (1993a) "Lawlikeness," *Nous* 27: 1–21.

——— (1993b) "When Would Natural Laws Have Been Broken?" *Analysis* 53: 262–69.

——— (1994) "Scientific Realism and Components: The Case of Classical Astronomy," *The Monist* 74: 111–27.

——— (1996) "Laws of Nature, Cosmic Coincidences, and Scientific Realism," *Australasian Journal of Philosophy* 74: 614–38.

——— (1998) "Salience, Supervenience, and Sellars," *Philosophical Studies*, in press.

——— (1999) "Calibration and the Epistemological Role of Bayesian Conditionalization," *Journal of Philosophy* 96: 294–324.

Laudan, Rachel (1987) *From Mineralogy to Geology* (Chicago: University of Chicago Press).

Leavitt, Henrietta (1912) "Periods of 25 Variable Stars in the Small Magellanic Cloud," *Harvard College Observatory Circular* no. 173.

Lewis, David (1973) *Counterfactuals* (Cambridge, Massachusetts: Harvard University Press).

——— (1983) "New Work for a Theory of Universals," *Australasian Journal of Philosophy* 61: 343–77.

——— (1986a) *On the Plurality of Worlds* (Oxford: Blackwell).

——— (1986b) *Philosophical Papers* vol. 2 (Oxford: Oxford University Press).

——— (1994) "Humean Supervenience Debugged," *Mind* 103: 473–90.

Loeb, Leonard (1934) *The Kinetic Theory of Gases* (New York: McGraw Hill).

Loewer, Barry (1976) "Counterfactuals with Disjunctive Antecedents," *Journal of Philosophy* 73: 531–37.

Lowry, T.M. (1926) *Historical Introduction to Chemistry* (London: Macmillan).

Loyd, Robert C.; Gray, Elmer; and Shipe, Emerson (1971) "Effect of Freezing on Hydrocyanic Acid Release from Sorghum Plants," *Agronomy Journal* 63: 139–40.

Lyttle, Terrence (1991) "Segregation Distortion," *Annual Review of Genetics* 25: 511–57.

MacArthur, Robert H. (1972) *Geographic Ecology* (Princeton, New Jersey: Princeton University Press).

Mackie, J.L. (1961) "The Sustaining of Counterfactuals," *Australasian Journal of Philosophy* 39: 283–86.

——— (1962) "Counterfactuals and Causal Laws," in *Analytic Philosophy*, ed. R.S. Butler (New York: Barnes and Noble), pp. 66–80.

——— (1963) "The Paradox of Confirmation," *British Journal for the Philosophy of Science* 13: 265–77.

——— (1973) *Truth, Probability, and Paradox* (Oxford: Clarendon Press).

Maxwell, James Clerk (1952) *Matter and Motion* (New York: Dover). (Originally published 1876)

Mayr, Ernst (1965) *Animal Species and Evolution* (Cambridge, Massachusetts: Harvard University Press).

——— (1969) *Principles of Systematic Zoology* (New York: McGraw-Hill).

McIlveen, Robin (1992) *Fundamentals of Weather and Climate* (London: Chapman and Hall).

McKay, Thomas, and van Inwagen, Peter (1977) "Counterfactuals with Disjunctive Antecedents," *Philosophical Studies* 31: 353–56.

McPherson, T. (1950) "Ramsey on Rules," *Analysis* 10: 85–92.

Mellor, D.H. (1980) "Necessities and Universals in Natural Laws," in *Science, Belief, and Behavior*, ed. D.H. Mellor (Cambridge: Cambridge University Press), pp. 105–125.

Mendeleev, Dmitri I. (1897) *The Principles of Chemistry*, trans. G. Kamensky, ed. A.J. Greenoway (London: Longmans).

Mill, John Stuart (1893) *A System of Logic*, 8th ed. (New York: Harper).

Millikan, Ruth Garrett (1993) "Explanation in Biopsychology," in *Mental Causation*, ed. John Heil and Alfred Mele (Oxford: Clarendon), pp. 211–32.

Molnar, George (1967) "Defeasible Propositions," *Australasian Journal of Philosophy* 45: 185–97.

Moore, G.E. (1962) *Common-Place Book 1919–1953*, ed. C. Lewy (London: Allen and Unwin).
Morton, Adam (1973) "If I Were a Dry Well-Made Match," *Dialogue* 12: 322–24.
Nagel, Ernest (1954) [Review of Toulmin, *The Philosophy of Science*], *Mind* 63: 403–12.
——— (1961) *The Structure of Science* (New York: Harcourt, Brace, and World).
Nash, Leonard K. (1956) *The Atomic-Molecular Theory* (Cambridge, Massachusetts: Harvard University Press).
Nernst, Walter (1904) *Theoretical Chemistry* (London: Macmillan).
Newton, Isaac (1971) *Sir Isaac Newton's Mathematical Principles of Natural Philosophy and His System of the World*, trans. Andrew Motte, rev. Florian Cajori, vol. 2 (Berkeley: University of California Press).
Ostriker, Jeremiah P. (1996) "Ostriker's Cosmic View from Nassau Hall," *Princeton*, Summer, p. 7.
Pap, Arthur (1958) "Disposition Concepts and Extensional Logic," *Minnesota Studies in the Philosophy of Science*, vol. 2, ed. H. Feigl and M. Scriven (Minneapolis: University of Minnesota Press), pp. 196–224.
Partington, J.R. (1949) *An Advanced Treatise on Physical Chemistry*, vol. 1 (London: Longmans, Green).
Pears, David (1950) "Hypotheticals," *Analysis* 10: 49–63.
Peirce, C.S. (1934) *Collected Papers*, vol. 5, ed. C. Hartshorne and P. Weiss (Cambridge, Massachusetts: Harvard University Press).
Pollock, John (1976) *Subjunctive Reasoning* (Dordrecht: Reidel).
Popper, Karl (1959) *The Logic of Scientific Discovery* (New York: Basic Books).
——— (1979) *Objective Knowledge: An Evolutionary Approach* (Oxford: Clarendon).
Putnam, Hilary (1975) "Philosophy and Our Mental Life," in *Mind, Language, and Reality: Philosophical Papers*, vol. 2 (Cambridge: Cambridge University Press), pp. 291–303.
——— (1993) "Aristotle after Wittgenstein," in *Modern Thinkers and Ancient Thinkers*, ed. R.W. Sharples (London: University College London Press), pp. 117–37.
Quine, W.V.O. (1960) *Word and Object* (New York: Wiley).
Ramsey, Frank P. (1931) "General Propositions and Causality," *The Foundations of Mathematics*, ed. R.B. Braithwaite (New York: Humanities Press), pp. 237–55.
Reichenbach, Hans (1938) *Experience and Prediction* (Chicago: University of Chicago Press).
——— (1947) *Elements of Symbolic Logic* (New York: Macmillan).
——— (1954) *Nomological Statements and Admissible Operations* (Dordrecht: North-Holland).
Remsen, Ira (1890) *Inorganic Chemistry* (New York: Henry Hall).
Rescher, Nicholas (1970) *Scientific Explanation* (New York: Free Press).
Ridley, Mark (1986) *Evolution and Classification* (New York: Wiley).
Rosenberg, Alexander (1985) *The Structure of Biological Science* (New York: Cambridge University Press).
——— (1987) "Why Does the Nature of Species Matter?" *Biology and Philosophy* 2: 192–97.
Rosencrantz, Roger D. (1977) *Inference, Method, and Decision* (Dordrecht: North-Holland).
——— (1981) *Foundations and Applications of Inductive Probability* (Atascadero, California: Ridgeview).

Ross, W.D. (1930) *The Right and the Good* (Oxford: Oxford University Press).
Ruse, Michael (1969) "Definitions of Species in Biology," *British Journal for the Philosophy of Science* 20: 97–119.
——— (1971) "Is the Theory of Evolution Different?" *Scientia* 106: 765–83.
——— (1973) *The Philosophy of Biology* (London: Hutchinson).
Ryle, Gilbert (1949) *The Concept of Mind* (New York: Barnes and Noble).
——— (1963) "'If', 'So,' and 'Because,'" in *Philosophical Analysis*, ed. M. Black (Englewood Cliffs, New Jersey: Prentice-Hall), pp. 302–18.
Salmon, Wesley C. (1963) "On Vindicating Induction," *Philosophy of Science* 30: 252–61.
——— (1989) "Four Decades of Scientific Explanation," *Minnesota Studies in the Philosophy of Science*, vol. 13, ed. Wesley C. Salmon and Philip Kitcher (Minneapolis: University of Minnesota Press), pp. 3–217.
Scheffler, Israel (1981) *The Anatomy of Inquiry* (Indianapolis: Hackett).
Schiffer, Stephen (1991) "Ceteris Paribus Laws," *Mind* 100: 1–17.
Schlick, Moritz (1931) "Die Kausalitat in der gegenwartigen Physik," *Die Naturwissenschaften* 19: 145–62.
——— (1949) "Causality in Everyday Life and in Recent Science," repr. in *Readings in Philosophical Analysis*, ed. H. Feigl and W. Sellars (New York: Appleton Century Crofts), pp. 515–33. (Originally published 1932)
Schwartz, Robert; Scheffler, Israel; and Goodman, Nelson (1970), "An Improvement in the Theory of Projectibility," *Journal of Philosophy* 67: 605–608.
Scriven, Michael (1955) [Review of Toulmin, *The Philosophy of Science*], *Philosophical Review* 64: 124–28.
——— (1959) "Truisms as the Grounds for Historical Explanations," in *Theories of History*, ed. Patrick Gardner (New York: Free Press), pp. 443–75.
——— (1961) "The Key Property of Physical Laws—Inaccuracy," in *Current Issues in the Philosophy of Science*, ed. H. Feigl and G. Maxwell (New York: Holt, Rinehart, and Winston), pp. 91–104.
Sellars, Wilfrid (1948a) "Concepts as Involving Laws and Inconceivable without Them," *Philosophy of Science* 15: 287–315.
——— (1948b) "Realism and the New Way of Words," *Philosophy and Phenomenological Research* 8: 601–34.
——— (1953) "Inference and Meaning," *Mind* 62: 313–38.
——— (1958) "Counterfactuals, Dispositions, and Causal Modalities," *Minnesota Studies in the Philosophy of Science*, vol. 2, ed. H. Feigl, M. Scriven, and G. Maxwell (Minneapolis: University of Minnesota Press), pp. 225–308.
——— (1963a) *Science, Perception, and Reality* (London: Routledge and Kegan Paul).
——— (1963b) "Abstract Entities," *Review of Metaphysics* 16: 627–671.
——— (1964) "Induction as Vindication," *Philosophy of Science* 31: 197–231.
——— (1965) "Scientific Realism or Irenic Instrumentalism," *Boston Studies in the Philosophy of Science*, vol. 2, ed. R. Cohen and M. Wartofsky (New York: Humanities Press), pp. 171–204.
——— (1968) *Science and Metaphysics* (London: Routledge and Kegan Paul).
——— (1977) "Is Scientific Realism Tenable?" *PSA 1976*, vol. 2 (Lansing, Michigan: Philosophy of Science Association), pp. 307–34.
——— (1989) *The Metaphysics of Experience*, ed. Pedro Amaral (Atascadero, California: Ridgeview).

Shapin, Steven (1994) *A Social History of Truth* (Chicago: University of Chicago Press).

Simpson, G.G. (1961) *Principles of Animal Taxonomy* (New York: Columbia University Press).

Sklar, Lawrence (1993) *Physics and Chance* (Cambridge: Cambridge University Press).

Skyrms, Brian (1980) *Causal Necessity* (New Haven, Connecticut: Yale University Press).

——— (1986) *Choice and Chance*, 3rd ed. (Belmont, California: Wadsworth).

Slote, Michael (1978) "Time in Counterfactuals," *Philosophical Review* 87: 3–27.

Smart, J.J.C. (1963) *Philosophy and Scientific Realism* (London: Routledge).

——— (1968) *Between Science and Philosophy* (New York: Random House).

——— (1985) "Laws of Nature and Cosmic Coincidences," *Philosophical Quarterly* 35: 272–80.

Sober, Elliott (1983) "Equilibrium Explanation," *Philosophical Studies* 43: 201–210.

——— (1984) *The Nature of Selection* (Chicago: University of Chicago Press).

——— (1988) "Confirmation and Law-Likeness," *Philosophical Review* 97: 93–98.

——— (1993) *Philosophy of Biology* (Boulder, Colorado: Westview Press).

Stace, Clive A. (1980) *Plant Taxonomy and Biosystematics* (London: Arnold).

Stanley, S.M. (1973) "An Explanation for Cope's Rule," *Evolution* 27: 1–26.

Strawson, P.F. (1952) *Introduction to Logical Theory* (London: Methuen).

Stuessy, Ted F. (1994) *Case Studies in Plant Taxonomy* (New York: Columbia University Press).

Suppes, Patrick (1966) "A Bayesian Approach to the Paradoxes of Confirmation," in *Aspects of Inductive Logic*, ed. J. Hintikka and P. Suppes (Amsterdam: North Holland), pp. 198–207.

Swartz, Norman (1985) *The Concept of Physical Law* (Cambridge: Cambridge University Press).

Swinburne, R.G. (1971) "The Paradoxes of Confirmation—a Survey," *American Philosophical Quarterly* 8: 318–30.

Swoyer, Chris (1982) "The Nature of Natural Laws," *Australasian Journal of Philosophy* 60: 203–23.

Teller, Paul (1983) "The Projection Postulate as a Fortuitous Approximation," *Philosophy of Science* 50: 413–31.

——— (1984) "Comments on Kim's Paper," *Southern Journal of Philosophy* 22 [suppl.]: 57–61.

Tooley, Michael (1977) "The Nature of Law," *Canadian Journal of Philosophy* 7: 667–98.

——— (1987) *Causation: A Realist Approach* (Oxford: Clarendon Press).

Toulmin, Stephen (1953) *The Philosophy of Science* (London: Hutchinson).

Underhill, Paco (1999) *What We Buy: The Science of Shopping* (New York: Simon and Schuster).

van der Waals, J.D. (1890) "The Continuity of the Liquid and Gaseous States of Matter," in *Physical Memoirs Selected and Translated from Foreign Sources*, vol. 1, trans. Richard Threlfall and John Adair (London: Taylor and Francis), pp. 333–496.

van Fraassen, Bas C. (1983) "Theory Confirmation: Tension and Conflict," in *Epistemology and Philosophy of Science: Proceedings of the Seventh International*

Wittgenstein Symposium, ed. P. Weingartner and M. Czermak (Vienna: Hoelder-Pichler-Tempsky), pp. 319–29.

——— (1989) *Laws and Symmetry* (Oxford: Clarendon Press).

——— (1997) "Belief and the Problem of Ulysses and the Sirens," *Philosophical Studies* 77: 7–37.

van Inwagen, Peter (1979) "Laws and Counterfactuals," *Nous* 13: 439–53.

van Spronsen, J.W. (1969) *The Periodic System of Chemical Elements* (Amsterdam: Elsevier).

Wambsganss, J.; Cen, R.; Ostriker, J.P.; and Turner, E.L. (1995) "Testing Cosmogonic Models with Gravitational Lensing," *Science* 268: 274–76.

Weber, Max (1949) "Objective Possibility and Adequate Causation," *Max Weber on the Methodology of the Social Sciences*, trans. and ed. Edward S. Shils and Henry A. Finch (Glencoe, Illinois: Free Press), pp. 164–88.

Weinert, Friedel (1995) "Laws of Nature—Laws of Science," in *Laws of Nature*, ed. Friedel Weinert (Berlin: de Gruyter), pp. 3–64.

Whewell, William (1840) *Philosophy of the Inductive Sciences* (London: Parker).

Wigner, Eugene P. (1967) *Symmetries and Reflections* (Bloomington: Indiana University Press).

Winsor, Mary (1991) *Reading the Shape of Nature* (Chicago: University of Chicago Press).

Wittgenstein, Ludwig (1958) *Philosophical Investigations*, 3rd ed., trans. G.E.M. Anscombe (New York: Macmillan).

——— (1978) *Remarks on the Foundations of Mathematics*, rev. ed., ed. G.H. von Wright, R. Rhees, and G.E.M. Anscombe, trans. G.E.M. Anscombe (Cambridge, Massachusetts: MIT Press).

Wolterstorff, Nicholas (1970) "On the Nature of Universals," in *Universals and Particulars*, ed. Michael Loux (Garden City, New York: Anchor Books), pp. 159–85.

Woodhead, Susan, and Bernays, Elizabeth (1977) "Changes in Release Rates of Cyanide in Relation to Palatability of Sorghum to Insects," *Nature* 270: 235–36.

Woodward, James (1992) "Realism about Laws," *Erkenntnis* 36: 181–218.

——— (1995) "Causation and Explanation in Econometrics," in *On The Reliability of Economic Models*, ed. Daniel Little (Dordrecht: Kluwer), pp. 9–61.

Woolley, R.G. (1978) "Must a Molecule Have a Shape?" *Journal of the American Chemical Society* 100: 1073–78.

Zilsel, Edgar (1942) "The Genesis of the Concept of Physical Law," *Philosophical Review* 51: 245–79.

Index

9 780195 331332

Printed in the United States
By Bookmasters